Series Editor

Prof. Dr. Michael J. Parnham
PLIVA
Research Institute
Prilaz baruna Filipovica 25
10000 Zagreb
Croatia

In Vivo Models of Inflammation

Douglas W. Morgan
Lisa A. Marshall

Editors

Birkhäuser Verlag
Basel · Boston · Berlin

Editors

Douglas W. Morgan, Ph.D.
Senior Group Leader
Cancer Research
D-47J AP9
Abbott Laboratories
Abbott Park, IL 60064
USA

Lisa A. Marshall, Ph.D.
Director and Head
Department of Immunology
P.O. Box 1539
SmithKline Beecham Pharmaceuticals
King of Prussia, PA 19406
USA

A CIP catalogue record for this book is available from the Library of Congress, Washington D.C., USA

Deutsche Bibliothek Cataloging-in-Publication Data
In vivo models of inflammation / ed. by Douglas W. Morgan ; Lisa A.
Marshall (ed.). - Basel ; Boston ; Berlin : Birkhäuser, 1999
(Progress in inflammation research)
ISBN 3-7643-5876-9 (Basel...)
ISBN 0-8176-5876-9 (Boston)

The publisher and editor can give no guarantee for the information on drug dosage and administration contained in this publication. The respective user must check its accuracy by consulting other sources of reference in each individual case.

Printed on acid-free paper produced from chlorine-free pulp. TCF ∞
Cover design: Markus Etterich, Basel
Printed in Germany
ISBN 3-7643-5876-9
ISBN 0-8176-5876-9

9 8 7 6 5 4 3 2 1

Contents

List of contributors

Karen M. Anderson, SmithKline Beecham Pharmaceuticals, Cardiovascular Pharmacology, UW 2510, 709 Swedeland Road, King of Prussia, PA 19406-0939, USA; e-mail: Karen_M_Anderson@sbphrd.com

Jill Bishop, Biology Department, Roche Discovery Welwyn, Welwyn Garden City, Herts. AL7 3AY, UK

Michael R. Briggs, Department of Cardiovascular Pharmacology, UW2511, SmithKline Beecham Pharmaceuticals, 709 Swedeland Road, King of Prussia, PA 1906, USA; e-mail: Michael_R_Briggs@stphrd.com

Ginny L. Bumgardner, Department of Surgery, The Ohio State University, Columbus, OH 43210-1228, USA

Richard P. Carlson, Department of Immunological Disease Research, Abbott Laboratories, D4NB, AP-9, 100 Abbott Park Road, Abott Park, IL 60064-350, USA

Richard W. Chapman, Schering-Plough Research Institute, 2015 Galloping Hill Road, Kenilworth, NJ 07033-0539, USA

Tai-Ping Fan, Department of Pharmacology, University of Cambridge, Tennis Court Road, Cambridge, CB2 1Q5, UK

Anne Flanigan, Division of Gastroenterology and Hapatology, Krancer Center for IBD Research, Allegheny University of the Health Sciences, Philadelphia, PA 19102-1192, USA

Stephen H. Gavett, Pulmonary Toxicology Branch, USEPA, Health Effects Research Laboratory, Research Triangle Park, NC 27711, USA

Ronald P. Gladue, Department of Immunology, Central Research Division, Pfizer Inc., Box 1189, Eastern Point Road, Groton, CT 06340, USA

David S. Grass, DNX Transgenic Sciences, 301 B College Road East, Princeton, NJ 08540, USA; e-mail: david.grass@chrysalisintl.com

Anna K. Greenham, Biology Department, Roche Discovery Welwyn, Welwyn Garden City, Herts., AL7 3AY, UK

Howard C. Hughes, SmithKline Beecham Pharmaceuticals, LAS, UW2620, 709 Swedeland Rd., King of Prussia, PA 19406-0939, USA;
e-mail: Bud_Hughes@SBPHRD.com

Jeffrey R. Jackson, Department of Oncology Research, SmithKline Beecham Pharmaceuticals, 709 Swedeland Rd., King of Prussia, PA 19406, USA

Peer B. Jacobson, Department of Immunological Disease Research, Abbott Laboratories, D4NB, AP-9, 100 Abbott Park Road, Abbott Park, IL 60064-3500, USA; e-mail: peer.b.jacobson@abbott.com

Leo A.B. Joosten, Dept. of Reumatology, University Hospital Nijmegen, Geert Grooteplein 8, 6500 HB Nijmegen, The Netherlands

Andrea Keane-Myers, Laboratory of Molecular Immunology, Schepens Eye Research Institute, Harvard Medical School, Boston, MA 02114, USA

Douglas Kuperman, Johns Hopkins School of Hygiene and Public Health, 615 N. Wolfe St., Baltimore, MD 21205, USA

E. Jonathan Lewis, Biology Department, Roche Discovery Welwyn, Welwyn Garden City, Herts. AL7 3AY, UK; e-mail: jon.lewis@roche.com

Kenneth N. Litwak, SmithKline Beecham Pharmaceuticals, LAS, UW2620, 709 Swedeland Rd., King of Prussia, PA 19406-0939, USA

Sreekant Murthy, Division of Gastroenterology and Hepatology, MA 131, Allegheny University of the Health Sciences, Hahnemann Division, Broad and Vine, Philadelphia, PA 19102-1192, USA; e-mail: murthys@auhs.edu

Sandhya S. Nerurkar, Department of Cardiovascular Pharmacology, UW2510, SmithKline Beecham Pharmaceuticals, 709 Swedeland Rd., King of Prussia, PA 1906, USA; e-mail: Sandhy_S_Nerurkar@sbphrd.com

Charles G. Orosz, Departments of Surgery, Pathology & Medical Microbiology/ Immunology, The Ohio State University, Columbus, OH 43210-1228, USA

Michael P. Seed, Panceutics, P.O. Box 11358, Swindon, Wilts., SN3 4GP, UK

William M. Selig, Hoechst Marion Roussel, Inc., Route 202-206, P.O. Box 6800, Bridgewater, NJ 08807-0800, USA; e-mail: william.selig@hmrag.com

Kenneth M. Tramposch, Bristol-Myers Squibb Pharmaceutical Research Institute, 100 Forest Avenue, Buffalo, NY 14213, USA;
e-mail: ken_tramposch@ccmail.BMS.com

David C. Underwood, Department of Pulmonary Pharmacology, SmithKline Beecham Pharmaceuticals, 709 Swedeland Road, King of Prussia, PA 19406, USA;
e-mail: David_C_Underwood@sbphrd.com

Wim B. van den Berg, Department of Rheumatology, University Hospital Nijmegen, Geert Grooteplein 8, 6500 HB Nijmegen, The Netherlands;
e-mail: w.vandenberg@reuma.azn.nl

M. Elaine Wakely, Departments of Surgery, Pathology & Medical Microbiology/ Immunology, The Ohio State University, Columbus, OH 43210-1228, USA

Elora J. Weringer, Department of Immunology, Central Research Box 1189, Pfizer Inc., Eastern Point Road, Groton, CT 06340, USA; e-mail: elora_j_weringer@groton.pfizer.com

Marsha A. Wills-Karp, Johns Hopkins School of Hygiene and Public Health, 615 N. Wolfe St., Baltimore, MD 21205, USA; e-mail: mkarp@welchlink.welch.jhu.edu

James D. Winkler, Department of Oncology Research, UW-532, SmithKline Beecham Pharmaceuticals, 709 Swedeland Road, P.O.Box 1539, King of Prussia, PA 19406, USA; e-mail: James_D_Winkler @SBphrd.com

Preface

The purpose of this volume in the series *Progress in Inflammation Research* is to provide the biomedical researcher with a description of the state of the art of the development and use of animal models of diseases with components of inflammation. Particularly highlighted are those models which can serve as *in vivo* correlates of diseases most commonly targeted for therapeutic intervention. The format is designed with the laboratory in mind; thus it provides detailed descriptions of the methodologies and uses of the most significant models. Also, new approaches to the development of future models in selected therapeutic areas have been highlighted. While emphasis is on the newest models, new information broadening our understanding of several well-known models of proven clinical utility is included. In addition, we have provided coverage of transgenic and gene transfer technologies which will undoubtedly serve as tools for many future approaches. Provocative comments on the cutting edge and future directions are meant to stimulate new thinking. Of course, it is important to recognize that the experimental use of animals for human benefit carries with it a solemn responsibility for the welfare of these animals. The reader is referred to the section on current regulations governing animal use which addresses this concern.

To fulfill our purpose, the content is organized according to therapeutic areas with the associated models arranged in subcategories of each therapeutic area. Concepts presented are discussed in the context of their current practice, including intended purpose, methodology, data and limitations. In this way, emphasis is placed on the usefulness of the models and how they work. Data on activities of key reference compounds and/or standards using graphs, tables and figures to illustrate the function of the model are included. The discussions include ideas on a given model's clinical correlate. For example, we asked our contributors to answer this question: How does the model mimic what is found in human clinical practice? They have answered this question in many interesting ways.

We hope the reader will find the information presented here useful for his or her own endeavours investigating processes of inflammation and developing therapeutics to treat inflammatory diseases.

October, 1998

Douglas W. Morgan
Lisa A. Marshall

Comparison of adjuvant and streptococcal cell wall-induced arthritis in the rat

Richard P. Carlson and Peer B. Jacobson

Department of Immunological Disease Research, Abbott Laboratories, D4NB, AP-9, 100 Abbott Park Road, Abbott Park, IL 60064-3500, USA

Introduction

The goals of this chapter are to discuss rat adjuvant and streptococcal cell wall-induced arthritis, both as animal models of chronic inflammation, and as relevant and predictive models of human rheumatoid arthritis (RA). The information which follows is intended to (1) provide the investigator with a detailed methodology for these two chronic models of inflammation, and (2) to review the pathogenesis and therapeutic interventions for which these models have historically been used to further our understanding of RA and its potential regulation by novel drugs. While these animal models have been extensively studied for over 30 years, the following information provides a rationale to better understand the similarities and differences of these models to RA. Furthermore, this chapter will describe how new technologies, such as magnetic resonance imaging (MRI), can be integrated with these "old" models to better predict clinical efficacy with the many new biological and immunomodulatory drugs currently in development.

Historical background

Stoerk et al. [1] were the first to observe adjuvant induced rat polyarthritis (AA) as we know it today. This adjuvant disease was the result of injecting extracts of spleen as an emulsion in Freund's complete adjuvant into the dorsal skin of rats. The initial observations of chronic polyarthritis in the joints of both front and rear paws were extended by Pearson [2], who reported that adjuvant disease could be induced by injecting an emulsion of Freund's adjuvant (without using minced tissue from spleen or muscle) into the skin in the posterior cervical region. However, by adding macerated muscle to the adjuvant, a greater incidence and severity of arthritic joints was observed. In 1959, Pearson and Wood [3] demonstrated that an intradermal injection of an emulsion containing a pure wax fraction of acid-fast bacilli labeled "Wax D, Strain Canetti" (Dr. E. Lederer, Institute Physico-Chemique, Paris) into the

In Vivo Models of Inflammation, edited by D.W. Morgan and L.A. Marshall

posterior cervical region was more effective on a dry weight basis when compared to other strains like *M. plei* and *M. butyricum*.

In 1959, Pearson and Wood reported a personal communication by B.H. Waksman that rat adjuvant arthritis could be successfully induced by intradermally injecting heat-killed tubercle bacilli in light mineral oil into the foot pad. This has subsequently been the method of induction of AA for the last 39 years. The first inhibitory effect of an antiinflammatory agent (hydrocortisone injected at 2 mg/kg) in a developing model was described by Pearson and Wood [3]. The early experiments for determining inhibitory activity were based upon an assessment of the severity of the disease by totaling the score of each individual paw and tail [3]. Newbould [4] initiated the first prophylactic approach by using the model to screen different classes of antiinflammatory drugs. His group induced the disease with a submaximal dose of killed tubercle in liquid paraffin (0.25 mg/rat) and measured paw swelling (antiinflammatory activity) with a micrometer to evaluate changes in paw diameters, and weight gain/loss for safety or toxicity. Winter and Nuss [5] and Glenn et al. [6] were probably the first to document the effects of nonsteroidal antiinflammatory drugs (NSAIDs) and steroids in the established disease model. Investigators who have used AA for drug testing and studying pharmacological mechanisms of action include Glenn et al. [6], Ward and Cloud [7], Winter and Nuss [5], and Graeme et al. [8].

It is important to note that Winter and Nuss [5] advanced the use of the semiautomatic mercury plethysmograph (originally developed by Dr. Gordon Van Arman, Wyeth Laboratories, 1961) to accurately evaluate the effects of experimental compounds on joint inflammation in a rapid and reproducible manner. Other innovative ideas for evaluating the severity of arthritis included measuring changes in paw temperature [9, 10] analyzing grip responses on a vertical screen [10], and using a rotarod technique to measure grip strength for determining improvement in disabled limbs and analgesic activity [11]. However, the measurement of paw edema by semiautomated plethysmography has been the method of choice for evaluating antiinflammatory drug activity for the last 31 years. Other parameters that are routinely used today to evaluate disease activity in AA include: measurement of acute phase proteins [6, 10, 12, 13], measurement of histological changes of the joints by noninvasive methods such as x-ray analysis [14], and recently, by MRI [15, 16]. Important reviews to read on this subject include those by Van Arman [17], Mohr and Wild [18], Billingham and Devies [19], Owen [20], Billingham [21], Lewis et al. [22], Rook [23] and Billingham [24].

The initiation of the rat streptococcal cell wall model (SCW) of arthritis began with the experiments of Schwab et al. [25], who found that the injection of covalent complexes of peptidoglycan and polysaccharide (PG-PS) from group A streptococci into rabbit skin produced nodular lesions, and induced remitting inflammation in other tissues. Subsequently, other groups demonstrated rheumatic-like lesions following PG-PS injection in a model of mouse endocarditis [26], synovitis in rabbit knee joints injected with PG-PS [27], and an erosive arthritis in rats (e.g. polyarthritis with reported cycles of remission and exacerbation of paw inflammation over several months) following a single i.p. injection of PG-PS (10S PG-PS model) [28, 29]. Wilder et al. [30], Spitznagel et al. [31], Ridge et al. [32], Wells et al. [33], Wahl et al. [34] and Sartor et al. [35, 36] have shown the phlogistic properties of streptococcal PG-PS in several tissues in rats (e.g. joints, uvea, liver and intestine). PG-PS from cell walls of normal intestinal bacteria or from two strains of bacteria prominent in patients with Crohn's disease or rheumatoid arthritis (RA) were shown to be arthropathogenic in the model by Severijnen et al. [37, 38]. Schwab [39] has very adequately reviewed the nomenclature, structural details, and relevant biological properties of PG-PS moieties associated with the induction of intestinal inflammation and polyarthritis.

In 1985, Esser et al. reported the development of a more synchronous and predictable monoarthritis in female rats, by injecting a small amount of PG-PS intraarticularly (i.a.) into the tibiotalar joint followed by an i.v. injection with a larger dose of PG-PS i.v. 1–6 weeks later [40]. Currently, most investigators inject 100P PG-PS i.v. 21 days after the i.a. injection for very predictable results and at a substantial reduction in cost compared to the 10S PG-PS SCW model. In addition, the use of a small quantity of 100 PG-PS by the i.v. route greatly reduces the granulomatous reaction in the liver, which is a potential liability associated witht the 10S PG-PS SCW model.

The close similarities of rat AA and SCW-induced polyarthritis in rats will be discussed in the following sections with regard to their pathophysiology, strain sensitivities, and their usefulness in discovering novel drugs and biologicals for the treatment of rheumatoid arthritis and advanced osteoarthritis in man.

Pathogenesis of rat adjuvant and streptococcal cell wall-induced polyarthritis

The pathology of these two induced chronic arthritic diseases is based upon events that occur in the uninjected joints, which begin only a few hours after the injection (subplantar for AA; i.p. for 10S PG-PS SCW) of the peptidoglycan antigen from mycobacterial/streptococcal cell walls. In AA, the antigen is presented to the host following a s.c. injection into the hindpaw or tail of a suspension of mycobacterium in light mineral oil, which soon after becomes systemically distributed via the lymphatics and circulating mononuclear cells. Antigen distribution and presentation occurs by similar mechanisms in the AA and SCW polyarthritis model, with the peptidoglycan moving rapidly in a few hours to the distal joints (ankles and wrists). This process has been described in both models in the synovial tissue using labeled mycobacterium [41, 42] and streptococcal cell wall fragments [43, 44]. However, following the i.p. injection of the 10S PG-PS

SCW material in saline, the inability of the reticuloendothelial system (RES), liver and spleen to systemically clear the antigen results in marked hepatic and splenic granulomas along with deposits in the sinusoids [27]. These "antigen depots" in the liver constitute a primary mechanism for the prolonged and progressive arthritis which is characteristic of the SCW polyarthritis model, and to a much lesser extent in AA.

The earliest recognizable event following the arrival of antigen is the deposition of fibrin along the synovial membrane and within the joint capsule – an event designed to wall off a bacterial infection. However, this defensive response is not a limiting step in these models, and fibrin continues to be laid down as antigen is continuously being presented from the depots (lymph nodes in AA; liver and spleen in SCW). Fibrin degradation products, chemokines/cytokines, and complement-platelet interactions generated near the synovial membrane recruit macrophages, neutrophils and lymphocytes into the joint space and surrounding tissues [45]. The continual arrival of antigen by activated $CD4^+$ T cells and monocytes drives this T cell mediated disease by a number of interdependent mechanisms to progressively erode the joint synovial tissue, cartilage and finally the bone. Th1 lymphocytes ($CD4^+$) are highly responsible for initiating the immune events that lead to the chronic arthritis which develops in both models. Verification for this is that anti-$CD4^+$ mAbs inhibit the developing disease, and sensitized T cells from spleens, and lymph nodes from animals with the active disease, can transfer the disease [46–48]. We have outlined in Table 1 the stepwise pathological sequalae involved in the pathogenesis of AA and SCW polyarthritis, and how these events compare to RA. Figure 1 also illustrates the likely occurrences of these pathologies during the time-course of AA. Many of the features outlined in Table 1 have been compiled from our experiences and modified from published reviews [2, 18, 22, 49–54]. It is quite surprising how similar these two models are to RA, but one has to realize that in rat AA and SCW arthritis, end stage ankylosis and ossification occurs 40-50 days post antigen while RA may take 30–50 years to reach a similar point. Another important difference between human RA and the AA and SCW arthritis models are the exacerbations or flares seen in RA which can occur many times over the long course of the disease. These flares probably enter the pathogenesis of the disease between the exudative and proliferative stages. Formation of intraarticular immune complexes

Figure 1
Time-course of rat adjuvant arthritis and general procedure associated with its implementation and use for evaluating disease progression and drug efficacy. See Methods for details. Data reprent mean hindpaw volumes (ml) ± SEM from seven separate 45 day studies using 10–16 rats/group. Rats were culled on day 15 as outlined under Methods; data prior to day 15 was generated using non-culled animals.

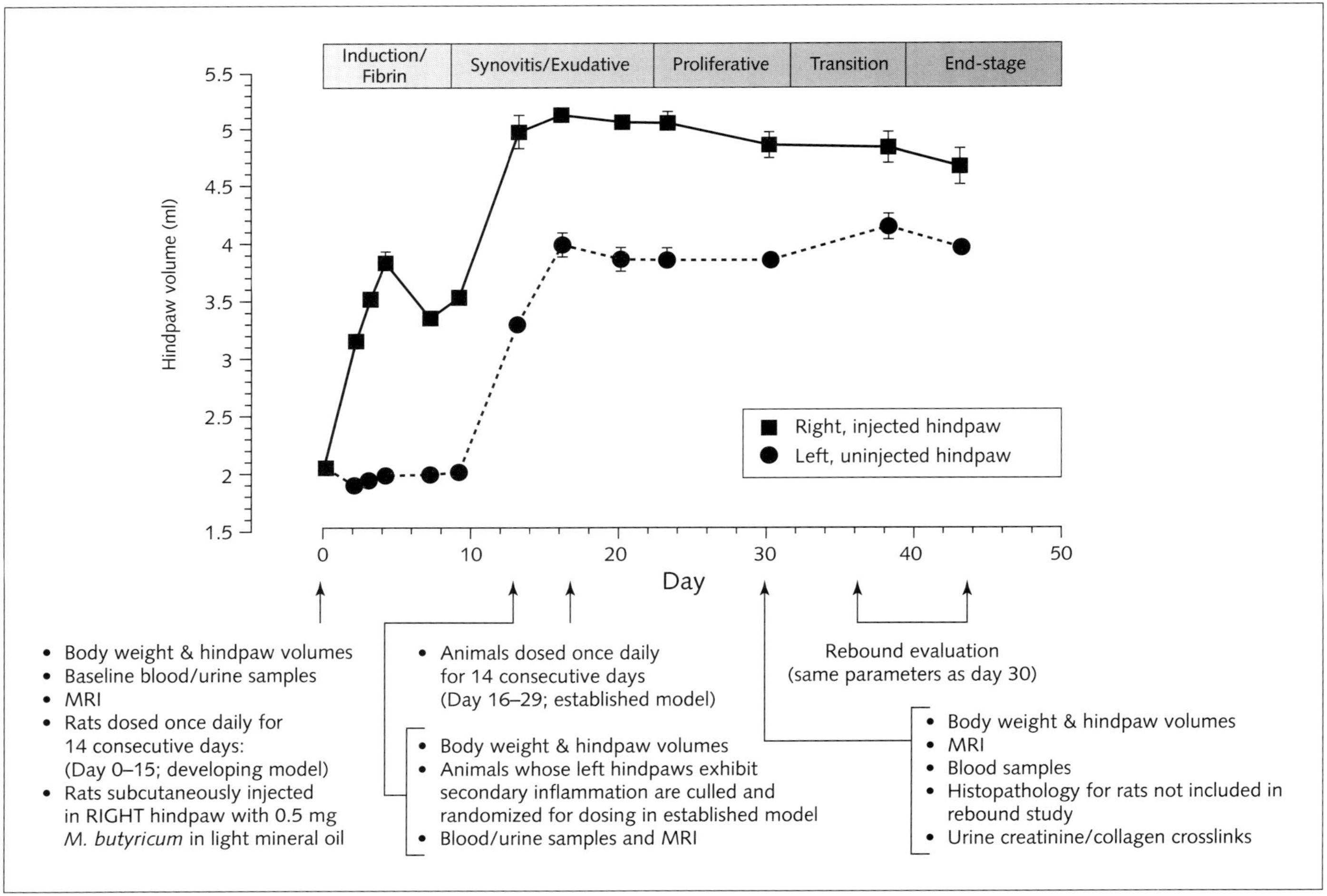
Induction/ Fibrin
Synovitis/Exudative
Proliferative
Transition
End-stage
Hindpaw volume (ml)
5.5
5
4.5
4
3.5
3
2.5
2
1.5
0
10
20
30
40
50
Day
Right, injected hindpaw
Left, uninjected hindpaw
• Body weight & hindpaw volumes
• Baseline blood/urine samples
• MRI
• Rats dosed once daily for 14 consecutive days: (Day 0–15; developing model)
• Rats subcutaneously injected in RIGHT hindpaw with 0.5 mg M. butyricum in light mineral oil
• Animals dosed once daily for 14 consecutive days (Day 16–29; established model)
• Body weight & hindpaw volumes
• Animals whose left hindpaws exhibit secondary inflammation are culled and randomized for dosing in established model
• Blood/urine samples and MRI
Rebound evaluation (same parameters as day 30)
• Body weight & hindpaw volumes
• MRI
• Blood samples
• Histopathology for rats not included in rebound study
• Urine creatinine/collagen crosslinks

Table 1 - Comparison of the pathogenesis of rat AA and SCW arthritis with human RA

Chronology	Disease characteristic	Rat (AA/SCW)	Human (RA)
Induction	antigen (AA: *M. butyr.*; SCW: PG-PS)	++	?
	rheumatoid factor genetic susceptibility	–	++
	sex difference	AA (–) SCW (++)	++
	environmental/gut flora/diet	++	+?
Fibrin	upregulated acute phase proteins	++	++
	fibrin deposition	++	++
Synovitis	continued organization of fibrin	++	++
	bursitis/tendonitis	++	++
	chondrocyte clusters	++	++
	upregulation of cell adhesion molecules	++	++
	neutrophil influx into synovial space	++	+
	$CD4^+$ lymphocytes into synovial space	++	++
	expansion of joint capsule	++	++
	elevated intraarticular cytokines	++	++
	osteophytes	++	–
	albumin/globulin ratio inversion	++	++
	edema	++	++
Exudative	*Soft tissue*:		
	$CD4^+$ lymphocyte	++	++
	dendritic cells	?	++
	neutrophils	++	++
	macrophages	++	++
	plasma cells	–	++
	Synovial space:		
	neutrophils	++	++
	$CD4^+$ lymphocytes	+	++
	macrophages	+	+
	other:		
	periosteal reaction	+	+
	osteitis	+	+
	synovial villi	+/–	++
	bursitis/tendonitis	++	++
	chondrocyte clusters	++	?
	elevated intraarticular cytokines	++	++
	elevated acute phase proteins	++	++
	albumin/globulin ratio inversion	++	++
	elevated MMPs	++	++
	edema	++	++

Table 1 (continued)

Chronology	Disease characteristic	Rat (AA/SCW)	Human (RA)
Proliferative	fibroplasia	++	++
	angiogenesis	+	++
	pannus	++	++
	lymphoid follicles	–	+
	upregulated acute phase proteins	++	++
	albumin/globulin ratio inversion	++	++
	fibroplasia	++	++
Transition	soft tissue edema	+	+
	leukocyte influx	+	+
Quiescence	angiogenesis	NA	++
	soft tissue changes	NA	–/+
	bony changes	NA	–/+
End-stage	albumin/globulin inversion		
	elevated acute phase proteins	–	+
	chondrocyte clusters	++	– ?
	osteitis	++	++
	periosteal reaction with ankylosis	++	+
	chronic deformation	++	++

NA, not applicable; (–), not observed; (+/–) not predictable; (+), moderately observed; (++) commonly observed; (?), unknown

accompanied by complement activation and cytokine/chemokine generation, probably directs this neutrophil influx into the joint, leading to increases in vascular permeability, edema, and pain.

Many early events in the pathogenesis of RA happen rapidly, and the T cell mediated aspects may be over in 2–5 years, leading to the macrophage dominant phase of tissue destruction. The end stage of RA cannot be treated effectively with glucocorticoids or disease modifying antirheumatic drugs (DMARDs), leaving orthopedic surgery as an eventual option to unlock ankylosed joints. Therefore, the target of new therapies should be directed at inhibiting the rapid and progressive T cell phase in early RA. Our observations from the study of novel antiarthritic drugs and biologicals in the developing and early established models of rat AA and SCW arthritis, either alone or in combination, suggest that these models may prove to be useful and predictive of clinical efficacy in human RA.

Strain susceptibility

Table 2 lists the most frequently used rat strains for the induction of AA and SCW polyarthritis over the last 40 years. The MHC-Class II (RT^1), along with "loosely" and non-linked MHC-class genes, are important not only in determining a strain's susceptibility to the peptidoglycan antigen, but are also very important in determining the incidence, severity, and duration of this chronic disease [55, 56]. Examples for this are shown in Table 2 where two different rat strains (Lewis and NIMRI Hooded) can be highly susceptible to the disease, but one has a short duration (e.g. in NIMRI Hooded rats where the disease undergoes spontaneous remission after day 21) while the other has a long duration of polyarthritis (e.g. Lewis rats where the chronic inflammation progresses to a permanent crippling state characterized by extensive ankylosis and fibrosis after day 35). Investigators who have done the majority of immunogenetic research on the above rats include: S.K. Anderle [57], J.R. Battisto [55], M.A. Griffiths [56], R.L. Wilder [58], E.M. Sternberg [59, 60], L. Klareskog [61], and I. Joosten [62].

It should also be noted that endogenous intestinal bacterial flora play an important role in the induction of the disease [63, 64]. For example, F344 rats in a conventional environment are resistant to the induction of chronic rat adjuvant and SCW arthritis, but not when raised in a germ-free environment [65]. In addition, during the last 10 years, Sternberg et al. [59, 60] and Wilder et al. [30, 58, 66] have shown the importance of the hypothalamic-pituitary-adrenal (HPA) axis on the induction of SCW arthritis in female Lewis and F344 rats. The high susceptibility of the arthritic disease in Lewis rats has been linked to a hypothalamic defect in the synthesis and release of corticotropic releasing hormone (CRH) and decreased sensitivity of the hypothalamus to inflammatory cytokines such a IL-1β [67].

There are no gender differences in susceptible strains for the induction of AA. However, only female Lewis or Sprague-Dawley rats have a high incidence and disease severity in the chronic phase of SCW arthritis when compared to the basically unresponsive males [66, 68]. The high female susceptibility in SCW arthritis has been associated with the estrogen levels in the blood and its inhibitory effect, especially on the hepatic reticuloendothelial system (Kupffer cells), on the clearance and detoxification of the PG-PS [68, 69]. This clearance defect in female rats allows for a higher continuous level of circulating PG-PS within the plasma compartment and mononuclear cells to reach the distal joints for initiating the primary lesion and the chronic secondary lesions beginning after day 10. However, Wilder has shown that PG-PS antigen reaches the same levels in the synovial membrane of distal joints in both susceptible female Lewis and resistant Fisher F344/N rats [44].

Our literature searches provide another possible reason for induction susceptibility of AA and SCW arthritis, especially in highly susceptible strains which produce long lasting chronic arthritis. We found that *Rattus rattus* and *norvegicus* had been in central China before the 12th century and afterwards in Europe (brought by

Table 2 - Strain susceptibility in the induction of rat adjuvant and streptococcal cell wall-induced arthritis

Rat adjuvant arthritis (*M. butyricum/tuberculosis* i.d. or s.c.)

Strain	Inbred (I) Outbred (O)	Sex	MHC-Class II RT1	Arthritis: Severity Low (L) Medium (M) High (H)	Arthritis: Duration Short (S') Long (L')
Lewis (Lew/N)	I	M/F	l	H	L
Sprague Dawley (SD)	O	M/F	b	H	L
Dark Agouti (DA)	I	M/F	av1	H	–
Wistar	O	M/F	u	H	L
Wistar Furth (WF)	O	M/F	u	M	L
Louvain (LOU/MN)	I	M/F	u	M	–
Holtzman (SD)	O	M/F	–	M	L
Long Evans (LE)	O	M/F	u	M	–
Brown Norway (BN)	I	M/F	n	L	L
Carworth (CFN)	O	M/F	–	M	–
NIMR Hooded	I	M/F	–	H	S
Lewis / (OLAC)	I	M/F	l	H	S
WAG/Rij	I	M/F	i	M	S
WAC/Ry	I	M/F	u	L	S
Buffalo (Buf/1)	I	M/F	u	M	–
Buffalo (Buf/2)	I	M/F	b	L/0	–
Fisher (F334/N)	I	M/F	lvl	0	–
August (AUC)	O	M/F	c	L	–
Wistar-King A (WKA)	I	M/F	k	–	–
SHR	I	M/F	k	L	L

Rat streptococcal cell wall-induced arthritis polyathritis

Strain	Inbred (I) Outbred (O)	Sex	MHC-Class II RT1	Arthritis: Severity Low (L) Medium (M) High (H)	Arthritis: Duration Short (S') Long (L')
Lewis (Lew/N)	I	F	l	H	L
Sprague Dawley (SD)	O	F	b	H	L
Dark Agouti (DA)	I	F	lvl	H	L
Wistar	O	F	u	M	–

Strain	Inbred (I) Outbred (O)	Sex	MHC-Class II RT1	Arthritis: Severity Low (L) Medium (M) High (H)	Arthritis: Duration Short (S') Long (L')
Brown Norway (BN)	I	F	n	M	–
Buffalo (Bual)	I	F	u	L/0	–
Buffalo(Bufn)	I	F	b	L/0	–
Fisher (F334/N)	I	F	lvl	L/0	–
SHR	I	F	k	L	L

a: S = recovers spontaneously after 21 days
b: L = joints undergo ankylosis >30 days with ossification >50 days

traders), and by living in close proximity to humans, were exposed to tuberculosis and streptococcal infections, thereby developing a resistance to these living organisms [70]. This natural immunity (which probably resides in CD4$^+$ lymphocytes and memory T cells) to mycobacteria and streptococcal cell wall has been carried over by immunoresponsive genes to the new inbred and outbred strains derived over the last 150 years from the original wild type rats. This is supported by the fact that Sprague-Dawley and Wistar rats are highly resistant to being infected by live *Mycobacterium tuberculosis*, and survive for 360–400 days after i.v. inoculation [71]. In contrast, these same strains are the most susceptible to the peptidoglycan which produces a severe and long duration polyarthritis. Aksetijevich et al. [72] have advocated this hypothesis in a slightly different way by speculating that the less robust HPA of Lewis rats makes them more susceptible to autoimmune diseases but more resistant to viral or bacterial infections, while the opposite scenario occurs in F344 rats. Since the susceptible strains are highly geared to launch an aggressive T cell mediated response to small amounts of the peptidoglycan which arrive at distal joints by CD4$^+$ lymphocytes, monocytes and macrophages, the interesting yet unanswered question is why does this "homing" of CD4$^+$ (TH1) cells and resident macrophages for these distal joints (ankle and wrists) occur? One can speculate, however, that these weight bearing/actively mobile joints may have small amounts of proteoglycan present from the breakdown/turnover of cartilage in the joint which primes the upregulation of adhesion proteins (e.g. VLA4) on post capillary venules, allowing the homing of the CD4$^+$ T cells and activated macrophages which are carrying the peptidoglycan antigen.

In partial support of this theory, humans with RA have the most severe pathology in the wrist and ankle joints. However, susceptibility of this disease in man likely depends upon several factors including HLA-DR subtypes and non-MHC linked genes, HPA defects, and microbial flora which allow unknown antigen(s) to perpetuate the induction and progression of the disease over time. Due to these similarities, AA and SCW arthritis will be important animal models in the future to help understand the pathogenesis of RA and to develop novel therapeutic agents (e.g. gene therapy and biologicals) with reduced side-effect profiles.

History of drug therapy and current antiinflammatory and immunomodulatory agents

The rat developing AA model was first used by Newbould in 1963 [4] at Imperial Chemical Industries (Zeneca, UK) as a screening model for the discovery of drugs which affect T cell mediated and macrophage driven chronic inflammation. This model has subsequently been used to identify orally active antiinflammatory drugs such as standard steroids (e.g. prednisolone), NSAIDs like phenylbutazone and aspirin, and other early DMARDs like chloroquine and i.m. gold thiomalate. In 1966, Merck [5] demonstrated the usefulness of the developing AA in marketing indomethacin as a drug which had superior potency compared to aspirin and phenylbutazone (e.g. oral ED_{50} for indomethacin in the developing model was about 0.5 mg/kg compared to 200 and 50 mg/kg, respectively, for aspirin and phenylbutazone [5]). However, while the gastroirritational side-effects for indomethacin were observed at doses one- to twofold above the ED_{50} doses in the acute rat carrageenan model of inflammation, gastric and intestinal lesions were not revealed in the developing AA model until doses exceeded the ED_{50} by six fold. Many laboratories have shown that indomethacin dosed at 3–10 mg/kg, p.o., for a few days in developing AA, would kill all the rats due to intestinal perforation. In the 1970s, SmithKline French (SmithKline Beecham), Pfizer, and Syntex used the model extensively to promote the development of auranofin, an oral gold compound, piroxicam, the first one-dose-a-day NSAID in man, and naproxen, which is now sold as an over the counter drug (Aleve™). Also, in the 1980s, etodolac and diclofenac were routinely evaluated in the rat adjuvant model, and today these drugs are still selling in the hundreds of millions of dollars.

Presently, pharmaceutical companies are testing novel NSAIDs called selective cyclooxygenase-2 (COX-2) inhibitors like celecoxib ([73]; Searle) and MK-966 ([74]; Merck), and both companies are expected to file New Drug Applications (NDAs) for these novel agents with the FDA by the end of 1998. COX-2 inhibitors have primarily been evaluated in established AA (see Tab. 3), possibly because our own research at Abbott has shown that this class of drugs is more potent in treating established rather than developing AA. The reduction in gastrointestinal side-effects in man associated with COX-2 inhibitors compared to standard mixed COX-1/2 NSAIDs has been remarkably positive; however, only the analgesic activity of celecoxib and MK-966 in RA and osteoarthritis have been reported.

Table 3 - Relative potencies of standard antiinflammatory drugs in rat adjuvant arthritis

Drug class	**Developing**		**Established**	
	Daily dosing regimen: Days 0–18		Daily dosing regimen: Days 12–29	
	Oral ED_{50} (mg/kg)[a]	**Ref.**[b]	**Oral ED_{50} (mg/kg)**[a]	**Ref.**[b]
Non-selective NSAIDs				
Indomethacin	0.6; 0.3; 0.4	[75–77]	0.3; 1.0; 0.3; 0.1	[15, 76, 78, 79]
Ibuprofen	44;100	[77]; (WA)[c]	75; 65	[80, 81]
Diclofenac	1.3; 0.9	[76, 77]	1.1	[76]
Naproxen	2.6; 8	[76]; (WA)[c]	8.9; 7; ≈7	[76, 82, 83]
Etodolac	≈1.5; 1.1; ≈1	[76, 84]; (WA)[c]	1.2; 0.5; 1.1	[76, 81]; (WA)[c]
Oxaprozin	175; 150	[77, 85]	ND	
Selective COX-2 Inhibitors				
Nimesulide	ND		1.6; 0.2; 0.2	[15, 80, 86]
NS-398	ND		3.3	[I 5]
SC58125	ND		0.2; 0.1	[15]; (WA)[c]
Celecoxib	35% @ 1.0 mpk	(ABT)[d,e]	0,3; 0.6	[73][f]; (ABT)[d,g]
Glucocorticoids				
Prednisolone	10; ≈20	[10, 14]	0.3; 0-3	[80]; (ABT)[d]
Methylprednisolone	5	(WA)[c]	ND	
Dexamethasone	0.008; <0.1	[5; 79]	0.005	(ABT)[d]
Immunosuppressants				
Methotrexate	0.1; ≈0.1; 0.2	[10, 87]; (WA)[c]	inactive @ 2	(WA)[c]
Cyclosporin A	2.4; >l	[88[h], 89]	>5	[88]
Tacrolimus (FK-506)	ND		2.1; 5.6	[16, 90]
Sirolimus (Rapamycin)	2; 0.3; 0.1	[88[h], 91[i]]; (ABT)[d,e]	4.5; 10; 1.2	[16, 88]; (ABT)[d,e]
Mycophenolate	30	(WA)[c]	inactive @ 60 mpk	(WA)[c,h]
Leflunomide	≈6	[92]	ND	
p38 MAP kinase/Cytokine inhibitors				
FR133605	63% @ 32 mpk	[79]	ND	
TAK 603	70% @ 3 mpk	[93]	ND	
SB203580	≈30	[94]	ND	
DMARDS				
d-penicillamine	inactive @ 200 mpk; inactive @ 100 mpk; 37% @ 100 mpk; inactive @ 50 mpk	[95, 79, 89, 96]	potentiated inflammation @ 50 mpk; inactive @ 100 mpk	[96, 97]

Drug class	Developing		Established	
	Daily dosing regimen: Days 0–18		Daily dosing regimen: Days 12–29	
	Oral ED_{50} (mg/kg)[a]	Ref.[b]	Oral ED_{50} (mg/kg)[a]	Ref.[b]
gold thiomalate	50% @ 3 mpk, (i m.); ≈30% @ 10 mpk, i.m.	[89, 96]	≈0.5 (i.m.); ≈30% @ 10 mpk, i.m.	[96, 98]
auranofin	not active @ 10 mpk; 60% @ 3 mpk, i.m.	[79, 89]	ND	

ND, limited or no data available; a, based upon historical data using arthritic index or paw volume for measuring disease severity using Lewis or Sprague Dawley rats, and oral dosing with 0.5% methylcellulose (MC), 0.5% Tween-80/MC, or 0.5% gum tragacanth vehicle suspensions unless otherwise noted; b, references are presented in order for each ED_{50} or note listed for that drug in either developing or established AA; c, unpublished data from Wyeth-Ayerst Laboratories, Princeton, NJ; d, unpublished data from Abbott Laboratories, Abbott Park, IL, using the enclosed protocols under Methods; e, dosed using EPC vehicle (20% EtOH, 30% propylene glycol, 50% D5W with 2% cremophor EL; f, orally dosed b.i.d.; g, dosed using 100% PEG400 vehicle; h, orally dosed MWF (6 doses); i, phosal 50 PG in 1% Tween 80 diluted 1:50 in dd H_2O.

Caution should be noted with regard to the interpretation of data with nonselective NSAIDs and COX-2 specific inhibitors in AA. In 2–3 week oral dosing studies with these inhibitors in AA, nearly complete inhibition of hindpaw edema, and the underlying soft tissue and bony changes can be achieved, even 1–2 weeks after drug therapy has stopped. The major reason for this excellent efficacy is that elevated prostaglandins are very important in driving the pathological changes observed in the soft tissue and bone [21]. Unfortunately, NSAIDs do not have this "curative," or disease-modifying effect in RA [21].

The newer immunosuppressants like sirolimus (rapamycin) and tacrolimus (FK-506) have been tested in the developing and established models, individually and in combination. As shown in Table 3, the oral potency (and efficacy) for sirolimus (ED50s) using a 3-day dosing regimen is greater in the developing model. Inhibitory effects in both models can be obtained using 6–7 doses on an every-other-day dosing regimen (Monday, Wednesday and Friday), achieving results comparable to 14-day dosed, q.d., regimens. The alternate day dosing regimens with these macrolides produced good inhibitory effects with a decrease in side-effects [88].

Combination studies with tacrolimus and sirolimus in established AA [16], and cyclosporin A and sirolimus in the developing AA model [91], have shown additive inhibitory effects, but not synergistic effects as observed in transplantation animal models [99]. The use of a two pronged attack (e.g. tacrolimus inhibiting early activation at T cell receptors through the inhibition of calcineurin and the transcription of IL-2 and the expression of IL-2 receptors, and sirolimus blocking the proliferation of lymphocytes at the late G1 phase of the cell cycle by preventing the sirolimus effector protein from interacting with the p70 ribosomal S6 kinase or possibly autophosphorylation of PI3 kinase) would be a very exciting combination to try using alternate day, low-dose regimens for treating recalcitrant RA and preventing flares. Also, the prolonged efficacy of sirolimus in AA after dosing cessation may help reduce the accumulation of drug and therefore, reduce immunosuppressive-related side-effects including infection and other compromised immune surveillance functions [16, 88, 91]. Overall, these animal models will help demonstrate to the clinicians a better and safer treatment for complicated autoimmune diseases like RA.

By surveying the ED_{50} values in Table 3 for several classes of antiinflammatory drugs using both prophylactic and therapeutic dosing regimens in developing and established AA, respectively, one can see the excellent reproducibility (e.g. NSAIDs and immunosuppressants) generated by different drug companies and research institutions over the last 15–20 years (using the male Lewis and Sprague Dawley rats along with an adjuvant injection of 0.4–0.6 mg/rat in light mineral oil). This reproducibility of AA has allowed drug companies the ability to assess the strengths of their compounds relative to the competition. This model is also very helpful for predicting possible immune-type toxicities in 3–6 month toxicological studies.

Many biologicals which have been used in AA, especially in the developing model, have demonstrated a diversity of immune-related mechanisms that are responsible for the pathogenesis of the chronic inflammation in the distal joints. As shown in Table 4, the importance of $CD4^+$ T cells, cell adhesion molecules and cytokines have been delineated in AA by several investigators [24, 89, 114], including the increased efficacy observed when combination therapy was employed. Combinational therapy of biological agents with the newer immunosuppressants (tacrolimus, sirolimus, and mycophenolate) will probably be the therapeutic regimens in the near future for RA. Presently, many combination therapies are being evaluated in the clinic, including methotrexate, corticosteroid and sulfasalazine, methotrexate and gold, methotrexate and hydroxychloroquine, and the favorable reports on these studies include less toxicity and greater efficacy than single dosing regimens for the individual drugs [115]. Recently, Enbrel™, the soluble TNF receptor p75 Fc fusion protein, and methotrexate have been used in combination in RA to produce greater efficacy than methotrexate alone [116, 117]. Single or combination therapies may also be effectively evaluated in AA using topical formulations [118, 119]. This approach not only targets/concentrates the drug at a particular

Table 4 – Activities of biological agents in rat adjuvant arthritis

Biological	Developing Variable dosing regimen: Days 0–18		Established Daily dosing regimen: Days 12–29	
	Activity	Ref.	Activity	Ref.
Anti T cell				
Anti-CD4 mAb or OX35 (IgG_{2a}) or W3/25 (IgG_1)	++	[100]	ND	
Anti TCR mAb;R73 (aβTCR)	ND		++	[101]
Cytokine antagonists				
Anti-TNFα mAb	+	[102]	ND	
Anti-IL-1β mAb+ anti-IL-1α mAb	–	[102]	ND	
Anti-TNFα mAb+ anti-IL-1β mAb + anti-IL-1α mAb	+	[102]	ND	
Anti-IL-2R mAb	–	[100, 103]	ND	
Cell adhesion/Neutrophil antagonists				
Anti VLA4 mAb + anti CD18	++	[104]	ND	
Anti-VLA4 mAb	+/++	[104, 105]	ND	
Anti CD18 mAb	+	[104]	ND	
Anti ICAM-1 mAb (1A29)	++	[106]	ND	
Anti P-selectin mAb	++	[107][a]	ND	
Anti E-selectin mAb	–	[107][a]	ND	
Anti P-selectin mAb + anti E-selectin mAb	++	[107][a]	ND	
Anti P-selectin mAb + anti E-selectin mAb + anti L-selectin mAb	++	[107]	ND	
Anti neutrophil mAb	++	[108]	ND	
Anti CD2 mAb (OX34)	++	[109]	++	[109]
Cytokine				
rIFNγ	+/–	[110, 111]	+/–	[110]
MHC II antagonist				
Anti-MHC class II	++	[100]	ND	
Heat shock proteins				
Mycobacterial HSP 65	++	[112]	ND	
Nonapeptide HSP 65	++	[113]	ND	
Collagen				
Oral tolerance/Chicken type II collagen	++	[113]	–	[113]

ND, limited or no data available; –, inactive; +, 30–50% inhibition; ++, >50% inhibition; a, based upon neutrophil and monocyte influx into tibiotalar joint on day 14 post adjuvant, at which time antibodies were given prior to labeled leukocytes.

Table 5 - Relative potencies of standard antiinflammatory drugs in rat streptococcal cell wall arthritis

Drug class	**Monoarticular (100P)**		**Polyarticular (10S)**	
	Daily dosing regimen: 48–72 h post i.v. challenge		Daily dosing regimen: 21–28 days post i.p. injection	
	Oral ED_{50} (mg/kg)	**Ref.**	**Oral ED_{50} (mg/kg)**	**Ref.**
Non-selective NSAIDs				
indomethacin	0.2	(ABT)[a]	ND	
Selective COX-2 inhibitors				
nimesulide	0.3	(ABT)[a]	ND	
Glucocorticoids				
dexamethasone	<0.1	(ABT)[a]	≈0.1 mpk, s.c.	[123]
Immunosuppressant				
methotrexate	ND		20% @ 0.125 mpk	[32]
cyclosporin A	ND		>80% @ 25 mpk	[32, 124]
DMARDs				
d-penicillarnine	ND		inactive @ 200 mpk	[32]
gold thioglucose	ND		inactive @ 5 mpk, s.c.	[32]

a, unpublished data from Abbott Laboratories using protocols described under Methods; ND, data not available

joint/site, but may also reduce the metabolic liability and side-effects associated with systemic administration of the drug. In the future, AA will also be used to assess the efficacy of gene therapy after local or systemic administration into joints. These therapies will be used to evaluate the modulation of the many different lymphocyte populations, antigen presentation cells, and inflammatory cytokines/mediators involved in this disease [120, 121].

The SCW model, especially the 10S polyarthritis model, has not been used extensively by drug companies to screen antiinflammatory drugs due to its variable arthritic response and impaired liver function. However, Wahl et al. [122] has shown using the purified 10S (PG-PS) that a very steady arthritic response can be observed from day 20–30 during the established disease. However, very few drugs during the past 20 years have been used in 2–3 week daily regimens in the developing phase of the 10S polyarthritis model, making it difficult to compare reproducibility among laboratories. Reported ED_{50} values for this model are listed in Table 5. Also, unlike for AA, no publications comparing several classes of antiinflammatory drugs in the systemic 10s polyarthritis model could be found after an

Table 6 - Activities of biological agents in rat streptococcal cell wall arthritis

Biologicals	Monoarticular (100P)		Polyarticular (10S)	
	Daily dosing regimen: 48–72 h post i.v. challenge		Daily dosing regimen: 21–28 days post i.p. injection	
	Activity	**Ref.**	**Activity**	**Ref.**
Anti-T cell				
Anti CD4 mAb W3/25 (IgG1)	ND		++	[125]
Anti-TCR mAb (R73)	ND		++	[126]
Cell adhesion inhibitors				
Anti-ICAM mAb	+	[127]	ND	
Anti-P-selectin mAb (PB 1.3 (Cytel))	++	[127]	ND	
Cytokine & chemokine antagonists				
Anti-IL-1α mAb	++	[127]	ND	
rhIL-1ra	++	[128]	ND	
Anti-TNFα mAb	++	[127]	ND	
Anti-MIP-2 mAb	–	[127]	ND	
Anti-MCP-1 mAb	++	[129]	ND	
Anti-TGFβ-1 mAb	ND		++	[130]
Anti-IL-4 mAb	+	[131]	ND	
Anti-IL-10 mAb	–	[131]	ND	
Anti-IFNγ mAb	–	[131]	ND	
Heat shock proteins/Cytokines				
mycobacterial HSP 65	ND		++	[132]
TGFβ-1	ND		++	[133]
rL-4	ND		++	[134]
rIFNγ	ND		++	[135]

ND, limited or no data available; –, not active; + 30–50% inhibition; ++ >50% inhibition

extensive search of the literature. A factor which likely contributes to this phenomenon is that the material costs for running the 10S PG-PS polyarthritis model as a screen can be exorbitant.

The 10S PG-PS polyarthritis model has primarily been used by academic scientists to test biological agents on disease mechanisms (Tab. 6) and on the effects of the HPA axis following the induction of SCW arthritis in Lewis and Fisher 344 rats [44, 60]. In contrast, the monoarticular 100P PG-PS SCW model is now being used by more investigators in academia and the pharmaceutical industry because of the increased reproducibility of the arthritic response after i.v. challenge, and because the joint pathology involving cartilage and bony changes is more moderate than observed with the 10S polyarthritis SCW and AA models. (See Tabs. 5 and 6 for details.)

Methods

Rat adjuvant arthritis

General considerations

In our laboratory, we use male Lewis rats with an arrival weight of 150–170 g, from Charles River or Harlan Laboratories. Lewis rats can be difficult to obtain, depending upon the number ordered and the time of year, so when we order 250 animals for a typical adjuvant study, we order the animals at least a month prior to shipment to ensure delivery. If the model is used on a routine basis, a standing order with the vendor can also be initiated to prevent delays. We find it is better to run a large study of rats with the same weight specifications, to allow inclusion of all the proper controls (e.g. normal, mineral oil injected, adjuvant positive controls ± test drug vehicle), including pharmacological internal controls (standard drugs or lead compounds or both), and experimental agents over a wide range of doses. In terms of the number of rats/group, we use 10–16 rats/group depending on how many terminal procedures will be performed throughout the disease process, and how many animals are needed for rebound studies which measure carryover drug effects. A minimum of eight rats/group for paw edema is suggested for the established model to determine drug efficacy and ED_{50}s.

When the animals arrive at our laboratory, they are quarantined for one week in our barrier facility prior to use, and sentinel rats of the same strain, size, sex, vendor and source room are monitored for viral infections. Body weights typically increase during this period to 175–200 g at the beginning of the experiment. Small-

Figure 2
A. Changes in body weight in normal and AA rats over time. Data represent mean body weight changes from seven separate AA studies using 10–16 rats/group. B. Weight gain profile comparison between normal and AA rats over three successive, 2-week intervals. Data represent differences in mean body weight (g) ± SEM (day 15–day 0; day 30–day 15; and day 44–day 30) from weight data generated studies described in Figure 2A. Asterisks indicate significant weight changes compared to the first two weeks of the study (p≤0.005 using Dunnetts multiple comparison tests; NS indicates no significant difference from changes seen during the first two weeks).

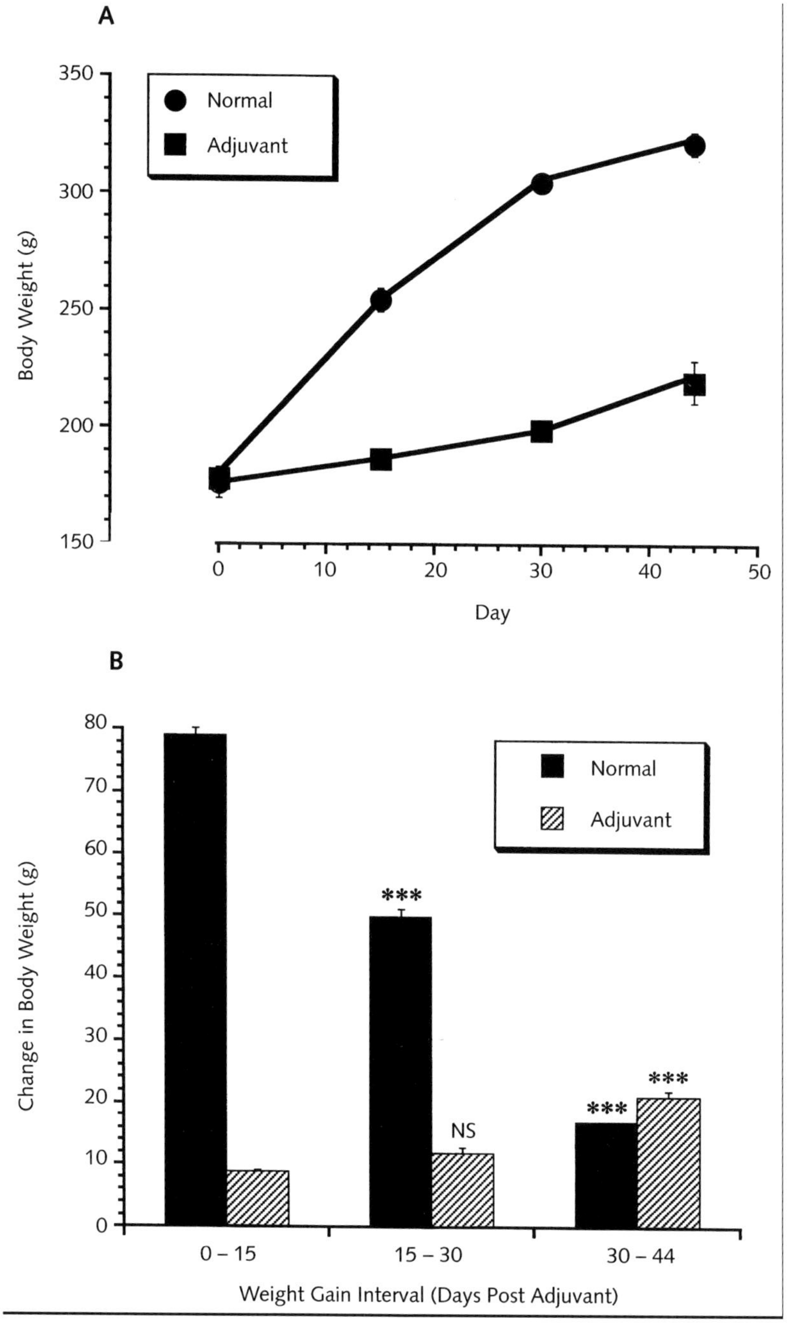
A
Normal
Adjuvant
Body Weight (g)
350
300
250
200
150
0
10
20
30
40
50
Day
B
Normal
Adjuvant
Change in Body Weight (g)
80
70
60
50
40
30
20
10
0
***
NS
***
***
0 – 15
15 – 30
30 – 44
Weight Gain Interval (Days Post Adjuvant)

er rats can be used, but it must be kept in mind that adjuvant rats gain very little weight over a 30-day period (< 10 g compared to 90–120g for a normal rat; see Fig. 2), and if any treatments are imposed upon these animals which affect weight, especially glucocorticoids and most immunosuppressives, the animals will actually lose weight compared to their starting weight at day 0. It is our practice, and insisted upon by our institutional animal care committee, that in the rare event an animal loses 20% of its original starting weight, the rat will be euthanized. It is also important to house chronically treated animals in the same area under similar conditions, along with 5–10 sentinel rats of the same age, sex, and strain for routine viral checks and blood analysis. Good laboratory procedures (including clean scrubs, jumpsuits, latex/nonlatex gloves, booties, safety glasses/goggles, and masks) should always be practiced for the investigators' safety (e.g. allergy prevention), the animals' welfare, and to ensure that the significant monetary and time investment of these chronic studies pays off and yields quality data. Additional concerns which should be addressed include regular inspections of water sources and elixirs for endotoxin contamination (E-Toxate limulus ameobocyte lysate test, Sigma, St. Louis, MO), adding double or triple bedding to the cages to minimize stress on diseased animals' joints, changing the bedding every other day, and importantly, making sure that as the disease progresses, that animal caretakers or investigators place food pellets directly on the bedding in consideration of reduced animal mobility.

A final word regarding the care of the animals. Most animal care workers/investigators who have not worked with this model may not be prepared for the morbidity associated with the physical appearance and progressive and destructive nature of this arthritic disease. The inflammatory responses in the paws are severe, with the enlarged paws at times developing skin ulcerations which can become infected. The rats also become quite irritated after the injection of adjuvant until about day 8, and fighting among animals in the cage is not uncommon during these first eight days following the injection. During this stressful period, what appears as a "bloody" discharge from the eyes and nares may be seen, along with a pinkish tint to the fur on the back. This condition is known as chromodacryorrhea, and is caused by the ocular secretion of a porphyrin-containing pigment from the Harderian glands, which are located deep within the orbits. The functional role of this secretion is not known, but chromodacryorrhea is a normal response to stress, especially under conditions of a chronic disease such as AA.

As the disease progresses beyond day 10, the animals' hindlimbs become immobile and they drag themselves around the cage by their front limbs. Between days 10 and 20, during the fulminating and rapidly developing inflammatory stage of the disease, the animals exhibit numerous pain responses (e.g. constant squeaking, ruffled yellowish/pinkish fur, and increased sensitivity to handling) and every measure which can reduce the stress on these animals should be taken (e.g. frequently changed double bedding, easily accessible food, water etc.). During the later chronic stage of the disease (day 30–45), the inflammatory response subsides (evident from decrease levels of circulating acute phase proteins and neutrophils), the joints become fibrotic, ankylosed, ossification occurs, and while the rats remain in a severely arthritic condition, they do not appear to be in as much pain or distress, and they begin to gain weight.

Adjuvant preparation and injection technique

The following procedures outline how the rat AA model is initiated in our laboratory as well as in other pharmaceutical companies. On day 0, male Lewis rats (150–170 g, Harlan Laboratories, Indianapolis, IN) are numbered (1–5 permanent marker slashes near the tip of the tail if the rats are housed five to a cage; left and right ear punches or indwelling transmitters can be used as more permanent means of identification), individually weighed, and the right and left hindpaw volumes (in ml) are measured using a Buxco semiautomated water plethysmograph (Buxco Electronics, Inc., Troy, NY). The right hindpaw volumes are rarely used for final calculations (due to the complex inflammatory reaction in the injected paw), but we typically measure the right paw anyway because: 1) it only takes a few seconds more to dip the right hindpaw, and 2) if the right hindpaw volume data is available, it can be used later if more comparisons are required. The rats then receive a 0.1 ml subplantar (s.c.) injection in the right hindpaw with a light mineral oil (Sigma, St. Louis, MO) suspension containing 0.5 mg of heat-killed, desiccated *Mycobacterium butyricum* (Difco Laboratories, Detroit, MI); vehicle control animals receive a subplantar injection of light mineral oil in the right hindpaw. The adjuvant dose prescribed above, where 0.5 mg of *M. butyricum* causes 80–85% maximal severity, is based upon literature references and previously determined in-house dose-response curves.

The procedure for preparing the mycobacterial suspension in mineral oil is very critical for high incidence of disease and reproducibility. Therefore, all personnel preparing the adjuvant should be properly and similarly trained. The mycobacteria is accurately weighed, and 50 mg is placed in a 15 ml glass homogenizer tube. Ten ml of light mineral oil is then pipetted into the tube. It is important not to add the light mineral oil to the tube first, because when the mycobacteria is added, it sticks to the top of the oil-coated tube and will not be completely mixed in the final volume. Once the oil has been added to the mycobacteria, the preparation is homogenized using a T-Line laboratory homogenizer (Model # 103; Montrose, PA) with a motorized Teflon pestle. This procedure should be done under a hood with the investigator using heavy cotton gloves (glass homogenizer tubes can break in ones hand), goggles and an appropriate respirator mask; potentially dangerous aerosols are generated when using this procedure. While using this homogenizer on mid to high speed (at a setting between 60 and 90 on the homogenizer described above), carefully move the tube up and down to the bottom of the tube, thereby forming a very fine suspension of the mycobacteria. The glass homogenizing tube can become very warm due to a tight fitting pestle and/or friction of the oil, but should be tolerable for 10–15 up/down cycles. When finished, we combine all homogenization samples of adjuvant from the individual tubes (if more than 10 ml are made) into a 50 ml polypropylene centrifuge tube. This tube is then rapidly vortexed, afterwhich the adjuvant is rehomogenized, 10 ml at a time in one of the glass homogenizer tubes used previously, for 10 additional cycles, and then recombined back into the 50 ml centrifuge tube. This procedure will produce a superfine suspension of *M. butyricum* in light mineral oil which remains suspended for at least an hour. However, a small amount of the *M. butyricum* will settle out over time, but a vigorous shake or vortex will resuspend the material adequately. We have tried other procedures such as stirring the adjuvant overnight in the oil using a magnetic stirrer in a glass Erlenmeyer flask, autoclaving for 20 min, grinding in a mortar before adding to a stirring flask, or sonicating, but the homogenization procedure described above works well and provides consistent results.

For injections, we use a 1/2", 25 gauge beveled stainless steel disposable needle attached to a 1 cc disposable polypropylene syringe. The combined preparation of adjuvant is vigorously shaken by hand in the 50 ml centrifuge tube, drawn up into the syringe, followed by attachment of the needle, and air displaced by pushing a small amount of the suspension out of the needle. At this point, the rat should be gently wrapped in a blue disposable diaper/absorbent pad, snugly placed under the arm with the rat's right hindpaw and tail held firmly by the left hand (if you are right handed). The needle of the syringe is then inserted at the front of the hindpaw, bevel up, into one of the two middle foot pads (this helps to prevent oil from leaking out from the injection site when the needle is withdrawn), and then moved under the skin approximately 8–10 mm toward the heel. One hundred µl of the fine suspension are injected, and the needle is held in place for 1 or 2 sec before withdrawing to further prevent oil from leaking out the injection site. Approximately 200 rats can be injected in this manner by an experienced technician in less than 3 h. Within 2 h after the injection, an acute inflammatory reaction is apparent and the response progresses as described under the *Pathogenesis* section of this review. Figure 1 illustrates the edema time-course in rat adjuvant arthritis, and the times at which various inflammatory parameters are measured and treatments initiated. Figures 3 and 4 illustrate the MRI/histopathological correlates for disease progression. Some laboratories have reported using subdermal tail injections to induce rat adjuvant arthritis, but the variability and necessity of measuring both hindpaws for edema evaluation makes the tail injection approach less desirable.

While few pharmaceutical or academic laboratories utilize MRI as an approach/technique for monitoring disease progression and therapeutic efficacy of experimental agents, MRI has distinct advantages over traditional inflammatory measurement parameters or other imaging techniques such as x-ray. In the context of arthritis research in small animals (rats and mice), MRI can noninvasively image arthritic joints in three dimensions over time in live animals. This feature provides important insight for the investigator regarding the state of the disease prior to the initiation of drug therapy, and importantly, after the cessation of drug administration, where the disease-modifying potential of the therapeutic intervention can be evaluated in the same animal over time. MRI provides three-dimensional anatomical/pathological details of bone and soft tissue (especially with advanced imaging techniques such as fat suppression, fractal analysis, gadolinium contrast enhancement, etc.) which are unattainable by other imaging techniques, and which can readily be correlated to and confirmed by histological analysis at the end of the study. While MRI will not replace semiautomated water plethysmography as a rapid and inexpensive means of measuring changes in hindpaw volume, our experience at Abbott Laboratories has reinforced the idea that the apparent antiinflammatory effects of a drug as reflected through paw volume measurements need to be confirmed by several inflammatory parameters including neutrophil counts, acute phase proteins, body weight, histopathology, and MRI [15, 16, 136]. MRI data from these studies can provide a striking visual and quantitative demonstration of the changes in soft tissue and bone structure before, during and after drug treatment. These effects, unlike data from paw volume or even histopathologic analysis, can be correlated with data obtained later during clinical trials, where MRI is being used more routinely as a diagnostic tool in RA.

Culling/Dosing protocols: Developing vs. established

As discussed, the adjuvant model can be run in two modes – developing and established. The developing model, in which animals are dosed between day 0 and 15 and evaluated on day 16, is a less rigorous model for evaluating the effects of drugs on paw volume, but is more sensitive than the established model for looking at drug effects on afferent immunological mechanisms resulting in soft tissue inflammation and synovitis. The developing model is, however, inherently more variable than the established model since the animals are not culled prior to dosing. One procedural note regarding the developing model which should be practiced by technicians regardless of experience is to randomize the animals after injection of the adjuvant prior to dosing. This should be done in case one or more animals were not injected properly for one reason or another. In general, 10 rats/group are suggested as a minimal number of animals when using the developing protocol for the determination of ED_{50} values. If the developing model is used as a screen to detect ≥30% inhibition of control paw volume by some intervention, then 6–8 rats/group may suffice. Finally, a day 4 adjuvant injected right paw for evaluation of subacute inflammation has been utilized by some laboratories to determine antiinflammatory activity of some drugs, and to discriminate that there is no acute antiinflammatory activity as observed for immunosuppressants like rapamycin. However, this subacute inflammation in AA (< 5 days post adjuvant injection) is difficult to suppress in a dose-related manner and for evaluating drug potency.

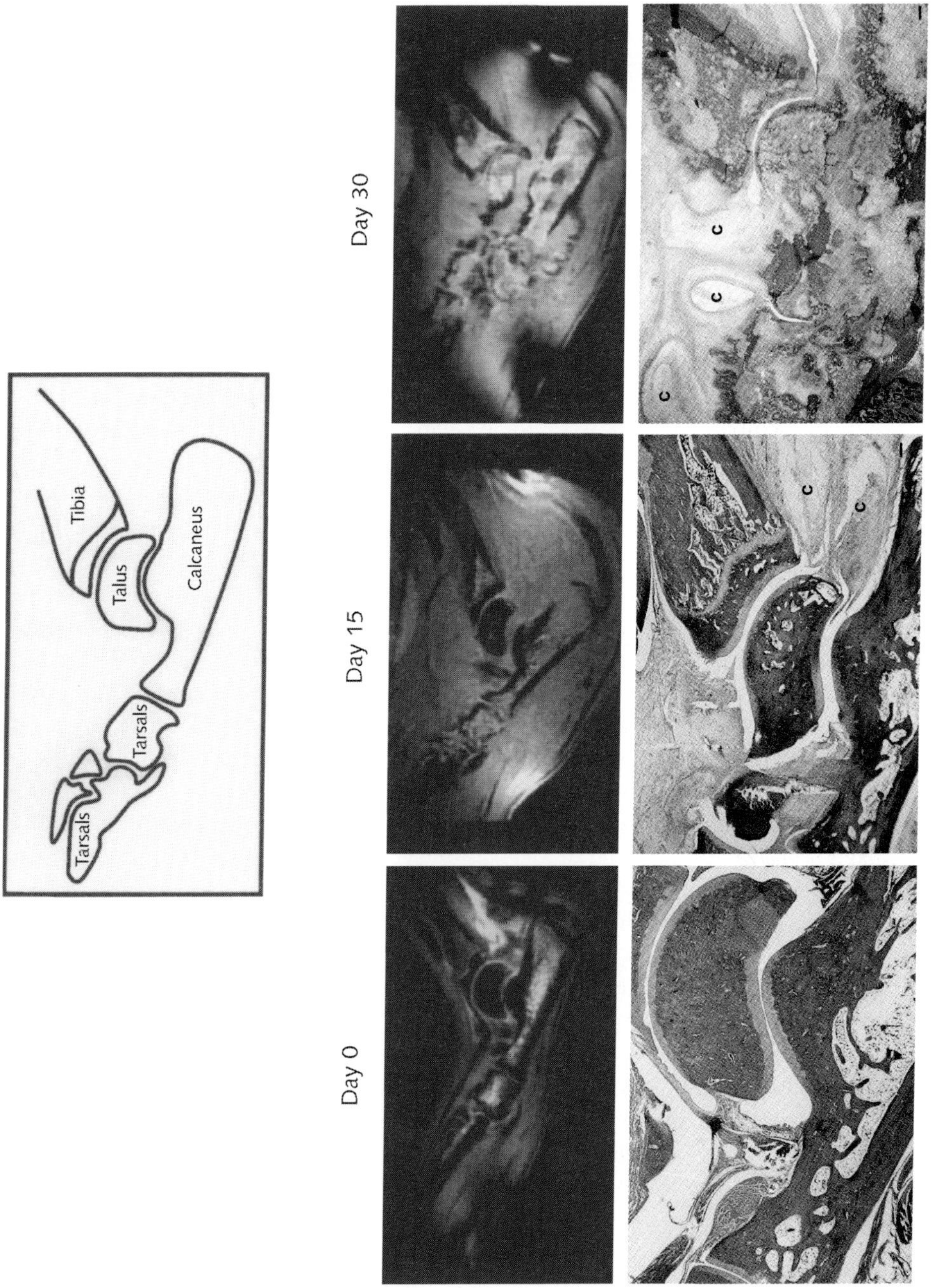
Tibia
Talus
Calcaneus
Tarsals
Tarsals
Day 0
Day 15
Day 30
c
c
c
c
c

The established model still exhibits pronounced soft tissue and synovial inflammation between day 16 and 30, but is accompanied by a marked progression of periosteal reactions, pannus formation, internal bone inflammation, fibrosis in the joints, and eventually, ankylosis (see Figs. 3 and 4). Using the established protocol, animals are initially weighed, hindpaws measured, and injected as described above, but immediate randomization following adjuvant injections is not necessary as in the developing model. Measurable swelling begins at approximately day 10 in the contralateral, uninjected left hindpaw. This process is very predictable and consistent, and is one of the most important reasons why AA is used extensively in drug discovery. On day 15, the rats are weighed again, and the left, uninjected hindpaw volumes are measured by semiautomated water plethysmography. There is no need to measure the right hindpaws at this time; left hindpaw volumes on day 15 are used only for selection/culling. Once the left hindpaws are measured, the edema is calculated (day 15 minus day 0 left hindpaw volumes), and the rats which exhibit a left hindpaw edema between 0.75 and 2.0 ml are randomized

Figure 3
Magnetic resonance imaging (MRI) with histopathological correlates depicting the pathogenesis of rat adjuvant arthritis. The cartoon illustrates the anatomy of a sagittal section through a rat hock, with the distal tibia, talus, calcaneus, and surrounding tissue as the primary landmarks used in our laboratory for disease evaluation. Day 0 represents a normal, untreated rat hock. Fifteen days after intradermal injection of Freund's complete adjuvant in the right hindpaw, significant soft tissue edema is noted in the uninjected left hindpaw (see Fig. 1 for time-course), including the presence of synovial cysts (distensions of the joint capsule); these are only apparent in t2 weighted or gadolinium-enhanced MRI images (not shown), but are evident in the histological photomicrograph (indicated by "C"). These cysts contain extensive numbers of neutrophils and mononuclear cells, with large accumulations of fibrin. Tibiotarsal joint space changes and fibrous connective tissue are somewhat variable at this time, and while periosteal proliferation is histologically evident at day 15 (judged from the presence of numerous osteoblasts lining immature osteoid), it is not seen in all rats. Between day 15 and 30, bony changes predominate the disease even as the earlier fulminating inflammatory response begins to subside. MRI and histology confirm that periosteal new bone formation is marked, especially at sites of tendon insertion (upper distal end of the tibia, and the lower distal end of the calcaneus). Periosteal proliferation is not seen in the talus, but internal bone inflammation in the talus is significant between day 30 and 45. Extensive invasion of the joint spaces with fibrous connective tissue and pannus is evident, leading to ossification and nearly complete ankylosis at day 45. Some synovial cysts remain on day 30 as seen in the photomicrograph (indicated by "C"), but in general are smaller, less numerous, and more fibrotic and organized in nature than day 15 cysts. Total magnification of histological sections is 10×, where the black bar represents 200 µm. Staining is by hematoxylin and eosin.

Normal

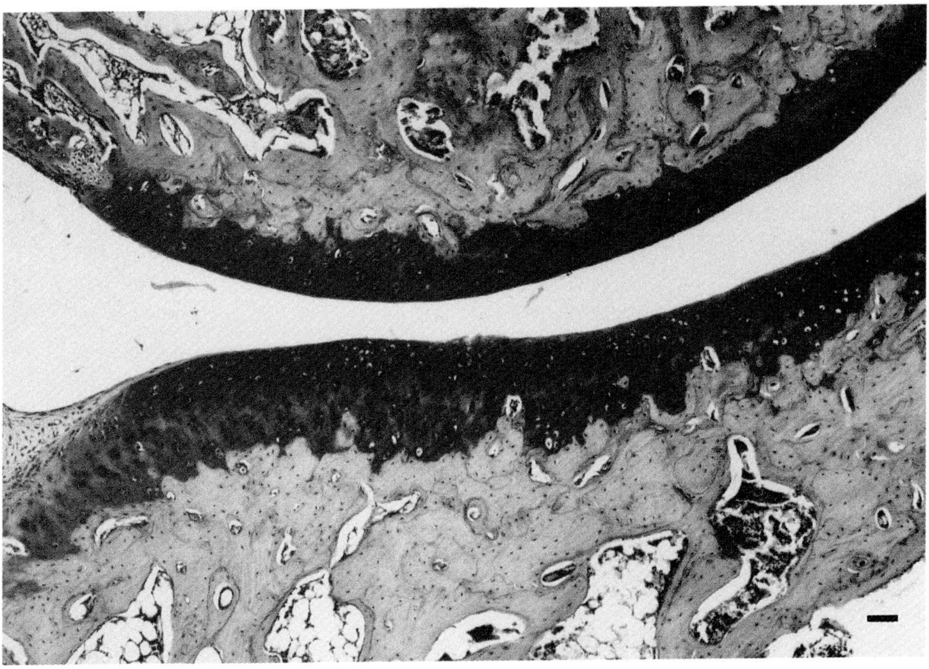

Day 30

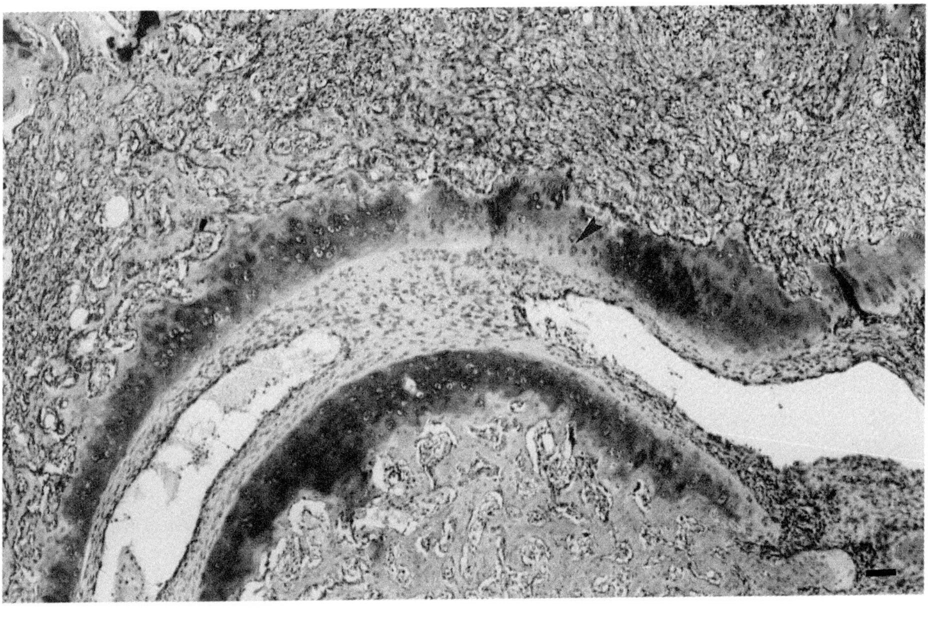

and used for the remainder of the study. The rats which do not exhibit an inflammatory response in the left hindpaw (<2% of total rats injected) and those which display a super-response to the antigen (>2ml of edema on day 15) are typically sacrificed following culling. These exclusion criteria result in approximately 25% of the starting number of animals being culled from the study (this should be factored into the initial animal order). The rats which remain are randomized and then renumbered on the tail (higher up with a different color or ear punched) and put into their new cages at five rats per cage. We have also found it helpful to double-check each cage as the rats are renumbered and moved into new cages to avoid confusion at the end when rats may be found "missing." Once the culling process has been completed, the animals are allowed to acclimate until the next day (16) when dosing begins (day 16-day 29). Rats should also be dosed at the same time of day throughout the dosing period. For example, glucocorticoids (e.g. prednisolone, dexamethasone) should be dosed in the afternoon when the natural corticosterone levels rise in rats (opposite to humans where cortisol peaks in early morning). This reduces the chances of interrupting the diurnal responses of the HPA axis, and simulates the way synthetic glucocorticoids are clinically prescribed for use. In both developing and established AA, rats are measured by semiautomated water plethysmography the day following the last dose, and then may be monitored for an additional period of time (7 and 14 days post drug cessation) to evaluate any rebound inflammation which occurs once the animals have been taken off the therapeutic regimen. In addition to hindpaw volume, a number of other biochemical and cellular parameters can be used to evaluate the immune status/inflammatory state of the animal; these are listed in Table 7. The use of one or more of these supplementary disease indicators is important, especially for evalu-

Figure 4

Sagittal histological sections of the tibiotalar articulation from a normal rat left hindpaw, and from the left uninjected hindpaw of a rat 30 days after a s.c. injection of Freund's complete adjuvant in the contralateral hindpaw. Sections have been stained with hematoxylin and eosin, followed by Safranin-O to identify glycosaminoglycans (GAG) in the articular cartilage. As seen in the normal joint, the joint space is clear of any fibrous connective tissue, GAG staining is of uniform thickness and intensity, and internal bone structure is normal. In contrast, the day 30 joint shows invasion of the joint space by fibrous connective tissue (pannus), and where it makes contact with the articular cartilage, significant erosion of the cartilage is evident and overall levels of articular cartilage GAG are also lower (less intense staining as indicated by arrowhead). Finally, the internal structure of the tibia and talus in the day 30 adjuvant joint is characterized by extensive loss of normal architecture, replacement by immature (woven) bone, and infiltrating neutrophils and mononuclear cells. Similar histopathology has been previously published [53, 137–139]. Final magnification is 40x, with the black bar representing 50 μm.

Table 7 - Supplementary inflammatory measurement parameters in rat adjuvant arthritis

Parameter	**Description**	**Comments**
complete blood count (CBC)	circulating neutrophils, lymphocytes, monocytes, eosinophils, basophils, platelets	30 µl 0.5 M EDTA/470 µl blood; 100–200 µl required for analysis using Abbott Cell-Dyn 3500
fibrinogen	acute phase protein	50 µl 0.105 M sodium citrate/450 µl blood; 100 µl plasma required/sample; analyzed on a Sysmex CA-1000 using fibrinogen kits from Sigma Chemical, St. Louis, MO
haptoglobin	acute phase prolein	5–50 µl serum required/well; Kallestad immunoradial diffusion kit from Sanofi Diagnostics, Chaska, MN
serum chemistry	liver enzymes, albumin, globulin, creatinine, electrolytes, glucose, etc.	100–200 µl serum required; use mottled-top Vacutainer tubes which contain silica activator; samples analyzed on an Abbott Spectrum
urine chemistry	electrolytes, protein, creatinine, etc.	100–200 µl urine from a 24 h urine collection; all urine analytes need to be normalized to urine creatinine; samples analyzed on an Abbott Spectrum
pyridinoline and deoxypyridinoline	urinary collagen crosslinks; bone resorption markers	100 µl urine from 24 h urine collection; normalized to urine creatinine; Metra Biosystems, Mountain View, CA
rat-specific ICTP and PIIINP	serum teliopeptides; bone formation/resorption markers	100 µl serum; Incstar, Stillwater, MN
corticosterone	endogenous corticosteroid; measure of adrenal function/HPA inte~rity	50 µl plasma (30 µl 0.5 M EDTA/470 µl blood); RIA for rat/mouse corticosterone from Amersham; RIA does not cross-react with prednisolone
ACTH	adrenocorticotropic hormone; secreted from anterior pituitary; measure of HPA integrity; negatively regulated by corticosterone/glucocorticoids	50 µl plasma (30 µl 0.5 M EDTA/470 µl blood); IRMA assay for rat/mouse/human ACTH from Nichols Institute, San Juan Capistrano, CA

Parameter	Description	Comments
IL-6, TNFα, IL-1β	inflammatory cytokines	plasma or synovial exudate; rat specific cytokine ELISAs available from R&D Systems, Endogen, and others
lymphoid tissues;	thymus, adrenal, spleen adrenal glands	wet weights (normalized to 100 g body weight); adrenals enlarge in absence of exogenous glucocorticoids; thymic involution observed as a consequence of elevated corticosterone; splenomegaly due to granulomatous reactions to heat-killed mycobacteria
histopathology	uninjected left hock	tarsal/metatarsal joint of left leg harvested, stripped of skin, placed in 10% buffered formalin for 24 h, followed by double distilled water rinse and storage; bone then decalcified, sectioned and stained (hematoxalin and eosin; safranin-o for glycosamino-glycans in cartilage. Brief fixation time is sufficient and preserves delicate immunohistochemical and histological features (e.g. synovial cysts)
magnetic resonance imaging (MRI); X-ray	uninjected left hock	if available, MRI provides valuable three-dimensional information re soft tissue and bone, prior to and after drug treatment; good dental x-rays also useful but less sensitive than MRI

ating antiinflammatory activity of drugs, because paw volume or edema is not always indicative of changes in immune function, joint disease, and pharmacological efficacy.

Streptococcal cell wall-induced arthritis

General considerations

Many of the comments regarding experimental procedures made in the previous section for AA are applicable to this and other animal models of inflammation. However, some general points need to be made at this time regarding which SCW model is chosen for routine use. The 10S model of systemic SCW-induced arthritis is probably the easier of the two models to run in less experienced hands (the other being the 100P model requiring intraarticular and i.v. injections). However, the 10S PG-PS model is the less predictable of the two SCW models, the most severe in terms of systemic involvement of the antigen and the severity of the disease, and is without question the most expensive.

The 10S model is initiated by a single i.p. injection, and the increase in hindpaw volume (edema) is monitored either subjectively by using an arbitrary arthritic index, or can be quantitated with a high precision caliper or semiautomated water plethysmography. While i.p. injections are a convenient and expedient way of introducing nonsoluble or soluble drugs/antigens into animals, if they are not done correctly and consistently, numerous adverse events can result. These include: (1) development of a counter irritant response which can result in the diminution of some distal inflammatory response, (2) variable rates of absorption in the abdomen depending upon the size of the animal and location of injection, and finally, (3) if the intestines, bladder, or other organs are punctured during the i.p. injection, peritonitis can rapidly ensue. A procedure which helps guard against this is to hold or lay the rat on its back at a 45° vertical position allowing the intestines and organs to fall away from the ventral surface. The injection should then be made close to the leg at the base of the peritoneal cavity.

Cost has been mentioned as a potential limiting factor in choosing between the 10S and 100p models. Stimpson and Schwab provide a detailed procedure for generating and processing ones own SCW preparations [63], but we find it much easier to purchase the material from commercial sources (Lee Laboratories, Grayson, GA). The current 1998 SCW prices for running the 10S PG-PS model using 100, 200 g rats would be nearly $ 5,400.00 (@ $ 18.00/mg 10S PG-PS; 300 mg required), but only $ 175.00 (@ $ 50.00/mg 100P PG-PS; 3.5 mg required) for the 100P model. This is approximately 30 times less expensive than the 10S model to run for the same number of rats! For comparison, the total cost for heat-killed desiccated *Mycobacterium butyricum* in an adjuvant arthritis study of the same size is roughly $ 7.00 (Difco: $ 87.00/600mg *M. butyricum*; 50 mg required).

The final choice regarding what SCW model is chosen depends not only on issues such as cost, but what type of disease is required for study. As discussed in previous sections, the 10S model promotes pathological changes in multiple tissues as well as the joints, is not as predictable as rat AA or the monoarticular 100P PG-PS SCW model in terms of secondary inflammatory responses, and has comparable histopathology in the bone to rat AA. The 100P model, on the other hand, is basically a delayed-type hypersensitivity response localized to the joint, whose pathogenesis is very predictable, and limited in duration and severity. Both SCW models, however, have not been used to a significant extent for drug discovery in academics and industry, and available information is limited in terms of the pharmacological and antiarthritic effects of classical antiinflammatory and immunosuppressive drugs in either system.

10S PG-PS systemic SCW model

As previously mentioned, the 10S PG-PS SCW model is very simple to initiate. The 10S PG-PS material purchased from Lee Laboratories is formulated as a stock solution in pyrogen-free saline, usually between 3–6 mg rhamnose/ml. According to Lee Laboratories' recommendations, as well as most published methods, the disease is initiated by injecting i.p. approximately 5 μg rhamnose from 10S PG-PS preparations/g body weight in a final injection volume of saline between 0.5 and 1 ml, using a 1 cc disposable polypropylene syringe with a 1/2", 25 g needle. The diluted SCW material should be briefly sonicated for no more than 1–2 min, in a glass tube/vial immersed in a Branson sonicating water bath; this procedure helps form a uniform suspension of SCW material. Rats do not need to be anesthetized for the i.p. injection procedure if it is done correctly. A vehicle control group should also be included which receive an i.p. injection of pyrogen-free saline. Rats should be weighed weekly throughout the study. Both hindpaw volumes should be measured every other day for the first 8–10 days when the model is first being developed in a laboratory so that the two phases of the disease can be clearly delineated (the group mean of the averaged right and left hindpaw volumes for each animal on a given day are typically used for disease and drug effect evaluation); thereafter, measurements can be taken every week depending upon the endpoints being evaluated. It should also be noted that too much handling (weighing, measuring paw volumes, blood samples) can induce added stress to the animals, increasing plasma corticosterone levels which may result in decreased inflammatory and immunologic responses.

Drug treatment regimens may be administered once the time-course of the disease has been established. For example, drugs may be administered from day 0, approximately 1 h prior to i.p. injection, through the end of the assay (e.g. day 30 or beyond), or they may be administered just prior to and during the secondary inflammatory response (from day 10–29). If a therapeutic rather than a prophylactic/developing protocol is used for drug therapy, culling of the rats followed by randomization may be possible between day 6 and 10 based upon those animals which respond during the first 5-days. We have found that > 70% of the rats injected i.p. with 10S PG-PS using the aforementioned conditions exhibit an average hindpaw volume (left + right hindpaw volumes/2) typically ranging between 2.3 and 2.6 ml by day 5. The timing and severity of the secondary phase of the disease can be somewhat variable depending upon the lot number of PG-PS, even when purchased from the same vendor. Edema (increased paw volume) can be measured by clinical score evaluations, high precision calipers, or more effectively with semiautomated water plethysmography. As previously mentioned, there is greater variability in the hindpaw volumes (e.g. day 30) among rats, even after culling, in the 10S PG-PS model compared to the rat AA model, but we have not seen significant differences between the right and left hindpaw volumes in the same animal in the 10S model. An example of the edema time-course seen in one representative experiment is shown in Figure 5, where typical day 30 changes in paw volume (after culling) can range from 0.75–2.5 ml. In Figure 6, the variability of the 10S PG-PS model is illustrated by two MRI images of randomly chosen right hindpaws from two rats injected i.p. on day 0 with 10S PG-PS, and which had similar day 5 mean hindpaw volumes (2.45 and 2.52 ml). While the MRI images show marked differences in soft tissue changes (edema) between the two rats, periosteal new bone formation and internal bone inflammation are clearly advanced in both rats.

The cellular and biochemical measurement parameters listed in Table 7 for AA may be applied to the SCW model. Other potential indices of disease severity which may be of interest to monitor include expression of adhesion molecules on leukocytes and in venules of the synovial membrane [127, 140], cell influx into the synovial space [127], glycosaminoglycan levels in the articular cartilage of the talus [141, 142] and granuloma formation in the liver [34].

100P PG-PS monoarticular SCW model

The 100P PG-PS material purchased from Lee Laboratories is formulated as a stock solution in pyrogen-free saline (approximately 5–8 mg rhamnose/ml). While the

Figure 5
Time-course of rat 10S-PG-PS SCW polyarthritis and general procedures associated with its implementation and potential use for evaluating disease progression and drug efficacy. See Methods for details. Data points represent the group mean ± SEM for averaged individual hindpaw volumes (for each individual rat, average hindpaw volume (ml) = left + right hindpaw volume /2; n=10 rats/data point). Rats were culled on day 8 as outlined under Methods; data prior to day 8 was generated using nonculled animals.

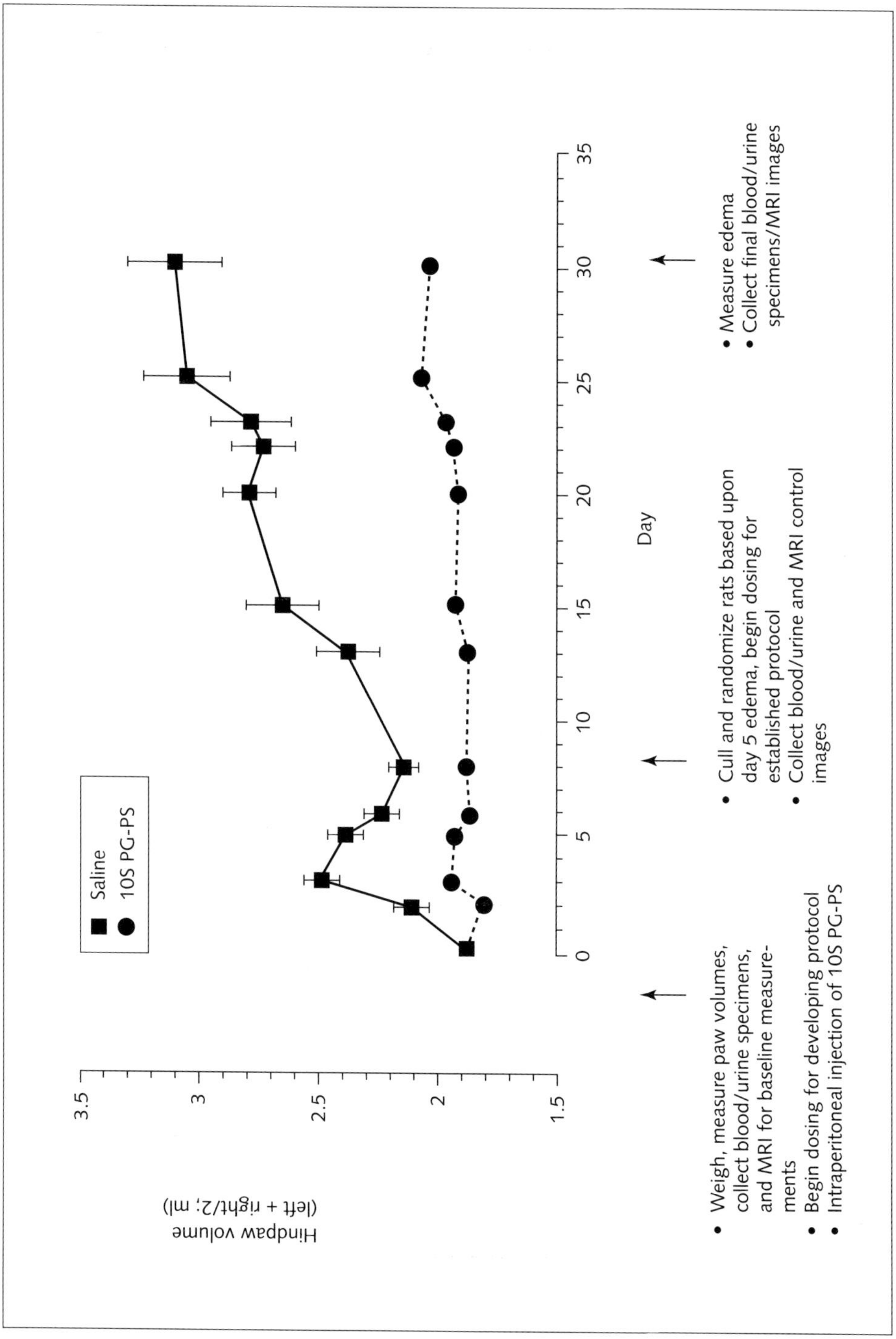
Hindpaw volume
(left + right/2; ml)
3.5
3
2.5
2
1.5
Saline
10S PG-PS
Day
0
5
10
15
20
25
30
35
• Weigh, measure paw volumes, collect blood/urine specimens, and MRI for baseline measurements
• Begin dosing for developing protocol
• Intraperitoneal injection of 10S PG-PS
• Cull and randomize rats based upon day 5 edema, begin dosing for established protocol
• Collect blood/urine and MRI control images
• Measure edema
• Collect final blood/urine specimens/MRI images

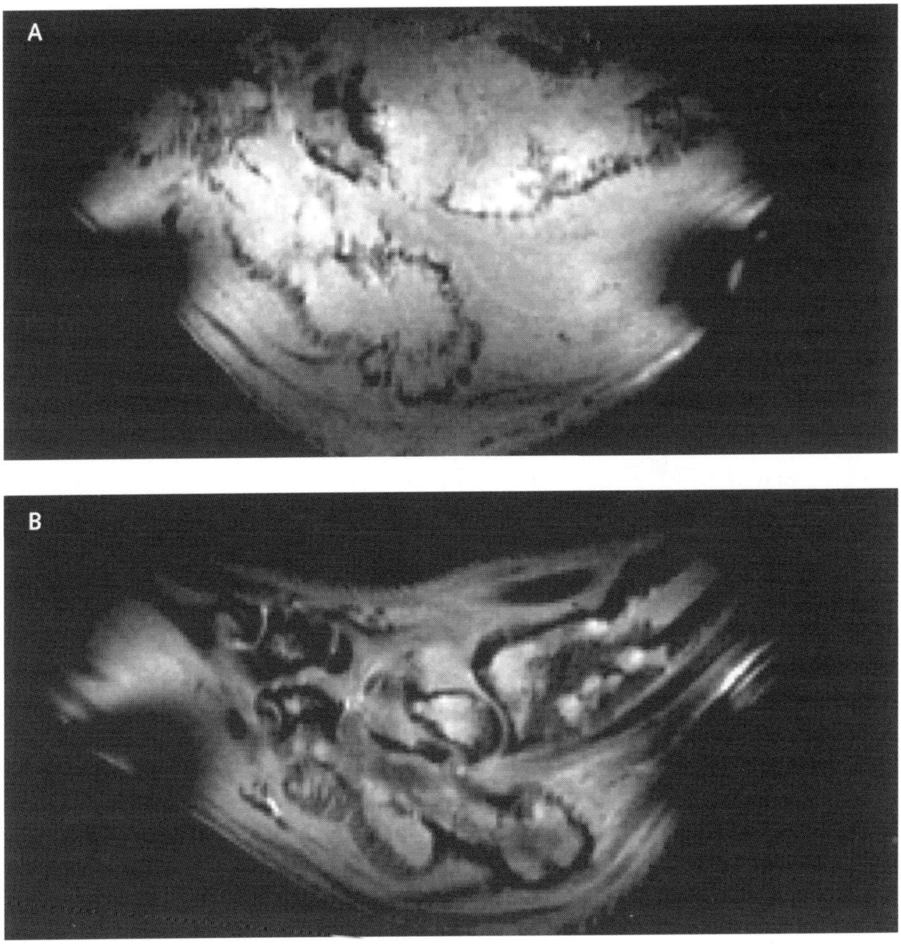

Figure 6
MRI images of two left hindpaws from rats injected i.p. 30 days previously with 10S PG-PS. Left hindpaw volumes of these two rats were not significantly different on day 5 when rats were culled, but by day 30 significant differences are observed in overall soft tissue swelling (edema), and periosteal new bone formation. While both rats display features of advanced arthritis, the histopathology depicted in panel A is clearly more severe. These images also demonstrate the important point that while paw volume is commonly used to measure anti-inflammatory activity or disease progression, paw volume alone does not reflect the underlying pathology. Furthermore, when used in the absence of other disease measurement parameters (acute phase proteins, leukocyte count, body weight, cytokine levels, histopathology, etc.), paw volume alone may grossly misrepresent the disease state or the effect of a drug in this model. These observed variations in MRI/histopathology among control 10S PG-PS rats are not seen in rat adjuvant arthritis on day 30.

techniques used in this model are not difficult once the necessary skills are learned, practice is necessary prior to initiating a full study. The first procedure requiring skill is the intraarticular injection into the tibiotalar region above the calcaneus (see cartoon of rat hock anatomy in Fig. 1). This procedure should be carried out under light anesthesia. Animals may be anesthetized with anesthetics such as ketamine/rompum or pentobarbital, but we find that the use of a gas anesthesia system (Omni Medical Equipment, Pleasanton, CA) in conjunction with an isoflurane/O_2 inhalation mixture is ideal. Using this approach, rats rapidly become anesthetized once they are placed in a large aquarium-type chamber saturated with isoflurane/O_2. Another advantage of this procedure is that the rats can safely remain in the chamber for an extended period of time ($\geq$ 1 h) if necessary. An additional piece of equipment worthy of investment includes a nose mask/tubing which extends from the saturation chamber and can be placed over the rat once it is removed from the chamber so that anesthesia can be continued during surgical or technical procedures. Alternatively, if the intraarticular injections can be completed within 30 s/rat, a 40%:60% mixture of CO_2/O_2 is a very safe and effective means of anesthetizing the rats. Animals recover more quickly from this gas mixture than with the isoflurane/O_2.

The stock 100P PG-PS is first diluted in pyrogen-free saline to the appropriate concentration (5 µg rhamnose units of 100P PG-PS/joint in a total injection volume of 20 µl). One method for accurately and reproducibly injecting 20 µl of the 100P PG-PS material into the tibiotalar joint space is to use a 20 µl adjustable micropipette (e.g. Eppendorf, Rainin, Gilson brands) with a 1/4", 27 g disposable needle snugly fitted over a standard disposable polypropylene pipette tip. Alternatively, a well calibrated 50 or 100 µl glass Hamilton repeating syringe with a tip modified to accept a 1/4", 27 g disposable needle will also work. Other devices may be used for injection, but overall they should be capable of injecting 10–20 µl volumes in an accurate and reproducible way. Care should also be taken to expel all air bubbles from the syringe and/or needle. The most important aspect to the intraarticular injection is how the needle is inserted into the joint. This is a blind injection, and a "feel" for the technique must be developed. Many investigators inject directly through the Achilles tendon into the joint. We find that this can occasionally cause an unnecessary secondary inflammation unrelated to the antigen-induced inflammatory response in the joint. The method we have used is to lay the anesthetized rat on its side and then flex the ankle/paw towards the tibia so that the proximal portion of the calcaneus is directed away from the tibia (see drawing of rat hock in Fig. 3). The needle is inserted into the hock on either side of the Achilles tendon, above the calcaneus, and then redirected towards the middle of the joint approximately 5 mm (almost to the hub of the needle described above). Occasionally you will hit the bone (talus or distal tibia) if the needle is inserted too far, and adjustments should be made for insertion distance. If done correctly, one can feel or sense a "pop" following penetration of the joint capsule/synovial space. The needle

the susceptibility of inbred and outbred rats to arthritis induced by cell walls of group A streptococci. *Infect Immun* 25: 484–490

58 Wilder RL (1988) Streptococcal cell-wall-induced arthritis in rats: an overview. *Int J Tiss React* 10: 1–5

59 Sternberg EM, Young III WS, Bernardini R, Calogero AE, Chrousos GP, Gold PW, Wilder RL (1989) A central nervous system defect in biosynthesis of corticotropin-releasing hormone is associated with susceptibility to streptococcal cell wall-induced arthritis in Lewis rats. *Proc Natl Acad Sci USA* 86: 4771–4775

60 Sternberg EM, Glowa JR, Smith MA, Calogero AE, Listwak SJ, Aksentijevich S, Chrousos GP, Wilder RL, Gold PW (1992) Corticotropin releasing hormone related behavioral and neuroendocrine responses in Lewis and Fischer rats. *Brain Res* 570: 54–60

61 Klareskog L, Wigzell H (1988) Immune reactions in the rheumatoid synovial tissue. In: JA Goodacre, WC Dick (eds): *Immunopathogenetic mechanisms of arthritis*. MTP Press Limited, Lancaster, 143–157

62 Joosten I, Wauben MHM, Holewijn MC, Reske K, Pedersen LO, Roosenboom CFP, Hensen EJ, van Eden W, Buus S (1994) Direct binding of autoimmune-disease related T-cell epitopes to purified Lewis rat MHC class-II molecules. *Int Immunol* 6: 751–759

63 Stimpson SA, Schwab JH (1989) Chronic remittent erosive arthritis induced by bacterial peptidoglycan-polysaccharide structures. In: JY Chang, AJ Lewis (eds): *Modern methods in pharmacology: pharmacological methods in the control of inflammation*. Alan R. Liss, Inc., New York, 381–394

64 van den Broek MF, van Bruggen MC, Koopman JP, Hazenberg MP, van den Berg WB (1992) Gut flora induces and maintains resistance against streptococcal cell wall-induced arthritis in F344 rats. *Clin Exp Immunol* 88: 313–317

65 Kohashi O, Kuwata J, Umehara K, Uemura F, Takahashi T, Ozawa A (1979) Susceptibility to adjuvant-induced arthritis among germfree specific-pathogen-free, and conventional rats. *Infect and Immun* 26: 791–794

66 Wilder RL, Calandra GB, Garvin AJ, Wright KD, Hansen CT (1982) Strain and sex variation in the susceptibility to streptococcal cell wall-induced polyarthritis in the rat. *Arth Rheum* 25: 1064–1072

67 Sternberg EM, Hill JM, Chrousos GP (1989) Inflammatory mediator-induced hypothalamic-pituitary-adrenal axis activation is defective in streptococcal cell wall arthritis susceptible Lewis rats. *Proc Natl Acad Sci USA* 86: 2374–2378

68 Allen JB, Blatter D, Calandra GB, Wilder RL (1983) Sex hormonal effects on the severity of streptococcal cell wall-induced polyarthritis in the rat. *Arth Rheum* 26: 560–563

69 Shear HL, Roubinian JR, Gil P, Talal N (1981) Clearance of sensitized erythrocytes in NZB/NZW mice. Effects of castration and sex hormone treatment. *Eur J Immunol* 11: 776–780

70 Robinson R (1979) Rat: Rat taxonomy. In: PL Altman, DD Katz (eds): *Inbred and*

genetically defined strains of laboratory animals. Part 1: Mouse and rat. Federation of American Societies for Experimental Biology, Bethesda, 233–350
71 Tobach E, Bloch H (1955) A study of the relationship between behavior and susceptibility to tuberculosis in rats and mice. *Adv Tuberc Res* 6: 62–89
72 Aksetijevich S, Whitfield HR, Young WS, Wilder RL, Chrousos GP, Gold PW, Sternberg EM (1992) Arthritis-susceptible Lewis rats fail to emerge from the stress hyporesponsive period. *Dev Brain Res* 65: 115–118
73 Penning TD Talley JJ Bertenshaw SR Carter JS Collins PW Docter S Graneto MJ Lee LF Malecha JW Miyashiro JM, et al. (1997) Synthesis and biological evaluation of the 1,5-diarylpyrazole class of cyclooxygenase-2 inhibitors: identification of 4-[5-(4-methylphenyl)-3-(trifluoromethyl)-1H-pyrazol-1-yl]benzenesulfonamide (SC-58635, celecoxib). *J Med Chem* 40: 1347–1365
74 Schwartz J, Mukhopadhyay S, McBride K, Jones T, Adcock S, Sharp P, Dedges J, Grasing K et al. (1998) Antipyretic activity of a selective cyclooxygenase (COX)-2 inhibitor, MK-966. *Am Soc Clin Pharmacol Ther* 63: 167
75 Carlson RP, Datko LJ, Chang J, Nielsen ST, Lewis AJ (1984) The antiinflammatory profile of (5H-dibenzo[a,d]-cyclohepten-5-ylidene)acetic acid (WY-41,770), an agent possessing weak prostaglandin synthetase inhibitory activity that is devoid of gastric side-effects. *Agents and Actions* 14: 654–661
76 Inoue K, Fujisawa H, Montonaga A, Inoue Y, Kyoi T, Ueda F, Kimura K (1994) Anti-inflammatory effects of etodolac: comparison with other non-steroidal anti-inflammatory drugs. *Biol Pharm Bull* 17: 1577–1583
77 Mukherjee A, Hale VG, Borga O, Stein R (1996) Predictability of the clinical potency of NSAIDs from the preclinical pharmacodynamics in rats. Inflamm Res 45: 531–540
78 Liyanage SP, Currey HLF (1972) Failure of oral d-penicillamine to modify adjuvant arthritis or immune response in the rat. *Ann Rheum Dis* 31: 521
79 Yamamoto N, Sakai F, Yamazaki H, Kawai Y, Nakahara K, Okuhara M (1996) Effects of FR133605, a novel cytokine suppressive agent, on bone and cartilage destruction in adjuvant arthritic rats. *J Rheum* 23: 1778–1783
80 Tanaka K, Shimotori T, Makino S, Aikawa Y, Inaba T, Yoshida C, Takano S (1992) Pharmacological studies of the new antiinflammatory agent 3-formylamino-7-methylsulfonylamino-6-phenoxy-4H-1-benzopyran-4-one. 1st communication: antiinflammatory, analgesic and other related properties. *Arzneim Forsch* 42: 935–944
81 Neuman RG, Wilson BD, Barkley M, Kimball ES, Weichman BM, Wood DD (1987) Inhibition of prostaglandin biosynthesis by etodolac. I. Selective activities in arthritis. *Agents and Actions* 21: 160–166
82 Calhoun W, Gilman SC, Datko LJ, Copenhaver TW, Carlson RP (1992) Interaction studies of tilomisole, aspirin, and naproxen in acute and chronic inflammation with assessment of gastrointestinal irritancy in the rat. *Agents and Actions* 36: 99–106
83 Ackerman NR, Rooks WH, Shott L, Genant H, Maloney P, West E (1979) Effects of naproxen on connective tissue changes in the adjuvant arthritic rat. *Arth Rheum* 22: 1365–1374

84 Weichman BM (1989) Rat adjuvant arthritis: a model of chronic inflammation. In: JY Chang, AJ Lewis (eds): *Pharmacological methods in the control of inflammation.* Alan R. Liss, Inc., New York, 363–380
85 Lewis AJ, Carlson RP, Chang J, Gilman SC, Nielsen S, Rosenthale ME, Janssen FW, Ruelius HW (1983) The pharmacological profile of oxaprozin, an antiinflammatory and analgesic agent with low gastrointestinal toxicity. *Curr Therap Res* 34: 777–794
86 Swingle KF, Moore GI, Grant TJ (1976) 4-nitro-2-phenoxymethanesulfonanilide (R-805): a chemically novel anti-inflammatory agent. *Arch Int Pharmacodyn* 221: 132–139
87 Connolly KM, Stecher VJ, Danis E, Pruden DJ, LaBrie T (1988) Alteration of interleukin-1 production and the acute phase response following medication of adjuvant arthritic rats with cyclosporin-A or methotrexate. *Int J Immunopharmacol* 10: 717–728
88 Carlson RP, Hartman DA, Tomcheck LA, Walter TL, Lugay JR, Calhoun W, Sehgal SN, Chang JY (1993) Rapamycin, a potential disease-modifying antiarthritic drug. *J Pharmacol Exp Ther* 266: 1125–1138
89 Connolly KM, Stecher VJ, Danis E, Pruden DJ, LaBrie T (1988) Alteration of interleukin-1 activity and the acute phase response in adjuvant arthritic rats treated with disease modifying antirheumatic drugs. *Agents and Actions* 25: 94–105
90 Hara M, Sugawara S, Hirose S, Kondo H, Irimajiri S, Uchida S, Hashimoto H, Kashiwazaki S (1995) Immunological and therapeutical effects of FK506 on patients with rheumatoid arthritis. *Inflamm Res* 44: S248
91 Carlson RP, Hartman DA, Ochalski SJ, Zimmerman JL, Glaser KB (1998) Sirolimus (Rapamycin, Rapamune) and combination therapy with cyclosporin A in the rat developing adjuvant arthritis model: correlation with blood levels and the effects of different oral formulations. *Inflamm Res* 47: 339–344
92 Bartlett RR, Schleyerback R (1985) Immunopharmacological profile of a novel isoxazol derivative, HWA 486, with potential antirheumatic activity. I. Disease modifying action on adjuvant arthritis of the rat. *Int J Immunopharmac* 7: 7–18
93 Ohta Y, Fukuda S, Baba A, Nagai H, Tsukuda R, Sohda T, Makino H (1996) Immunomodulating and articular protecting activities of a new anti-rheumatic drug, TAK-603. *Immunopharm* 34: 17–26
94 Badger AM, Bradbeer JN, Votta B, Lee JC, Adams JL, Griswold DE (1996) Pharmacological profile of SB 203580, a selective inhibitor of cytokine suppressive binding protein/p38 kinase, in animal models of arthritis, bone resorption, endotoxin shock and immune function. *J Pharmacol Exp Ther*ap 279: 1453–1461
95 Gouret C, Mocquet G, Raynaud G (1976) Use of Freund's adjuvant arthritis test in anti-inflammatory drug screening in the rat: value of animal selection and preparation at the breeding center. *Lab Animal Science* 26: 281–287
96 Arrigoni-Martelli E, Bramm E (1975) Investigations on the influence of cyclophosphamide, gold sodium thiomalate and D-penicillamine on nystatin oedema and adjuvant arthritis. *Agents and Actions* 5: 264–267
97 Dunn CJ, Prouteau M, Delahaye M, Purcell T, Branceni D (1984) Prolonged treatment

with D-penicillamine: effects on adjuvant arthritis in the rat. *Agents and Actions* 14: 269–273

98 Finkelstein AE, Ladizesky M, Borinsky R, Kohn E, Ginsburg I (1988) Antiarthritic synergism of combined oral and parenteral chrysotherapy. I. Studies in adjuvant-induced arthritis in rats. *Inflammation* 12: 373–382

99 Morris RE, Meiser BM, Wu J, Shorthouse R, Wang J (1991) Use of rapamycin for the suppression of alloimmune reactions *in vivo*: schedule dependence, tolerance induction, synergy with cyclosporine and FK-506, and effect on host-versus-graft and graft-versus-host reactions. *Transplant Proc* 23: 521–524

100 Billingham MEJ, Hicks CA, Carney SL (1990) Monoclonal antibodies and arthritis. *Agents and Actions* 29: 77–87

101 Yoshino S, Schlipkoter E, Kinne R, Hunig T, Emmrich F (1990) Suppression and prevention of adjuvant arthritis in rats by a monoclonal antibody to the αβ T cell receptor. *Eur J Immunol* 20: 2805–2808

102 Issekutz AC, Meager A, Otterness I, Issekutz TB (1994) The role of tumour necrosis factor-alpha and IL-1 in polymorphonuclear leucocyte and T lymphocyte recruitment to joint inflammation in adjuvant arthritis. *Clin Exp Immunol* 97: 26–32

103 Fergusson KM, Osawa H, Diamanstein T, Oronsky AL, Kerwar SS (1988) Treatment with an anti-interleukin 2 receptor antibody protects rats from passive but not adjuvant arthritis. *Int J Immunother* IV: 29–33

104 Issekutz AC, Ayer L, Miyasaka M, Issekutz TB (1996) Treatment of established adjuvant arthritis in rats with monoclonal antibody to CD18 and very late activation antigen-4 integrins suppresses neutrophil and T-lymphocyte migration to the joints and improves clinical disease. *Immunology* 88: 569–576

105 Seiffge D (1996) Protective effects of monoclonal antibody to VLA-4 on leukocyte adhesion and course of disease in adjuvant arthritis in rats. *J Rheum* 23: 2086–2091

106 Iigo Y, Takashi T, Tamatani T, Miyasaka M, Higashida T, Yagita H, Okumuru K, Tsukada W (1991) ICAM-1-dependent pathway is critically involved in the pathogenesis of adjuvant arthritis in rats. *J Immunol* 147: 4167–4171

107 Walter UM, Issekutz AC (1997) The role of E- and P-selectin in neutrophil and monocyte migration in adjuvant-induced arthritis in the rat. *Eur J Immunol* 27: 1498–1505

108 Santos LL, Morand EF, Hutchinson P, Boyce NW, Holdsworth SR (1997) Anti-neutrophil monoclonal antibody therapy inhibits the development of adjuvant arthritis. *Clin Exp Immunol* 107: 248–253

109 Hoffmann JC, Herklotz C, Zeidler H, Bayer B, Rosenthal H, Westermann J (1997) Initiation and perpetuation of rat adjuvant arthritis is inhibited by the anti-CD2 monoclonal antibody (mAb) OX34. *Ann Rheum Dis* 56: 716–722

110 Jacob CO, Holoshitz J, van der Meide P, Strober S, McDevitt HO (1989) Heterogeneous effects of IFN-gamma in adjuvant arthritis. *J Immunol* 142: 1500–1505

111 Nakajima H, Takamori H, Hiyama Y, Tsukada W (1991) The effect of treatment with recombinant gamma-interferon on adjuvant-induced arthritis in rats. *Agents and Actions* 34: 63–65

112 Feige U, Schulmeister A, Mollenhauer J, Brune K, Bang H (1994) A constitutive 65kDa chondrocyte protein as a target antigen in adjuvant arthritis in Lewis rats. *Autoimmunity* 17: 233–239

113 Zhang ZJ, Lee CSY, Lider O, Weiner HL (1990) Suppression of adjuvant arthritis in Lewis rats by oral administration of tÿpe II collagen. *J Immunol* 145: 2489–2493

114 Halloran MM, Szekanecz Z, Barquin N, Haines GK, Koch AE (1996) Cellular adhesion molecules in rat adjuvant arthritis. *Arth Rheum* 39: 810–819

115 Clegg DO, Dietz F, Duffy J, Willkens RF, Hurd E, Germain BF, Wall B, Wallace DJ, Bell CL, Sleckman J (1997) Safety and efficacy of hydroxychloroquine as maintenance therapy for rheumatoid arthritis after combination therapy with methotrexate and hydroxychloroquine. *J Rheum* 24: 1896–1902

116 Murray KM, Dahl SL (1997) Recombinant human tumor necrosis factor receptor (p75) Fc fusion protein (TNFR:Fc) in rheumatoid arthritis. *Ann Pharmacother* 31: 1335–1338

117 Brower V (1997) Enbrel's phase III reinforces prospects in RA. Nat Biotechnol 15: 1240

118 Niederer RR (1984) Topical anti-inflammatory activity of suprofen in the adjuvant arthritic rat. *Bull Tech/Gattefosse* 77: 43–45

119 Hiramatsu Y, Akita S, Salamin PA, Maier R (1990) Assessment of topical non-steroidal anti-inflammatory drugs in animal models. *Arzneim Forsch* 40: 1117–1124

120 Lopez-Guerrero JA, Lopez-Bote JP, Ortiz MA, Gupta RS, Paez E, Bernabeu C (1993) Modulation of adjuvant arthritis in Lewis rats by recombinant vaccinia virus expressing the human 60-kilodalton heat shock protein. *Infect Immun* 61: 4225–4231

121 Makarov SS, Olsen JC, Johnston WN, Schwab JH, Anderle SK, Brown RR, Haskill JS (1995) Retrovirus mediated *in vivo* gene transfer to synovium in bacterial cell wall-induced arthritis in rats. Gene Ther 2: 424–428

122 Wahl SM, Allen JB, Ohura K, Chenoweth DE, Hand AR (1991) IFN-gamma inhibits inflammatory cell recruitment and the evolution of bacterial cell wall-induced arthritis. *J Immunol* 146: 95–100

123 Sano H, Hla T, Maier JAM, Crofford LJ, Case JP, Maciag T, Wilder RL (1992) *in vivo* cyclooxygenase experession in synovial tissues of patients with rheumatoid arthritis and osteoarthritis and rats with adjuvant and streptococcal cell wall arthritis. *J Clin Invest* 89: 97–108

124 Yocum DE, Allen JB, Wahl SM, Calandra GB, Wilder RL (1986) Inhibition by cyclosporin A of streptococcal cell wall-induced arthritis and hepatic granulomas in rats. *Arth Rheum* 29: 262–273

125 van den Broek MF, van de Langerigt LGM, van Bruggen MCJ, Billinghan MEJ, van den Berg WB (1992) Treatment of rats with monoclonal anti-CD4 induces long-term resistance to streptococcal cell wall-induced arthritis. *Eur J Immunol* 22: 57–61

126 Yoshino S, Cleland L, Mayerhofer G, Brown R, Schwab J (1991) Prevention of chronic erosive streptococcal cell wall-induced arthritis in rats by treatment with monoclonal antibody against the T cell antigen receptor α,β. *J Immunol* 146: 1487–1489

127 Schimmer RC, Schrier DJ, Flory CM, Dykens J, Tung DKL, Jacobson PB, Friedl HP, Conroy MC, Schimmer BB, Ward PA (1997) Streptococcal cell wall-induced arthritis.

Requirements for neutrophils, P-selectin, intercellular adhesion molecule-1, and macrophage-inflammatory protein-2. *J Immunol* 159: 4103–4108
128 Schwab JH, Anderle SK, Brown RR, Dalldorf FG, Thompson RC (1991) Pro- and anti-inflammatory roles of interleukin-1 in recurrence of bacterial cell wall-induced arthritis in rats. *Infect Immun* 59: 4436–4442
129 Schrier DJ, Schimmer RC, Flory CM, Tung DK, Ward PA (1998) Role of chemokines and cytokines in a reactivation model of arthritis in rats induced by injection with streptococcal cell walls. *J Leuk Biol* 63: 359–363
130 Wahl SM, Allen JB, Costa GL, Wong HL, Dasch JR (1993) Reversal of acute and chronic synovial inflammation by anti-transforming growth factor beta. *J Exp Med* 177: 225–230
131 Schimmer RC Schrier DJ Flory CM Laemont KD Tung D Metz AL Friedl HP Conroy MC Warren JS Beck B, et al. (1998) Streptococcal cell wall-induced arthritis: requirements for IL-4, IL-10, IFN-γ, and monocyte chemoattractant protein-1. *J Immunol* 160: 1466–1471
132 van den Broek MF, Hogervorst EJM, van Bruggen MCJ, van Eden W, van der Zee R, van den Berg WB (1989) Protection against streptococcal cell wall-induced arthritis by pretreatment with the 65-kD mycobacterial heat shock protein. *J Exp Med* 170: 449–466
133 Brandes M, Allen JB, Ogawa Y, Wahl SM (1991) Transforming growth factor beta 1 suppresses acute and chronic arthritis in experimental animals. *J Clin Invest* 87: 1108–1113
134 Allen JB, Wong HL, Costa GL, Bienkowski MJ, Wahl SM (1993) Suppression of monocyte function and differential regulation of IL-1 and IL-1ra by IL-4 contribute to a resolution of experimental arthritis. *J Immunol* 151: 4344–4351
135 Allen JB, Bansal GP, Feldman GM, Hand AO, Wahl LM, Wahl SM (1991) Suppression of bacterial cell wall-induced polyarthritis by recombinant gamma interferon. *Cytokine* 3: 98–106
136 Jacobson PB, Morgan S, Wilcox D, Carlson R, Harris R, Nuss M (1998) A new spin on an old model: magnetic resonance imaging of the adjuvant arthritis rat. *in vivo* evaluation of disease progression by MRI with respect to standard inflammatory parameters and histopathology. *Arth Rheum; submitted*
137 Jones RS, Ward JR (1963) Studies on adjuvant-induced polyarthritis in rats. II. Histogenesis of joint and visceral lesions. *Arth Rheum* 6: 23–35
138 Weichman BM, Chau TT, Rona G (1987) Histopathologic evaluation of the effects of etodolac in established adjuvant arthritis in rats: evidence for reversal of joint damage. *Arth Rheum* 30: 466–470.
139 Mohr W, Wild A, Wolf HP (1981) Role of polymorphs in inflammatory cartilage destruction in adjuvant arthritis in rats. *Ann Rheum Dis* 40: 171–176
140 Wilder RL, Case JP, Crofford LJ, Kumkumian GK, Lafyatis R, Remmers EF, Sano H, Sternberg EM, Yocum DE (1991) Endothelial cells and the pathogenesis of rheumatoid

arthritis in humans and streptococcal cell wall arthritis in Lewis rats. *J Cell Biochem* 45: 162–166

141 Schrier DJ, Flory CM, Finkel M, Kuchera SL, Lesch ME, Jacobson PB (1996) The effects of the phospholipase A2 inhibitor, manoalide, on cartilage degradation, stromelysin expression, and synovial fluid cell count induced by intraarticular injection of human recombinant interleukin-1 alpha in the rabbit. *Arth Rheum* 39: 1292–1299

142 Sano H, Forough R, Maier JA, Case JP, Jackson A, Engleka K, Maciag T, Wilder RL (1990) Detection of high levels of heparin binding growth factor-1 (acidic fibroblast growth factor) in inflammatory arthritic joints. *J Cell Biol* 110: 1417–1426

Murine collagen-induced arthritis

Wim B. van den Berg and Leo A.B. Joosten

Dept. of Rheumatology, University Hospital Nijmegen, Geert Grooteplein 8, NL-6500 HB Nijmegen, The Netherlands

Introduction

A major research goal in the field of arthritis is to unravel the pathogenesis of chronic arthritis and the concomitant joint destruction. A second, more practical goal is to define targeted therapies, selectively inhibiting the progression of destructive arthritis, yet leaving host defence mechanisms virtually intact. Although animal models are not ideal in terms of precise mimicry of human arthritic disease, they do reflect key aspects of their human counterparts and offer a useful approach to understand arthritic processes and to improve therapeutic treatment. As the previous chapter dealt with adjuvant-induced arthritis in rat the current chapter will deal with collagen induced arthritis in mouse.

Rheumatoid arthritis is characterized by chronic inflammation in the joints and progressive destruction of cartilage and bone. Histopathological features include immune complexes in the articular cartilage layers and variable amounts of macrophages and T cells in the synovium, accompanied by fibrosis and synovial hyperplasia. The disease is often considered as an autoimmune process, the articular cartilage being an intriguing component, since it is the victim but also a likely trigger of the disease. Arguments for this are based on the observation that destructive forms of RA tend to decline when the cartilage is fully destroyed. Moreover, total joint replacement often results in a complete remission of arthritis in that particular joint, without the need of concomitant synovectomy. This is compatible with cartilage components being joint specific autoantigens or cartilage tissue functioning as an avascular reservoir, retaining yet unidentified arthritogenic triggers. Models have been developed which have proved the arthritogenic potential of cartilage autoantigens such as collagen type II and proteoglycan [1–4]. More recently [5, 6], arthritogenic potential has been demonstrated for novel cartilage components such as collagen types IX and XI, cartilage derived oligomeric protein (COMP) and hyaline cartilage glycoprotein 39 (HC gp-39). These models all elude to the same principle: arthritis due to the loss of tolerance against a cartilage specific autoantigen.

In Vivo Models of Inflammation, edited by D.W. Morgan and L.A. Marshall

The present chapter will be confined to detailed discussion of key events in arthritis induced with collagen type II (CII). Collagen induced arthritis (CIA) is a widely accepted arthritis model, based on T cell and antibody mediated autoimmunereactivity against cartilage CII. The model is characterized by severe cartilage and bone erosions. Induction has been demonstrated in various strains of rats and mice, susceptibility showing tight genetic restriction. More recently, collagen arthritis has been induced in non-human primates as well. The model of collagen arthritis is highly suited to analyse principles of autoimmune disease expression as well as antigen-specific immunosuppression. Moreover, it can be used to study mechanisms and mediators involved in autoimmune cartilage and bone destruction. The following sections will mainly deal with features of collagen-induced arthritis in the mouse.

Induction of collagen arthritis

The model of CIA was first described in 1977, by Trentham and colleagues [1], as a coincidental finding in protocols to induce autoantibodies to purified collagen preparations. The initial observations indicated that arthritis was confined to sensitization with native collagen type II; denatured CII or native CI not showing arthritogenicity. The crucial element in this arthritis is the induction of immunity to foreign collagen type II, subsequently cross-reacting with homologous CII. Plain immunization with homologous instead of heterologous CII can also be used, but then much stronger immunization regimens are needed to override natural tolerance.

In Lewis rats a single immunization with CII in Freunds complete adjuvant at the base of the tail is sufficient to get full blown expression of a polyarthritis within 14 days. In mice the disease expression is more gradual, starting after 3–4 weeks in some animals, whereas a 10% incidence commonly takes 8–10 weeks. Both chicken as well as bovine collagen type II preparations are proper heterologous antigens to induce CIA in mice. Susceptible strains include DBA/1j mice and B10RIII mice, having the H-2q and H-2r haplotype, respectively. Dominant epitopes of the CII molecule are different for DBA/1j and B10RIII mice, consistent with the different haplotype [7, 8]. Male mice show higher susceptibility as compared to female mice. In general, we are using the DBA/1j male mouse and bovine CII to induce this model.

Bulk quantities of bovine CII can be isolated from articular cartilage slices, taken from a knee joint of 1–2 year-old cows, according to Miller and Rhodes [9]. In brief, proteoglycans are extracted from the cartilage with 4 M guanidiniumchloride in a neutral 0.05 M Tris buffer, 24 h at room temperature. After washing the cartilage is digested with pepsin (1 mg/ml in 0.1 M acetic acid) for 48 h at room temperature. The suspension is then centrifugated at 1000 g and the supernatant is adjusted to pH 7.4 with 2 M NaOH. Bulk protein is precipitated with addition of solid NaCl, reaching a final concentration of 20% NaCl and equilibration for 2 days at 4° C. After centrifugation at 27 000 g for 30 min, the pellet is resolved in 0.5 M acetic acid and dialysed against this solution, overnight at room temperature. The collagen is then resolved and remaining material is removed by centrifugation (10 min 2000 g). The collagen is then selectively precipitated at a final concentration of 5% NaCl, overnight at 4° C. Spinning down, resolution and precipitation at 5% NaCl is repeated twice to obtain a purer collagen preparation. Final dialysation is done against 0.05 M acetic acid and the preparation can be stored as such at –20° C, or lyophilized.

Proper, native CII preparations are poorly soluble in water, which provides a first, simple check on quality. Purity can be analysed by gel-electrophoresis. In addition, newly prepared CII batches are first screened in an ELISA, in comparison with former arthritogenic CII batches and the use of a standard set of anti-CII antibodies, obtained from a pool of arthritic mice. To obtain a defined solution for immunization, CII is slowly resolved in 0.05 M HAc in an overnight procedure at 4° C (concentration 2 mg/ml) and then emulsified in an equal volume of Freunds complete adjuvant, containing 2 mg/ml *Mycobacterium tuberculosis*, strain H37Ra. Mice are then immunized intradermally at the base of the tail with 100 µl emulsion (100 µg CII). At day 21 a booster injection is given with 100 µg CII in 100 µl PBS, administered intraperitoneally.

Onset of arthritis starts around day 25–28, often first affecting some digits of hind and fore paws, then spreading to multiple sites in the paw, including the ankle compartments (Fig. 1). When not heavily boosted the onset may be rather gradual, not even reaching a 100% incidence at 8 weeks and with limited numbers of joints affected (Fig. 2).

The model is a mixture of an immune complex disease and a delayed type hypersensitivity reaction in the joint. Although anti-CII antibodies alone are able to induce arthritis after passive transfer, high concentrations are needed, in particular of complement activating subclasses, recognizing multipe epitopes, and even then at best a transient arthritis occurs [10]. Passive transfer with bulk T cells or T cell clones also yielded poor disease expression [11]. Probably, antibodies are needed to bind to the cartilage surface, herein promoting further release of collagen epitopes upon complement fixation and the attraction of leucocytes. Attachment of granulocytes to the cartilage surface is a crucial element of CIA. Subsequent influx of anti-CII specific Th1 cells will than further drive the arthritic process. Of interest, the cytokine pattern in the lymphoid organs showed a dominant Th1 pattern after immunization with CII in FCA [12]. Susceptibility to CIA is enhanced when the amount of *Mycobacterium tuberculosis* within the FCA preparation is enhanced. Moreover, severe arthritis can also be induced upon immunization with CII in incomplete Freunds adjuvant (IFA), provided that the mice are then treated with recombinant IL-12 during the immunization period, herein strongly promoting a Th1 response [13]. Bacterial preparations are potent inducers of IL-12.

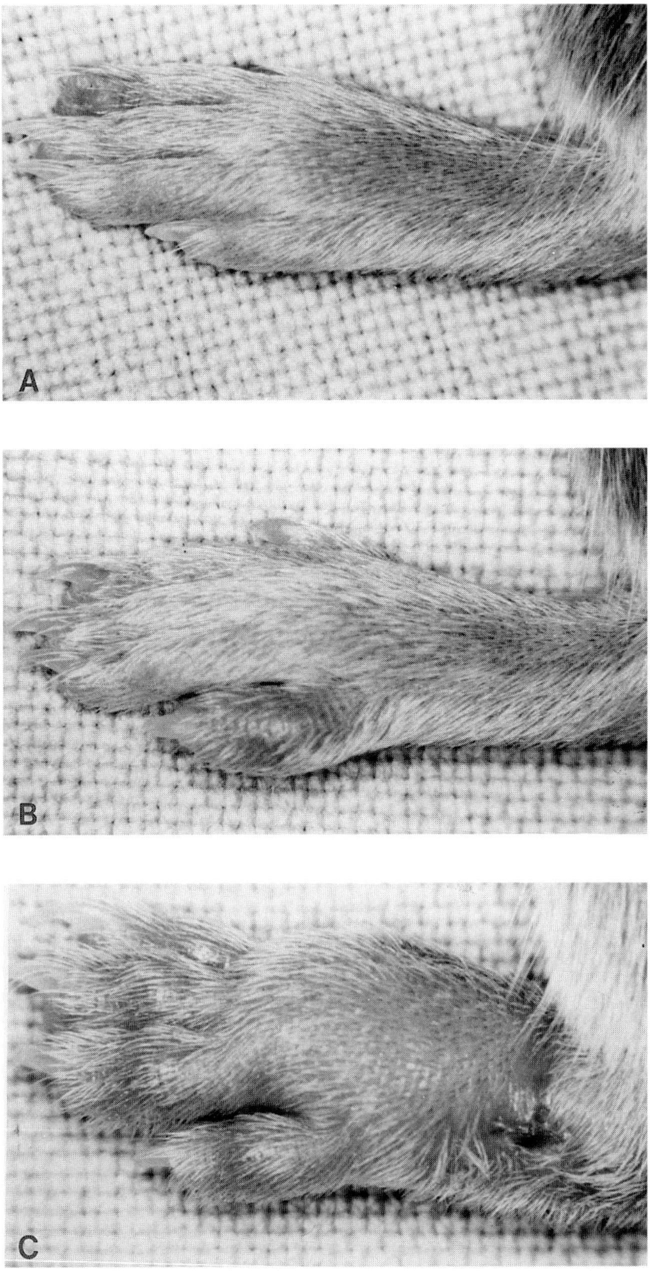

Figure 1
Macroscopic appearance of collagen arthritis, ranging from normal (A) to one affected (B) and full expression in the whole paw (C).

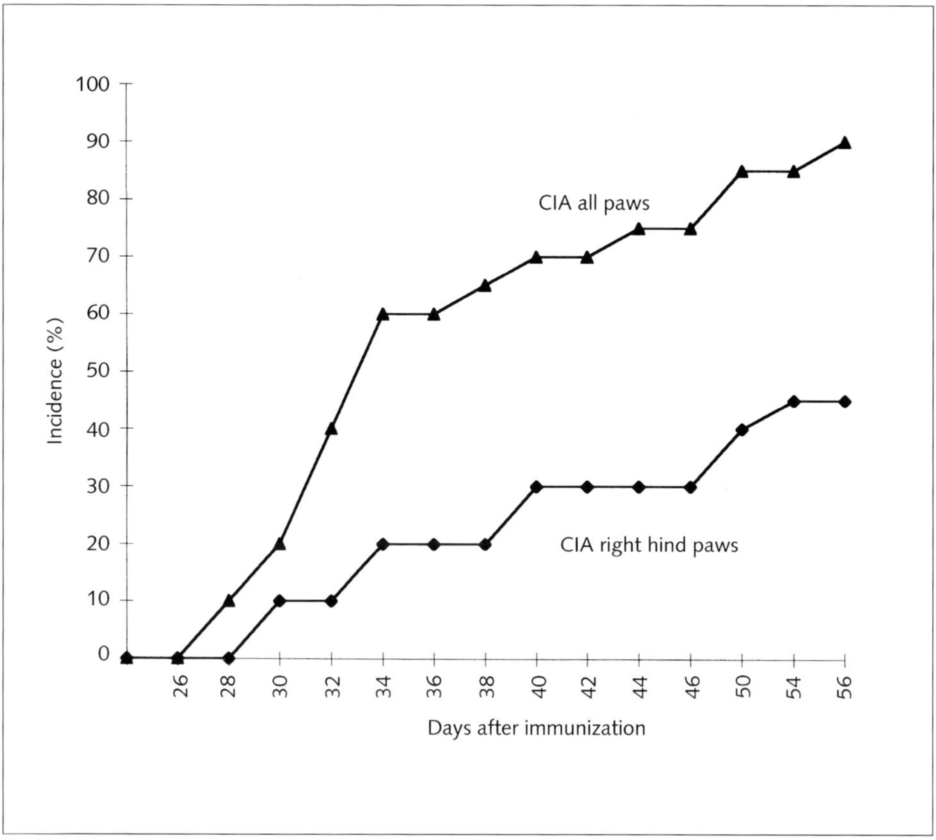

Figure 2
Incidence of CIA, as scored in one selected paw or in all paws. Note that the incidence in the right hind paw is still less than 50% after 50 days. Reflects a group of 20 mice.

Remarkably, high doses of IL-12 given during standard immunization with CII in FCA were shown to suppress the CIA, associated with a marked reduction in CII-specific antibodies. Although there is no doubt that Th1 reactivity is needed, the critical importance of high levels of anti-CII antibodies is further underlined in studies in susceptible and resistant mice strains. Additional IL-12 treatment during immunization with CII in FCA was shown to enhance CII-specific Th1 responses in C57Bl and B10.Q mice, but this protocol still failed to induce arthritis in these mice. Analysis of anti-CII antibody titers revealed that the levels remained markedly lower in these strains as compared to those found in susceptible DBA mice [14].

Expression of arthritis

Apart from the generation of adequate levels of complement fixing anti-CII antibodies as well as the presence of anti-CII specific Th1 cells, it is clear that expression of autoimmune arthritis depends on local conditions in joint tissues. Collagen arthritis shows a higher incidence in male as compared to female mice, whereas other models such as antigen or Zymosan induced arthritis show the opposite sex preponderance. Apart from the autoimmune character, the elicitation of the joint inflammation by direct injection into the knee joint is a major difference between these models and CIA. In collagen arthritis no local insults are given and arthritis develops "spontaneously". The male preponderance of CIA might be linked to the impact of hormones on this arthritic process, but it is tempting to suggest that the consistent fighting of male mice makes a major contribution as well, causing microtrauma in joint tissues and being crucial in triggering onset of arthritis. In line with this, it is also our experience that disease incidence is generally higher when the mice are housed in large groups instead of small groups.

Threatening autoimmune reactivity is generated by the immunization protocol, but precipitation of the autoimmune process in the joint is facilitated by nonspecific inflammation at such sites or systemic generation of proinflammatory mediators. It is long recognized that systemic administration of IL-1, shortly before onset of the disease, markedly accelerates CIA expression [15]. This seems related to activation of endothelium, facilitating influx of inflammatory cells. Moreover, IL-1 is a potent cartilage destructive mediator, causing loss of cartilage proteoglycans and herein denuding the autoimmune target in CIA, collagen type II in the articular cartilage. In addition, is was shown that local injection of TNFα or TGFβ potentiates CIA expression in the injected joint [16, 17]. TNF (tumor necrosis factor) is a pivotal proinflammatory cytokine in arthritis and an inducer of IL-1. TGFβ (transforming growth factor-β), although having immunosuppressive potential, is a potent chemoattractant. All of this fits with unmasking of dormant autoimmune reactivity by nonspecific attraction of inflammatory cells to the joints, including CII-reactive T cells, and amplification of the process by inflammation mediated exposure of autoimmune epitopes.

A single injection of LPS provides an elegant alternative for the acceleration of CIA by systemic administration of recombinant IL-1 [18]. In our standard protocol we give a booster immunization with 100 μg CII at day 21 and a single intraperitoneal injection of 10–40 μg LPS around day 28. This will not only greatly enhance the severity of the arthritis, but also synchronize the expression in a group of mice and enlarge the number of affected joints in one animal. Bear in mind that spontaneous expression of CIA after plain immunization often affects only a limited number of joints, such as one or two toes, complicating grading and histologic analysis. A critical prerequisite for the accelerated expression still remains the proper immunization with CII. In animals showing poor immunity to CII the collagen arthritis can-

Table 1 - Factors influencing expression of murine CIA

• isotype of the anti collagen type II antibodies
• degree of Th1 anti-CII reactivity
• acceleration with the cytokines IL-1, TNFα and TGFβ
• control by the suppressive cytokines IL-4 and IL-10 (TGFβ)
• nonspecific trauma in joint tissues promotes expression

not be accelerated, neither with a single nor with repeated LPS challenge. The mechanism behind CIA acceleration with LPS can be linked to the generation of TNF and IL-1, as well as marked IL-12 production. LPS induced acceleration can be blocked with antibodies against IL-12 and acceleration can be induced with systemic administration of recombinant IL-12 [19].

Apart from LPS induced cytokine generation it is conceivable that the LPS induced elicitation of IL-12 promotes Th1 generation. When applied shortly after the booster injection with CII it will influence the critical process of generation of cross-reactive T cell activity from heterologous CII to the homologous CII of the mouse, including the process of epitope spreading. Although collagen arthritis can be induced by immunization with a small CII fragment, containing the dominant epitope, the arthritis is less severe as compared to the one induced with the whole CII molecule, suggesting multiple epitope involvement in classic CIA [9].

In line with the notion that any inflammatory stimulus generating proinflammatory mediators such as TNF or IL-1 will accelerate CIA expression (Tab. 1), we have demonstrated that a systemic injection with Zymosan (yeast particles) highly accelerates CIA in DBA/1j mice [20]. Consistent enhancement of CIA incidence and severity was seen with a dose of 3 mg Zymosan, injected intraperitoneally. This injection induces a marked peritonitis and onset of accelerated CIA expression was noted after a few days. With increasing dosages of Zymosan the arthritis expression could not be further enhanced (Fig. 3). We even observed a delay in day of onset, apparently linked to a more prolonged, distracting inflammation in the peritoneal cavity.

Unilateral CIA

We recently developed a variant of the polyarthritic Zymosan induced CIA, by local injection of Zymosan in one knee joint [20]. This injection is given around day 25 after the first immunization with CII at day 0 and boosting at day 21. Upon local injection in nonimmunized DBA mice, Zymosan induces a transient arthritis, with reversible cartilage proteoglycan depletion. When injected in CII immunized DBA mice, a dose of 60 μg Zymosan is sufficient to accelerate expression of collagen

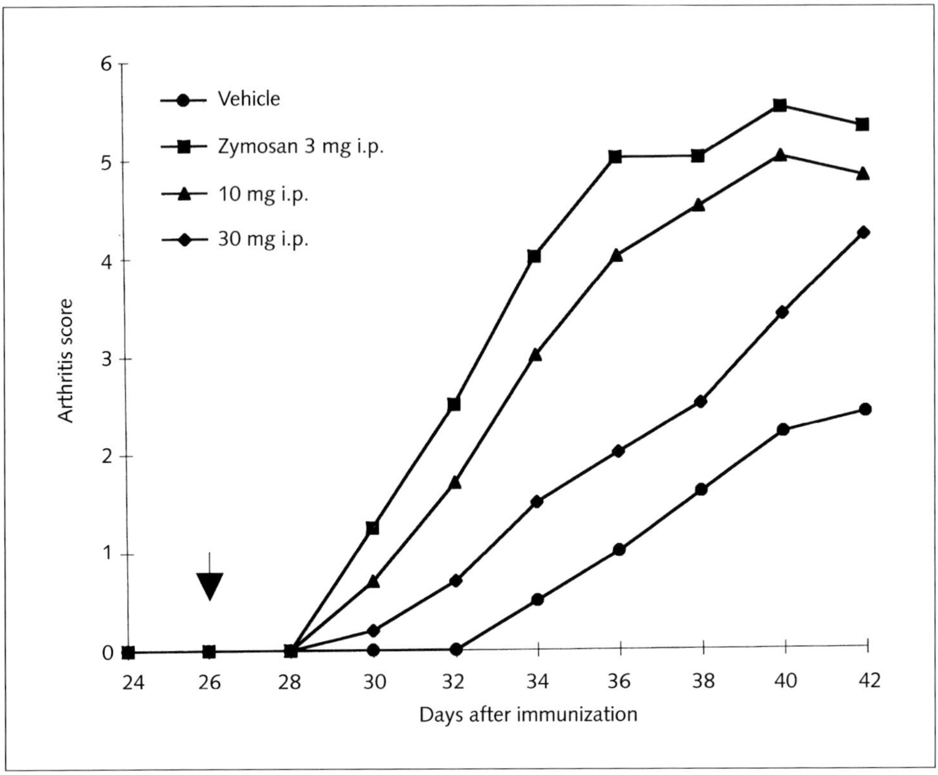

Figure 3
Accelerated expression of CIA and more severe arthritis after intraperitoneal injection of Zymosan at day 26. Reflects groups of 10 mice.

arthritis in that knee joint, as reflected by the characteristic, aggressive cartilage destruction and prolonged joint inflammation. When a higher dose is injected, 180 μg Zymosan, the expression of collagen arthritis is not restricted to the injected joint, but also extends to the ipsilateral ankle joint, whereas the expression in the other paws was not enhanced. With the 180 μg dose of Zymosan the accelerated and synchronized expression in the ankle reached an incidence of 90%, whereas the incidence dropped to 10–50% with lower Zymosan dosages. Anti-TNF treatment markedly reduced this spreading to the ipsilateral paw, whereas anti-IL-1 treatment fully prevented the expression in that ankle joint, making it likely that the Zymosan induced expression is related to TNF and IL-1, produced locally in the knee joint and diffusing to the ankle.

To obtain optimal advantage of a fully synchronized and localized expression model of CIA, it is essential that the initial immunization and boosting with CII is not too optimal, creating already a considerable number of mice with affected ankles around day 25. In daily practice, when aiming for this unilateral model, we apply normal amounts of *Mycobacterium tuberculosis* (1 mg) in the initial immunization, give no boosting with LPS and perform the CII boosting at day 21 in saline and not in Freunds adjuvant. Finally, we do a prescreening of the mice at day 25, discarding all animals having any sign of arthritis from the experiment and performing the local Zymosan acceleration in this negative-selection group. This approach highly reduces the variation normally encountered in the spontaneous polyarthritic CIA.

Arthritis score and histopathology

The course of CIA is routinely scored by macroscopic analysis of arthritic signs in peripheral joints. In general, mice are examined every other day, from day 25, to get an impression of the course of the disease. Clinical severity of arthritis is graded on a scale of 0–2 for each paw, according to changes in redness and swelling. At late stages of the disease ankylosis of the ankles can be included. The macroscopic score is expressed as a selective value in one paw or as a cumulative value for all paws, with a maximum of 8.

Histologically, the inflammation is characterized by a florid exudate in the joint space, containing numerous amounts of granulocytes, and a progressive destruction of the articular cartilage. Erosion of bone is pronounced, but periosteal new bone formation is seen as well. Bone marrow is affected, but markedly less as compared to the polyarthritic adjuvant disease. Apart from the high number of granulocytes in the joint cavity, the synovial tissue contains large numbers of macrophages and lymphocytes, but also in this compartment granulocytes are prominent in the first 2 weeks after onset. The most characteristic feature of collagen arthritis is the aggressive attack of the inflammatory process at the articular cartilage (Fig. 4). In this model heavy sticking of granulocytes at the cartilage surface is a common finding. Moreover, cartilage damage is not limited to loss of proteoglycans from the matrix, but in short periods of time roughening and deep erosions of the surface are consistently observed, further facilitating attachment of granulocytes at these sites.

In that sense granulocytes probably play an active role in the cartilage destruction, linked to sticking to anti-CII immune complexes in the surface layers. It is long recognized that granulocytes contain high amounts of elastase and cathepsin G, which are potent mediators of cartilage proteoglycan depletion in cocultures of cartilage and activated granulocytes, *in vitro*. When such cultures are done in the presence of full serum or synovial fluid, the destruction is limited, due to large amounts of high molecular weight enzyme inhibitors in these fluids. However, when the granulocytes are pelleted on the cartilage surface, heavy destruction is again noted, implying direct extrusion of enzymes in the matrix and escape from natural

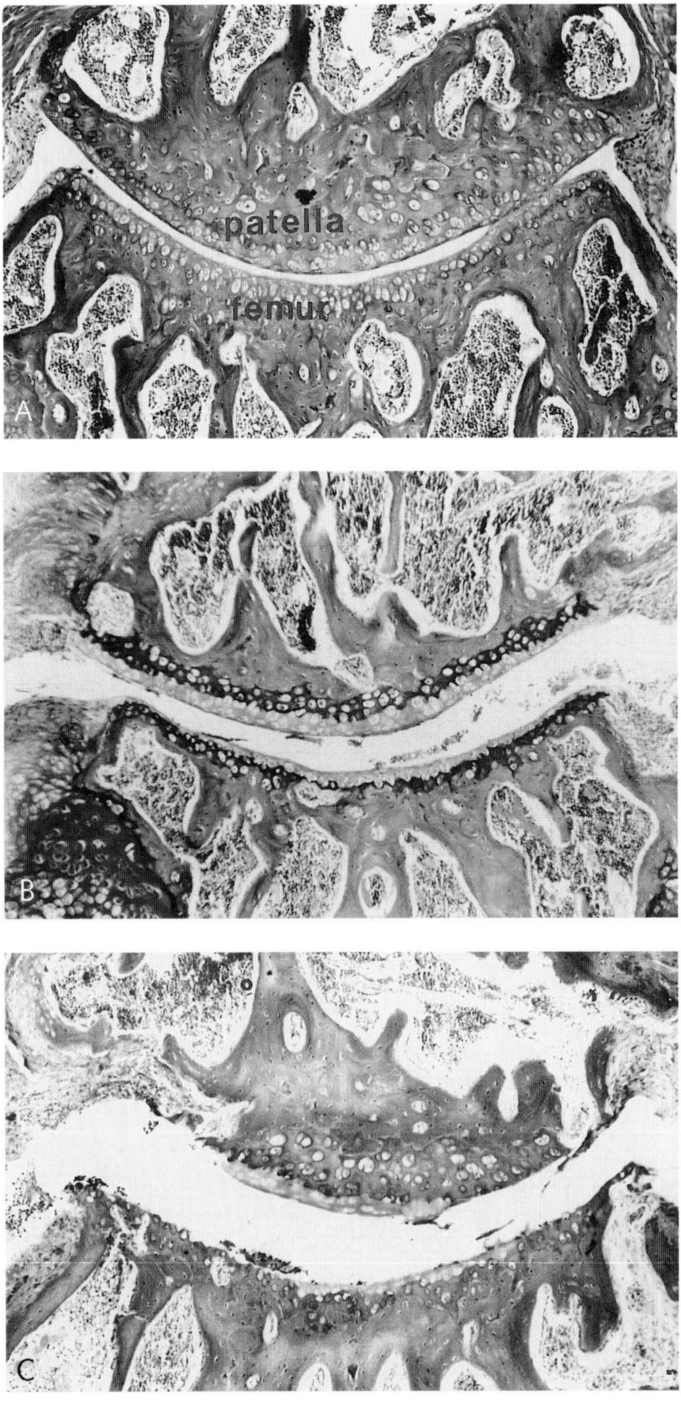
patella
femur
A
B
C

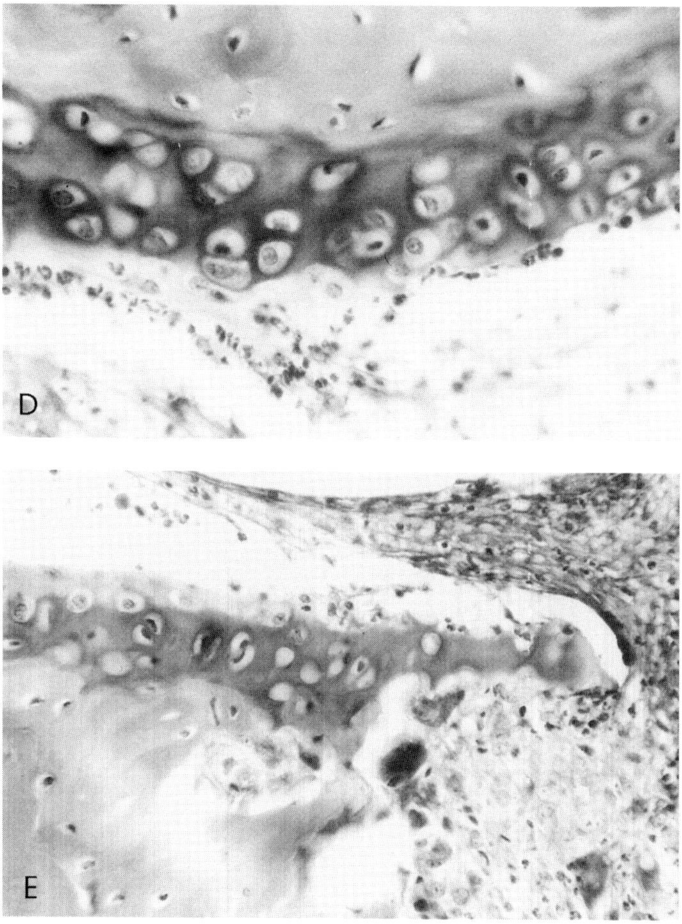

Figure 4
Characteristic histopathology of murine CIA in the area of the patella and opposite femur of the knee. Homogeneous Saffranin O staining of proteoglycans in normal cartilage surface layers (A). Depletion of proteoglycan in the surperficial cartilage but still intact surface in mBSA arthritis (B) and marked depletion as well as erosion in late stage CIA (C). Attachment of granulocytes and irregular cartilage surface in active CIA (D). Cartilage erosion but also bone erosion underneath by ingrowth of granulation tissue in CIA (E).

inhibitors once the cells are in full contact with the cartilage surface [21]. This process of pronounced sticking is seen *in vivo*, in the model of collagen arthritis, provided that the anti-CII antibody levels are high. In addition, we noted such sticking also in murine, methylated bovine serum albumin (mBSA)-induced arthritis, only when the joints are immobilized [22]. Apparently, the immune complex formation between antibodies and cationic mBSA, planted in the cartilage surface, is on its own insufficient to generate heavy sticking under normal movement conditions, but when the cells are allowed to settle under immobilized conditions, the interaction remains intact. Like the situation in collagen arthritis, this condition results in erosions of the surface and further settlement and digging into the roughened surface by the attached granulocytes. With respect to granulocyte involvement in cartilage destruction in human RA, it is clear that immune complexes can be found in the articular cartilage surface of a large number of patients. Whether the concentration is high enough to allow for consistent granulocyte attachment is yet unclear and considerable variation between different RA patients seems obvious. Apart from this process, cartilage erosion at the cartilage margins, linked to pannus overgrowth, is considered to make a significant contribution in RA patients.

At later stages of collagen arthritis, cartilage erosion at the margins and pannus formation is a prominent feature as well (Fig. 4). After a few weeks the model often progresses to complete loss of the whole cartilage, ending up in bone to bone contact and variable degree of ankylosis. This dramatic destruction of the cartilage reflects the directed autoimmune attack at the cartilage and the arthritis in the synovial tissue burns out in a particular joint, when the cartilage is fully destroyed. The synovium then displays a mixture of an immune infiltrate, macrophages and a pronounced fibrotic reaction.

The highly destructive character of collagen arthritis is also reflected in the massive occurrence of the proteoglycan breakdown neoepitope VDIPEN throughout the cartilage. This epitope is indicative for the involvement of metalloproteases, in particular stromelysin. In contrast to the lack of such epitopes in reversible cartilage proteoglycan depletion in Zymosan arthritis and the variable degree of these neoepitopes in murine antigen induced arthritis, only showing expression at particular sites of the cartilage displaying irreversible lesions, the expression is fast and much more pronounced in collagen arthritis (Fig. 5). Detailed studies in antigen induced arthritis made it clear that VDIPEN expression is linked to IL-1 driven processes [23] and colocalizes with collagen breakdown neoepitopes. All of this is compatible with a role of stromelysin in activation of collagenase, and a dominant role of this process

Figure 5
Expression of cartilage proteoglycan breakdown neoepitope VDIPEN. Note the absence in normal cartilage (A); local expression in mBSA arthritis (B); fully affected cartilage in CIA (C).

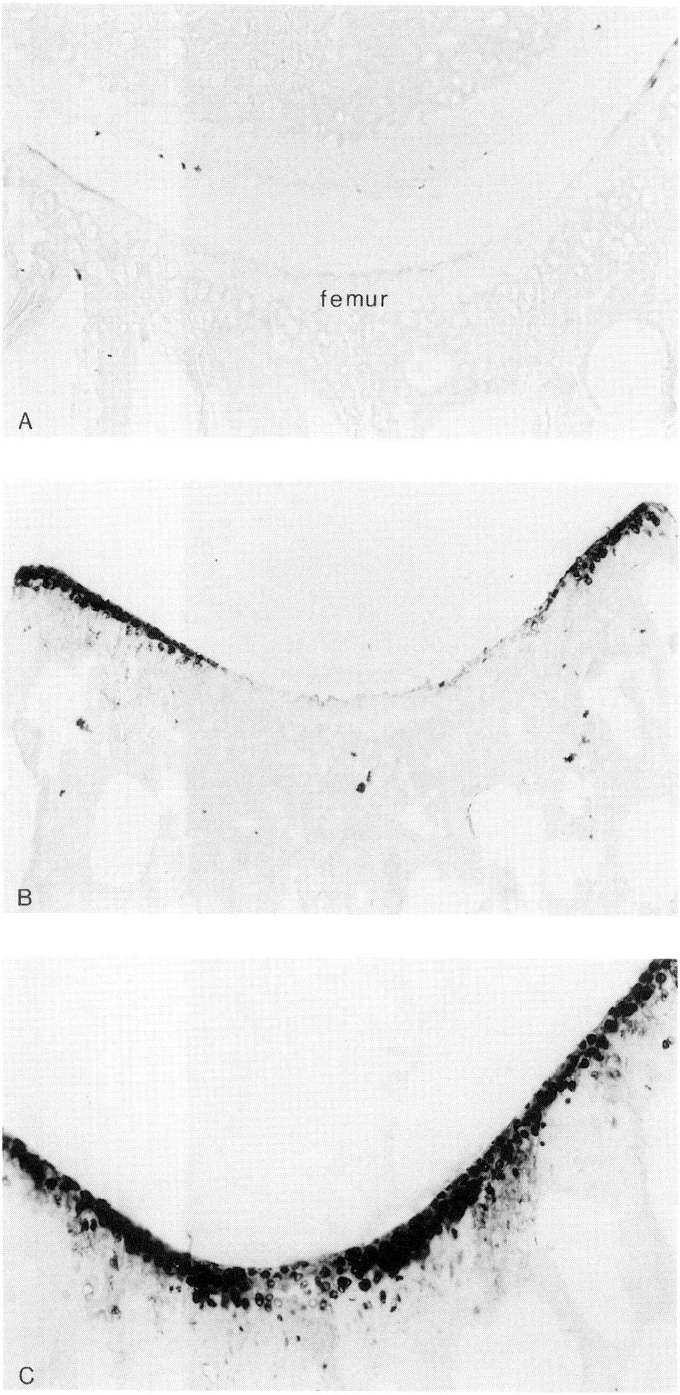
femur
A
B
C

in cartilage erosion in CIA. Stromelysin is produced in a latent form after activation of synovial cells or cartilage with IL-1, suggesting that further activation by granulocyte enzymes may contribute as well, elastase being a likely candidadte in this process. Recent studies with elastase inhibitors revealed efficacy in murine collagen arthritis [24].

Given the rapid development of the arthritic changes in this model, it is clear that consistent histologic scoring of the severity of the arthritis is seriously complicated by variable days of onset of arthritis in individual mice, variation in onset between different paws or even between digits in one paw. The latter variability furthermore asks for highly standardized semiserial sectioning of complicated joint structures of the whole paw. All of this flaws the design of proper drug studies. Attempts, discussed above, to synchronize expression in groups of mice, or perhaps even better, to precipitate the arthritis in a given joint by a local inflammatory insult, provide valuable improvements of applicability of the collagen arthritis model.

Although the classic macroscopic scoring of collagen arthritis is always done in paws, with additional analysis of histology of the ankle joints, it is our experience that there is a high correlation between occurrence of arthritis in the knee and the ankle. Since the standardized joint sectioning is much easier in the knee as compared to the ankle, histologic analysis should preferrably be done in the knee. We have carefully compared the characteristic histopathology in ankles and knees and the patterns of synovitis and cartilage destruction are very similar (Tab. 2). It should be noted that this is unlike the situation in adjuvant arthritis in the rat. The latter model of arthritis shows predominant expression in the ankles, whereas knee joints are rarely affected.

Table 2 - Histology of CIA. Comparison of affected knee and ankle joints

	Day 37*		**Day 42**	
	Knee joint	**Ankle joint**	**Knee joint**	**Ankle joint**
Infiltrate	1.5 ± 0.6	1.5 ± 0.3	1.3 ± 0.4	1.7 ± 0.4
Cartilage damage	1.0 ± 0.6	0.9 ± 0.4	1.2 ± 0.3	1.1 ± 0.2
Proteoglycan depletion	1.9 ± 0.8	1.6 ± 0.4	2.2 ± 0.5	1.9 ± 0.5

** Days after the first immunization with type II collagen. The values represent the mean ± SD of at least 20 knee or ankle joints. Histology was scored on a scale ranging from 0 to 3. Infiltrate is scored as the amount inflammatory cells in synovial tissue and joint cavity. Cartilage damage reflects surface erosions, proteoglycan depletion indicates loss of Saffranin O staining.*

Cytokine involvement

In line with a major role of the cytokine TNF in human RA, the onset of collagen arthritis is TNF dependent. Studies with neutralising anti-TNF antibodies or soluble TNF receptors revealed a major suppressive effect, when treatment was started shortly before onset of CIA [25, 26]. When the arthritis is fully expressed, subsequent blocking of TNF appeared only marginally effective, implying that TNF is crucial in onset but less so in propagation of arthritis. In clear contrast, blocking of IL-1 with neutralising antibodies or IL-1ra (receptor antagonist) markedly reduced severity of the arthritis [27], also when the arthritis was fully established (Fig. 6). Moreover, anti-IL-1 treatment markedly reduced cartilage damage. Elegant studies in IL-1β deficient mice showed full resistance to CIA induction and the critical importance of IL-1β was also emphasized by greatly reduced CIA in IL-1 converting enzyme (ICE)-deficient mice and efficacy of ICE inhibitors in CIA in normal mice. Recent studies in TNF-receptor knock-out mice revealed a lower incidence and a milder form of CIA in the absence of proper TNF-receptor interaction. However, once a joint was afflicted, the progression of arthritis in that joint was indistinguishable from that in wild type mice [28], again underlining the limited role of TNF in propagation and cartilage destruction. It emphasizes that TNF is helpful in acceleration of arthritis expression, but that TNF independent onset can occur as well. Although it is claimed in human RA that TNF is driving most of the IL-1 production and that TNF blocking would be sufficient to block the whole arthritic process, this is not found in collagen arthritis. Recent studies in murine streptococcal cell wall-induced arthritis also revealed major TNF dependence of initial joint swelling, but IL-1β dependence of cartilage destruction [29]. This was found using neutralising antibodies and confirmed in TNF and IL-1β knock-out mice. Again, TNF blocking did not sufficiently prevent IL-1 production.

These findings imply that anti-TNF treatment in RA patients would be beneficial when the disease is in fact a chronic process, due to repeated flares, with each acute exacerbation showing strong TNF dependency. Of interest, when expression of collagen arthritis is not highly stimulated by additional boosting or synchronizing injections with LPS or additional cytokines, onset of arthritis starts only in a small number of joints, with gradual involvement of additional joints with time. This creates a seemingly extended period of TNF dependency of the model, which is lost upon synchronization and speedy propagation to established arthritis in most joints.

An intriguing element in control of collagen arthritis expression is formed by the synovial lining cells. This layer consists of synovial fibroblasts and macrophages. When macrophages are selectively depleted from this layer by local injection of toxic liposomes and the subsequent process of engulfment of liposomes by these phagocytes and subsequent apoptotic cell death, such a joint becomes refractory to the onset of collagen arthritis [30]. Further analysis revealed that the lining macro-

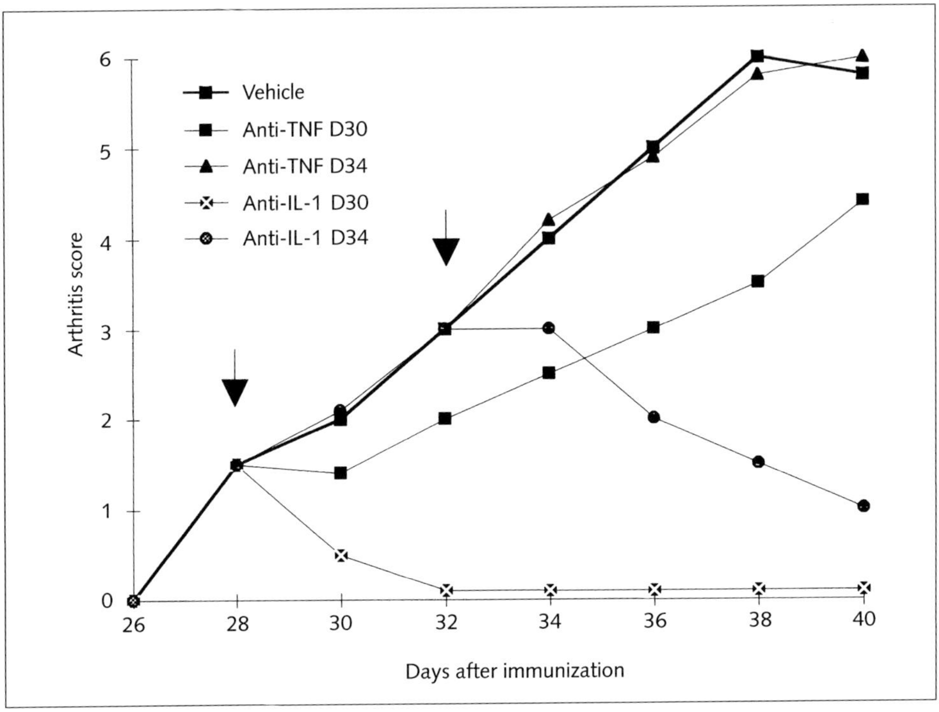

Figure 6
Collagen arthritis is treated with systemic administration of neutralizing antibodies against TNF or IL-1. Treatment was started in the various groups at either day 28 or 32. Anti-TNF is still effective shortly after onset, but not in established disease. In contrast, anti-IL-1 is highly effective, even in late arthritis.

phages are a major source of chemotactic factors, needed to direct the initiating leucocyte influx into the joint. TNF and IL-1 are potent inducers of chemokine production in lining cells, and when these recombinant cytokines are injected in a lining depleted, naïve joint, they do not induce leucocyte influx. In contrast, C5a is still fully capable to attract leucocytes in such a joint. This further establishes the TNF/IL-1 dependence, with an intermediate role of the lining cells, in collagen arthritis. Subsequent studies in other models revealed that immune complex (IC) arthritis was totally abolished in lining depleted joints, whereas a strong T cell driven arthritis was hardly affected. Moreover, IC arthritis showed strong IL-1 dependence, sharing this feature with collagen arthritis and further emphasizing that onset of collagen arthritis is more an immune complex phenomenon as compared to a T cell process.

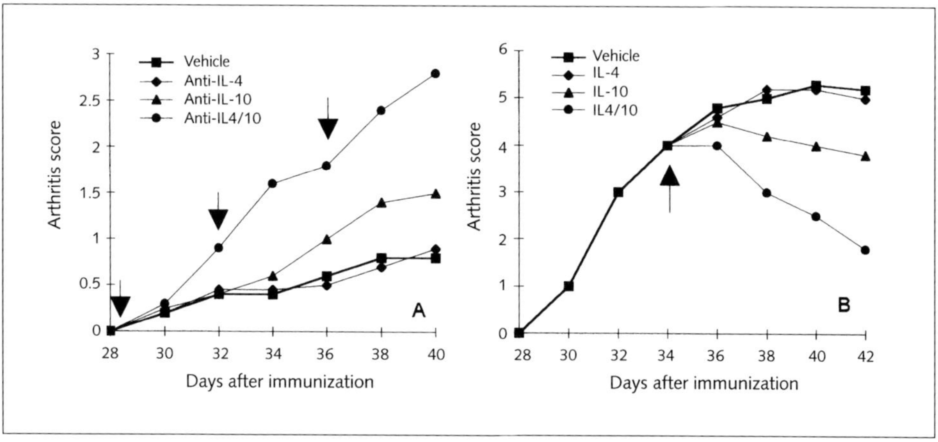

Figure 7
Treatment of CIA with systemic neutralizing antibodies (A) or recombinant IL-4 and/or IL-10 (B). Antibodies given at days 28, 32 and 36. Recombinant cytokines given daily, from day 34. For experiment (A), mice were selected at day 28, having no signs of arthritis. In (B), groups of mice are depicted which were challenged with LPS at day 28, to obtain high arthritis expression in the control mice. Spontaneous CIA expression is under the control of endogenous IL-10. Suppression of established arthritis is still possible with additional IL-4/IL-10.

Apart from a pivotal role of TNF and IL-1 in onset and propagation of CIA, regulation of the arthritis is exerted by the cytokines IL-4 and IL-10. These so-called modulatory cytokines have a critical impact on the arthritic process at various levels, including the control of the Th1/Th2 balance, inhibition of macrophage TNF/IL-1 production and stimulation of chondrocytes in the articular cartilage. Expression of collagen arthritis is under the control of endogenous IL-10 [31]. High levels are found in the synovial tissue and anti-IL-10 antibodies, given shortly before expected onset, enhance incidence and severity. Anti-IL-4 antibodies were without effect, in line with the difficulty to detect significant levels of IL-4 in the synovium, but combined anti-IL-4/IL-10 treatment promoted the strongest expression of CIA (Fig. 7). The opposite approach, i.e. treatment of CIA with systemically injected recombinant IL-4 or IL-10, revealed that IL-10 reduced the joint swelling in CIA, but more marked suppression, including reduced cartilage destruction was noted with the combination treatment with IL-4 and IL-10 [32]. Of interest, IL-10 is a potent reducer of macrophage TNF production, but IL-1 production was only suppressed with the combination of IL-10 and IL-4. Moreover, this combination also upregulated the IL-1ra/IL-1 balance, both in the synovium as well as in the cartilage.

A final cytokine deserving major attention is IL-12. This cytokine originates from macrophages and is produced after activation with bacterial components. IL-12 is a potent inducer of IFNγ and promotes Th1 generation and propagation. As stated above, IL-12 when given at the expected onset of CIA, greatly enhances incidence and severity and systemic anti-IL-12 treatment prevented LPS-accelerated CIA expression. However, when anti-IL-12 was applied in the established phase of CIA, it appeared poorly suppressive and upon interruption of anti-IL-12 treatment we noted a marked exacerbation. Moreover, late treatment with recombinant IL-12 suppressed instead of enhanced the arthritis, prolonged IL-12 treatment markedly enhanced IL-10 levels and the suppressive effect of IL-12 could be abrogated with anti-IL-10 [19]. This suggests a dual role of IL-12 in early and late disease. The potent induction of IL-10 reflects an intriguing feedback pathway to control for excessive and prolonged Th1 responses, but seriously hampers therapeutic targeting of IL-12 in autoimmune arthritis.

Applicability of the model

The model of collagen induced arthritis establishes that an autoimmune reaction to a cartilage component can lead to a chronic, destructive polyarthritis. Although it is far from accepted that collagen type II is a crucial antigen in human RA, the findings in the model may exemplify common principles in arthritis directed against cartilage autoantigens. The model is highly suitable and widely used to try to understand the immunoregulation in autoimmune arthritis and to identify ways to induce tolerance using peptide fragments, or to selectively target the T cell receptors involved in collagen epitope recognition. Detailed discussion of these topics goes beyond the scope of the present chapter and a recent review of Myers et al. [9] as well as the chapter of Lewis et al. in this volume is advised for further reading.

Apart from the immunoregulation, the model is highly suitable to try to understand the complex cytokine interplay in onset and propagation of arthritis and to identify therapies aimed at prevention of cartilage destruction. Examples of cytokine involvement are already discussed above. As mentioned before, the onset of collagen arthritis is an immune complex phenomenon. This stage shows high sensitivity to NSAIDs and, in fact, all therapies which will interfere with the initiating leucocyte influx will show efficacy. The more interesting part of the model is the established phase of the arthritis and the ongoing destruction of the articular cartilage.

When efficacy of drugs in onset of arthritis is investigated it should be realized that this stage is rather stress sensitive. Daily treatment by i.p. or oral injection may have a large impact at that stage and handling of mice should be done with great subtlety. We have often noted a significant suppression of arthritis onset with prophylactic daily vehicle treatment, whereas this effect was absent at later stages, when the arthritis is fully established. Treatment which is started after onset of arthritis

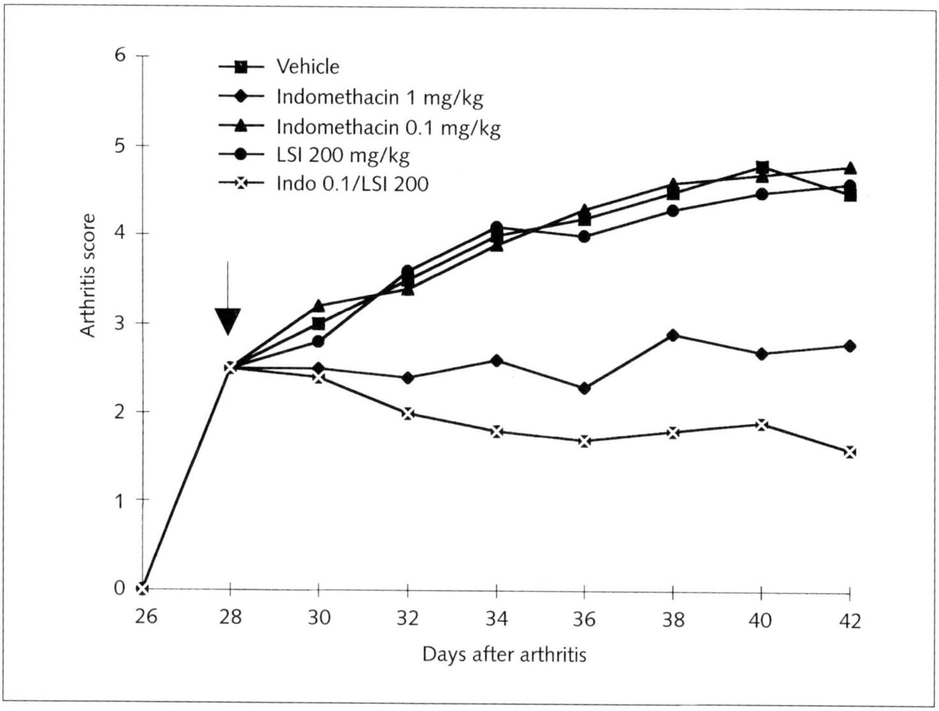

Figure 8
CIA mice received daily oral treatment with indomethacin and/or a leukotriene synthesis inhibitor (Bay W 5676), from day 28 for 14 consecutive days. The data represents groups of 10 mice. Note the marked synergy.

has the further advantage that grouping of the mice can be done by weighted randomization, creating a similar mean index of arthritis at the start of the various control and treatment groups. When randomization has to be done before onset, higher variation between groups is unavoidable and in general, the use of at least 10 mice per experimental group in such protocols is warranted. To illustrate the efficacy of some drugs in murine collagen arthritis a few examples will be discussed in more detail.

The onset of collagen arthritis is sensitive to treatment with indomethacin. A dose of 1 mg/kg significantly suppressed the macroscopic signs of arthritis. Intriguingly, when low dosages of indomethacin are used, sensitivity is lost. However, when treatment with indomethacin is combined with a leukotriene synthesis inhibitor, marked synergistic suppression was observed (Fig. 8). This clearly illustrates that both prostaglandins and leukotrienes are of importance at arthritis onset. The role of leukotrienes was also nicely illustrated in the poor induction of CIA in

Table 3 - Histology of CIA after treatment with either COI, LSI or combination of COI/LSI

Treatment	Dose	Infiltrate	Cartilage damage	Proteoglycan depletion
Vehicle	–	1.3 ± 0.7	1.5 ± 0.7	2.0 ± 0.7
COI	1 mg/kg	0.7 ± 0.6	0.8 ± 0.5	1.5 ± 0.8
COI	0.1 mg/kg	1.4 ± 0.9	1.6 ± 0.6	2.2 ± 0.9
LSI	200 mg/kg	1.5 ± 1.0	1.4 ± 0.8	2.1 ± 0.5
COI/LSI	0.1/200	0.6 ± 0.4	1.0 ± 0.3	1.2 ± 0.8

Treatment of arthritic mice was started at day 28 after immunization with type II collagen. Mice were injected i.p. twice a day with cyclooxygenase inhibitor (COI) indomethacin or leukotriene synthesis inhibitor (LSI) Bay W 5676 or the combination for 14 consecutive days. The data represent the mean ± SD of at least 10 mice per group. Histology was scored on a scale ranging from O to 3.

lipoxygenase deficient mice [33]. Of interest, cartilage destruction was also markedly reduced with the combination treatment (Tab. 3). It suggests that NSAIDs with a combined profile would be the better anti-arthritic drug. More recent interest focused on the dominant cox-1 and/or cox-2 inhibitory pattern of NSAIDs, further pinpointing the profiling of suitable NSAIDs.

A second example of therapeutic approaches in this model of arthritis is provided by the demonstration of synergy between steroids and IL-10. Steroids are potent suppressors of arthritis but their clinical application is seriously hampered by side-effects such as osteoporosis. It would be desirable to find ways to combine drugs at lower, nontoxic concentrations, yet retaining the beneficial effects. We found that prednisolone treatment suppressed collagen arthritis at a dose of 1–5 mg/kg per day, but dosages of 0.05 or 0.1 mg/kg were without effect. Interestingly, daily IL-10 treatment at a dose of 5 μg/mouse reduced macroscopic signs of swelling, but did not suppress other parameters of arthritis. When IL-10 was combined with the low dose prednisolone treatment, this resulted in marked suppression of arthritis (Fig. 9), including reduction in cartilage destruction (Tab. 4). The latter effect was also evident when COMP levels were measured in the serum. COMP is a marker of enhanced cartilage turnover [34] and its serum concentration raises from 4 μg/ml in control mice to 8 μg/ml in mice with active collagen arthritis. Moreover, we observed a straight correlation between COMP levels around day 40 and the degree of histologic cartilage damage at that stage. After treatment with the combination of low-dose prednisolone and IL-10, the COMP values were fully normalized.

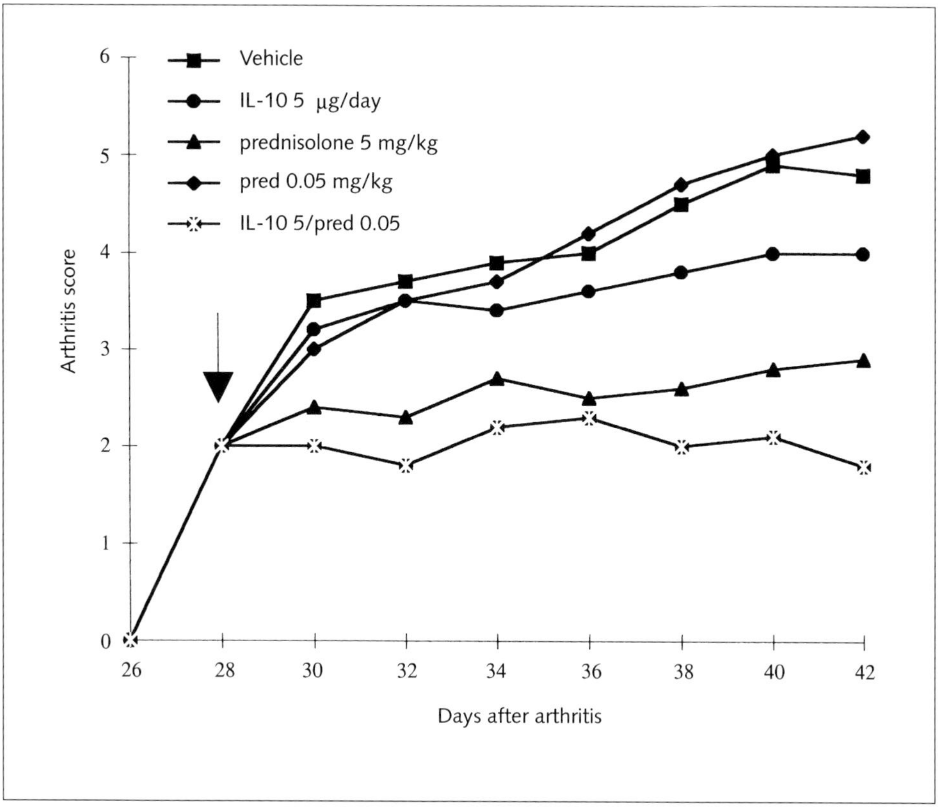

Figure 9
CIA mice received daily intraperitoneal treatment with prednisolone and/or murine IL-10. Note the synergy between low dose steroid and IL-10.

A final application of therapeutic manipulation in collagen arthritis is illustrated with a gene therapy approach with IL-1ra. Shortly before onset of collagen arthritis, we injected fibroblasts, transfected *in vitro* with a retroviral gene construct, containing IL-1ra, directly into the knee joint. Normal fibroblasts served as controls. The onset of arthritis was almost completely prevented in knee joints containing these IL-1ra producing cells. Intriguingly, this treatment also prevented the occurrence of arthritis in the ipsilateral paw, whereas it had no effect on arthritis in the other paws [35]. These findings underline the critical role of IL-1 in this model and furthermore support the applicability of gene transfer, allowing for treatment of a major joint, yet influencing arthritis expression in nearby, smaller joints.

Table 4 - Histology of CIA after treatment with IL-10, prednisolone or IL-10/prednisolone

Treatment	Dose	Infiltrate	Cartilage damage	Proteoglycan depletion	COMP (µg/ml)
Vehicle	–	1.7 ± 0.9	1.5 ± 1.0	2.3 ± 1.3	8.2 ± 0.8
IL-10	5 µg/day	1.3 ± 0.7	1.2 ± 0.6	2.0 ± 1.0	8.0 ± 0.7
prednisolone	0.05 mg/kg	1.9 ± 1.0	1.7 ± 0.9	2.1 ± 1.0	10.2 ± 1.8
IL-10/pred.	5/0.05	1.0 ± 0.8	0.8 ± 0.7*	1.5 ± 1.0*	4.8 ± 0.6*

*Treatment of arthritic mice was started at day 28 after immunization with type II collagen. Mice were injected i.p. twice a day with murine IL-10, prednisolone or the combination for 14 consecutive days. The data represent the mean ± SD of at least 10 mice per group. Histology was scored on a scale ranging from 0 to 3. Serum COMP levels were determined by ELISA, levels in normal sera amount 4.2 ± 0.6 µg/ml. *$p < 0.05$ Students' t-test compared to vehicle.*

Final remarks

Murine collagen arthritis is highly IL-1 dependent and is therefore an excellent model to screen novel drugs with IL-1 related activity. However, it should be realized that the model is rather sensitive to other drugs as well, including a range of NSAIDs. The latter effect is less pronounced in established CIA, whereas IL-1 sensitivity still remains and it might be suggested to use the more chronic phase for screening of IL-1 drugs. Another peculiar element of CIA is the dominant involvement of PMNs and the highly erosive character, with full destruction of the articular cartilage and major bone erosions.

Apart from the discussion that CII might not be a crucial autoantigen in RA, the model reflects autoimmune arthritis driven by a cartilage autoantigen and as such represents patterns probably holding for a range of cartilage arthritogens. The high scientific interest in immunomodulation studies in this model will soon provide further insight in therapeutic applicability of tolerance induction [36], be it in an antigen specific way or using the principle of bystander suppression. The latter stands for the suppression of inflammation by Th2 or Th3 cells, directed against a nonrelated antigen and causing suppression by local production of suppressive cytokines such as IL-4, IL-10 or TGFβ [37]. In that sense defined tolerization against cartilage-specific "antigens" might prove of therapeutic value. Collagen arthritis is only one of the many available arthritis models. In the pharmaceutical industry the first screening of drugs is often done in adjuvant arthritis, mainly related to ease of handling and historical reasons as discussed in Chapter 1. The choice for the one or the other model should be based on the desired drug profile and the characteristic arthritic aspects of the various models. Recent reviews on arthritis models [38–40] are suggested for further reading.

References

1 Trentham DE, Townes AS, Kang AH (1977) Autoimmunity to type II collagen: an experimental model of arthritis. *J Exp Med* 146: 857–868
2 Stuart JM, Townes AS, Kang AH (1982) Nature and specificity of immune responses to collagen in type II collagen-induced arthritis in mice. *J Clin Invest* 69: 673–683
3 Holmdahl R, Malmstrom V, Vuorio E (1993) Autoimmune recognition of cartilage collagens. *Ann Med* 25: 251–264
4 Glant TT, Mikecz K, Arzoumanian A, Poole AR (1987). Proteoglycan-induced arthritis in Balb/C mice. Clinical features and histopathology. *Arthritis Rheum* 30: 201–212
5 Cremer MA, Ye XJ, Terato K, Owens SW, Seyer JM, Kang AH (1994) Type XI collagen-induced arthritis in the Lewis rat. Characterization of cellular and humoral immune responses to native types XI, V, and II collagen and constituent a-chains. *J Immunol* 153: 824–832
6 Verheijden GFM, Rijnders AWM, Bos E, Coenen-de Roo CJJ, van Staveren CJ, Miltenburg AMM, Meijerink JH, Elewaut D, de Keyser F, Veys E, Boots AMH (1997) Human cartilage glycoprotein-39 as a candidate autoantigen in rheumatoid arthritis. *Arthritis Rheum* 40: 1115–1125
7 Wooley PH, Luthra HS, Stuart JM, David CS (1981) Type II collagen-induced arthritis in mice: I. major histocompatibility complex (I-region) linkage and antibody correlates. *J Exp Med* 154: 688–700
8 Myers LK, Rosloniec EF, Cremer MA, Kang AH (1997) Collagen-induced arthritis, an animal model of autoimmunity. *Life Sciences* 61: 1861–1878
9 Miller EJ, Rhodes RK (1982) Preparation and characterization of the different types of collagen. *Methods Enzymol* 82: 33–65
10 Stuart J, Townes A, Kang A (1984) Collagen autoimmune arthritis. *Annu Rev Immunol* 2: 199–218
11 Holmdahl R, Klareskog L, Rubin K, Larsson E, Wigzell H (1985) T lymphocytes in collagen IL-induced arthritis in mice. Characterization of arthritogenic collagen II-specific T-cell lines and clones. *Scand J Immunol* 22: 295–306
12 Mauri C, Williams RO, Walmsley M, Feldmann M (1996) Relationship between Th/Th2 cytokine patterns and the arthritogenic response in collagen-induced arthritis. *Eur J Immunol* 26: 1511–1518
13 Germann T, Szeliga J, Hess H, Störkel S, Podlaski F, Gately M, Schmitt E, Rüde E (1995) Administration of IL-12 in combination with type II collagen induces severe arthritis in DBA/1 mice. *Proc Natl Acad Sci USA* 92: 4823
14 Szeliga J, Hess H, Rüde E, Schmitt E, Germann T (1996) IL-12 promotes cellular but

not humoral type II collagen-specific Th1-type responses in C57Bl/6 and B10.Q mice and fails to induce arthritis. *Int Immunol* 8: 1221–1227

15 Killar LM, Dunn CJ (1989) Interleukin-1 potentiates the development of collagen-induced arthritis in mice. *Clin Sci* (Colch) 76: 535–538

16 Cooper WO, Fava RA, Gates CA, Cremer MA, Townes AS (1992) Acceleration of onset of collagen-induced arthritis by intra-articular injection of tumor necrosis factor or TGFβ. *Clin Exp Immunol* 89: 244–250

17 Santambrogio L, Hochwald GM, Leu GH, Thorbecke GJ (1993) Antagonistic effects of endogenous and exogenous TGFβ and TNFα on autoimmune diseases in mice. *Immunopharmacol Immunotoxicol* 15: 461–478

18 Caccese RG, Zimmerman JL, Carlson RP (1992) Bacterial lipopolysaccharide potentiates type II collagen-induced arthritis in mice. *Mediators Inflamm* 1: 273–279

19 Joosten LAB, Lubberts E, Helsen MMA, van den Berg WB (1997) Dual role of IL-12 in early and late stages of murine collagen type II arthritis. *J Immunol* 159: 4094–4102

20 Joosten LAB, Helsen MMA, van den Berg WB (1994) Accelerated onset of collagen-induced arthritis by remote inflammation. *Clin Exp Immunol* 97: 204–211

21 Schalkwijk J, van den Berg WB, Joosten LAB, van de Putte LBA (1987) Elastase secreted by activated polymorphonuclear leucocytes causes chondrocyte damage and matrix degradation in intact articular cartilage: escape from inactivation by alpha-1-proteinase inhibitor. *Brit J Exp Pathol* 68: 81–88

22 Van Lent PLEM, van den Bersselaar L, van de Putte LBA, van den Berg WB (1990) Immobilization aggravates cartilage damage during antigen-induced arthritis in mice. Attachment of polymorphonuclear leucocytes to articular cartilage. *Am J Pathol* 136: 1407–1416

23 Van Meurs JBJ, van Lent PLEM, Singer II, Bayne EK, van de Loo FAJ, van den Berg WB (1998) IL-1ra prevents expression of the metalloproteinase-generated neoepitope VDIPEN in antigen-induced arthritis. *Arthritis Rheum* 41: 647–656

24 Kakimoto K, Matsukawa A, Yoshinaga M, Nakamura H (1995) Suppressive effect of neutrophil elastase inhibitor on the development of collagen-induced arthritis. *Cell Immunol* 165: 26–32

25 Williams RO, Feldmann M, Maini RN (1992) Anti-tumor necrosis factor ameliorates joints disease in murine collagen-induced arthritis. *Proc Natl Acad Sci USA* 89: 9784–9788

26 Joosten LAB, Helsen MMA, van de Loo FAJ, van den Berg WB (1996) Anticytokine treatment of established type II collagen-induced arthritis in DBA/1 mice: a comparative study using anti-TNFa, anti-IL-1a/ß, and IL-1ra. *Arthritis Rheum* 39: 797–809

27 Van den Berg WB, Joosten LAB, Helsen MMA, van de Loo AAJ (1994) Amelioration of established murine collagen-induced arthritis with anti-IL-1 treatment. *Clin Exp Immunol* 95: 237–243

28 Mori L, Iselin S, Delibero G, Lesslauer W (1996) Attenuation of collagen induced arthritis in 55kDa TNF receptor type 1 (TNFR1) IgG1 treated and TNFR1 deficient mice. *J Immunol* 157: 3178–3182

29 Kuiper S, Joosten LAB, Bendele AM, Edwards CK III, Arntz OJ, Helsen MMA, van de Loo FAJ, van den Berg WB (1998) Different roles of TNFα and IL-1 in murine streptococcal cell wall arthritis. *Cytokine* 10: 690–702

30 Van Lent PLEM, Holthuysen AEM, van den Bersselaar LAM, van Rooijen N, Joosten LAB, van de Loo FAJ, van de Putte LBA, van den Berg WB (1996) Phagocytic lining cells determine local expression of inflammation in type II collagen-induced arthritis. *Arthritis Rheum* 39: 1545–1555

31 Kasama T, Strieter RM, Lukacs NW, Lincoln PM, Burdick MD, Kunkel SL (1995) IL-10 expression and chemokine regulation during the evolution of murine type II collagen-induced arthritis. *J Clin Invest* 95: 2868–2876

32 Joosten LAB, Lubberts E, Durez P, Helsen MMA, Jacobs MJM, Goldman M, van den Berg WB (1997) Role of IL-4 and IL-10 in murine collagen-induced arthritis. *Arthritis Rheum* 40: 249–259

33 Griffiths RJ, Smith MA, Roach ML, Stam EJ, Milici AJ, Scampoli DN, Eskra JD, Byrum RS, Koller BH, McNeish JD (1997) Collagen-induced arthritis is reduced in 5-lipoxygenase-activating CT: protein-deficient mice. *J Exp Med* 185: 1123–1129

34 Mansson B, Carey D, Alini M, Ionescu M, Rosenberg LC, Poole AR, Heinegard D, Saxne T (1995) Cartilage and bone metabolism in RA. Differences between rapid and slow progression of disease identified by serum markers of cartilage metabolism. *J Clin Invest* 95: 1071–1077

35 Bakker AC, Joosten LAB, Arntz OJ, Helsen MMA, Bendele A, van de Loo FAJ, van den Berg WB (1997). Prevention of murine collagen-induced arthritis in the knee and ipsilateral paw by local expression of human IL-1ra protein in the knee. *Arthritis Rheum* 40: 893–900

36 Myers LK, Seyer JM, Stuart JM, Kang AH (1997) Suppression of murine collagen-induced arthritis by nasal administration of collagen. *Immunol* 90: 161–164

37 Miossec P, van den Berg WB (1997) Th1/Th2 cytokine balance in arthritis. *Arthritis Rheum* 40: 2105–2115

38 Van den Berg WB (1998) Animal models of arthritis: Applicability. In: P Maddison, D Isenberg, P Woo, D Glass (eds): *Oxford textbook of arthritis*. Oxford University Press, Oxford, 559–573

39 Van den Berg WB, van den Broek MF, van de Putte LBA, van Bruggen MCJ, van Lent PLEM (1991) Experimental arthritis: Importance of T cells and antigen mimicry in chronicity and treatment. In: Kresina TF (ed): *Monoclonal antibodies, cytokines, and arthritis*. Dekker, New York, 237–252

40 van den Berg WB (1998) Role of T cells in arthritis: lessons from animal models. In: P Miossec, WB van den Berg WB, GS Firestein (eds): *T cells in arthritis*. Birkhäuser Verlag, Basel, 75–92

Joint and cartilage degradation

E. Jonathan Lewis, Jill Bishop and Anna K. Greenham

Biology Department, Roche Discovery Welwyn, Welwyn Garden City, Herts. AL7 3AY, UK

Introduction

Degradation of cartilage is recognised as a key event in arthritis, yet this is a difficult process to accurately quantify *in vivo*. In the clinic, joint space narrowing is used as a measure of cartilage loss because cartilage itself cannot be visualised by radiography. There are limitations to this technique because of the precision of alignment of the joint needed to achieve reproducible quantification [1, 2]. High definition images of cartilage in the joint can be produced by magnetic resonance imaging [3]. Quantification of cartilage loss would then require MRI systems which capture slices from the entire joint to perform volumetric analysis of the total joint cartilage. At the present time this can be achieved, but it is expensive and requires long imaging runs of many hours which would be impossible with conscious patients.

Surrogate markers of joint degradation have been evaluated and some of these markers show promise, but the structure of the degrading joint may limit the usefulness of the markers [4]. The measured synovial or plasma concentrations of a cartilage-related marker could arise from either a small volume of cartilage in a badly eroded joint or from a large volume of healthy cartilage which is only partially degraded.

In animal models of arthritis, degradation of cartilage has traditionally been quantified by histology [5, 6]. This provides a visual score of cartilage loss but realistically only offers an estimate of the change unless sequential sections are scored, which is time-consuming. Other techniques have been tried in animal models such as quantifying proteoglycan concentration using radiolabelled ligands [7, 8]. This technique requires living tissue and proteoglycan is only one measure of intact cartilage, which is composed of a network of collagen fibrils interspersed with large proteoglycan molecules to confer rigidity to the structure.

To address the issue of accurately measuring cartilage changes during inflammatory reactions we have developed/modified two animal models. The sponge/cartilage model was developed to study the effects of drugs on cartilage degradation by

In Vivo Models of Inflammation, edited by D.W. Morgan and L.A. Marshall

trifugation the sponge collapses into the sleeve and the exudate (maximum of 100 μl) forced out of the sponge. The exudate can then be examined for drug concentration levels, cytokines by ELISA assays or potentially for metalloproteinase levels if specific assays can be developed.

Cartilage matrix was analysed in a variety of ways. The proteoglycan content of cartilages was determined by incubating the cartilages in a papain mixture (0.5 ml of papain in buffer) at 65° C for 1 h and assaying using the Farndale method [15] which has been adapted for use on an automated centrifugal spectro-photometer [16]. Samples were diluted prior to assay: implanted controls 1/10, 1/100 or for non-implanted cartilages 1/1000. To analyse collagen content, the cartilage digests were hydrolysed in 1 M hydrochloric acid for 18 h and the hydrolysate diluted and assayed using an assay for hydroxyproline [17] again adapted for use on the Cobas-bio system [16].

Results

Time-course studies of the sponge/cartilage model have shown that implanted cartilages undergo a series of changes related to the inflammatory reactions (Fig. 1).

Sponges containing *Mtb* induce an inflammatory response which can be easily characterised by measuring the cellular infiltration into the sponge and by weighing the surrounding granuloma. The morphology of the infiltrating cells was determined histologically and shown to be composed of mainly polymorphonuclear leukocytes (PMNs). The number of infiltrating PMNs reached a maximum during the early phase of the inflammatory response (day 9) and in 'live' cartilage implants remained at that level until termination of the experiment (Fig. 1A).

Granulomas were formed by fibroblasts surrounding, but not invading, the sponge. Initially, the granulation tissue weight was large, but in time the weight of

Figure 1
Time-course related changes in the sponge/cartilage model.
Cellulose sponges containing 1 mg of Mtb *and a bovine nasal cartilage cylinders were implanted subcutaneously into the back of groups of five male rats, AHH/r strain. Cellular infiltration into the sponge is shown in Figure 1A. Cartilage which had been repeat freeze/thawed to killed chondrocytes is referred to as dead (■), and untreated as live (▲). Granulation tissue formed around the sponge as shown in Figure 1B. Symbol shows mean values of ten sponge/cartilages taken from the five rats. Changes in the composition of the implanted cartilage cylinders is shown as cartilage weight (Fig. 1C), proteoglycan levels as measured as GAG (Fig. 1D) and collagen as measured by hydroxyproline levels (Fig. 1E). All changes were statistically significant according to Student's* t-test *(unpaired, two-tailed) unless denoted by ns (non-significant).*

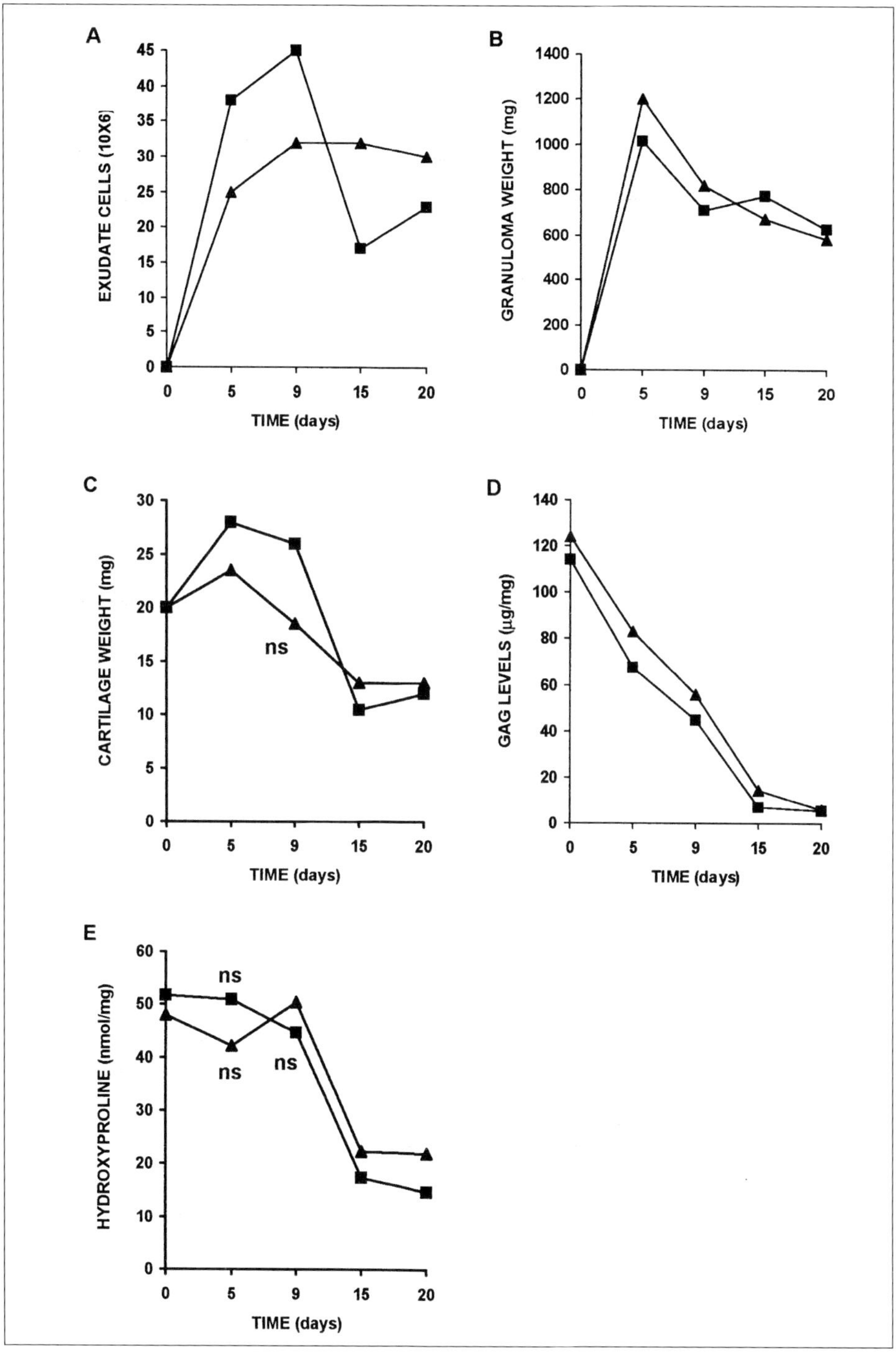
A
EXUDATE CELLS (10X6)
45
40
35
30
25
20
15
10
5
0
0
5
9
15
20
TIME (days)
B
GRANULOMA WEIGHT (mg)
1400
1200
1000
800
600
400
200
0
0
5
9
15
20
TIME (days)
C
CARTILAGE WEIGHT (mg)
30
25
20
15
10
5
0
ns
0
5
9
15
20
TIME (days)
D
GAG LEVELS (µg/mg)
140
120
100
80
60
40
20
0
0
5
9
15
20
TIME (days)
E
HYDROXYPROLINE (nmol/mg)
60
50
40
30
20
10
0
ns
ns
ns
0
5
9
15
20
TIME (days)

tissue decreased, possibly as a consequence of remodelling or due to gradual removal of the irritant from the sponge (Fig. 1B). On day 5, the granuloma enclosed a large volume of fluid (150–250 μl) with the same composition as plasma, but within 9 days a blood barrier was formed by the granulation tissue which prohibited the entry of large molecular weight proteins. This type of barrier is also seen in synovial joints [18]. Studies of the exudate have shown that levels of large plasma proteins (such as caeruloplasmin, with a molecular weight of 132 kD) are reduced relative to plasma levels. Thus, there are probably low levels of α_2 macroglobulin (molecular weight 800 kD), a naturally occurring inhibitor of matrix metalloproteinases (MMPs), within the exudate.

Time-course analysis of implanted cartilage has shown that soon after implantation cartilage weight increases initially before degradation occurs. In cartilages where the chondrocytes have been killed by repeat freeze thawing, there was a greater weight increase than in live cartilage (Fig. 1C). Weight gain has been shown to be related to increased water/fluid content, presumably as water replaces lost proteoglycan. Proteoglycan levels in the cartilage decrease rapidly with 50% loss by day 9 (Fig. 1D). Killed cartilages lose proteoglycan slightly faster than live cartilage, therefore proteoglycan loss is not due to MMPs synthesised by chondrocytes. Previous studies have shown that if the *Mtb* is replaced by the less potent irritants carrageenan or zymosan, there are smaller granulomas and less proteoglycan lost [16]. Proteoglycan loss is probably related to MMPs released from fibroblasts in the granuloma or from the infiltrating cells. The collagen content of cartilages, measured as hydroxyproline, increases in live cartilage on day 9 (Fig. 1E). This effect is possibly because of active synthesis of collagen by chondrocytes before degradation begins. Live and killed cartilages lose hydroxyproline from day 9 onwards, the maximum loss is achieved by day 14–18.

The effects of a number of drugs have been studied in this model (Tab. 1). The major use of this model has been to identify and develop MMP inhibitors such as Ro32-3555 (Trocade™) which is undergoing phase II/III clinical trials for the treatment of rheumatoid arthritis. These drugs show excellent protection of the cartilages which is obvious from the intact physical appearance of the treated cartilages compared to the degraded vehicle-control implanted cartilages [19]. The protective effect of Ro32-3555 is related to its actions as a collagenase selective inhibitor, since this compound lacks potent inhibitory activity against (human) gelatinases and stromelysin. Ro32-3555 dosed animals have smaller granulomas, but this effect was not dose-related to its protective effect on cartilages. Ro32-3555 did not inhibit cellular infiltration, but protected the cartilages from degradation as shown for the dose-related effects on cartilage weight loss, proteoglycan loss and hydroxyproline loss. The inhibition of proteoglycan loss was minor relative to the protection of collagen and was thought to be connected to protection of the collagen which traps the proteoglycan in the collagen matrix. Indomethacin and dexamethasone both protected the cartilages, but both drugs also inhibited cellular infiltration and granulo-

Table 1 - Effect of drugs in the sponge/cartilage model

Drug	Dose mg/kg	% Inhibition				
		Granuloma weight loss	Cellular infiltration	Cartilage weight loss	Proteo-glycan loss	Hydroxy-proline loss
Ro32-3555	25	26***	4	131***	8*	79***
	10	21***	0	82**	9*	48**
	5	25***	9	57	9*	40*
	2.5	20**	0	68**	5	17*
	1	19**	6	0	4	0
Aurothiomalate-	10	6	1	0	0	0
D-penicillamine	150	0	4	26	0	0
Cyclophosphamide	10	6	20***	36	0	0
Indomethacin	2	14**	31**	86***	0	69***
Dexamethasone	0.1	35***	50***	114***	0	60*

The drugs were administered daily for the duration of the experiments 14–16 days. Aurothiomalate was administered by intramuscular injection into the hind leg, dexamethasone administered subcutaneously, Ro32-3555 orally (twice/day), indomethacin, cyclophosphamide and d-penicillamine orally (once per day). Statistical significance of the measured parameters was determined by comparison to sponge/cartilages implanted in vehicle treated animals. Student's t*-test was used for this analysis and shown as* **p<0.05,* ****p<0.01 and* ****p<0.001. % Inhibition was determined relative to vehicle treated animals and also to non-implanted cartilages when appropriate. Group size was 10 animals per dose.*

ma formation thus reducing the release of MMPs by decreasing the numbers of inflammatory cells. The DMARDs tested, D-penicillamine, aurothiomalate and cyclophosphamide had no effect on cartilage protection.

Conclusions

This model has been routinely used to determine the *in vivo* activity of MMP inhibitors selected because of their activity against isolated enzymes and has provided reliable and reproducible results. The specific advantages of this model over other cartilage degradation models is that only one nasal cartilage is needed per experiment to provide cartilage samples for all the implants instead of having to use large numbers of donor rats or mice. The assays developed were accurate and repeatable and measured all the major matrix components cartilage i.e. proteoglycan, collagen and water content by weight. Degradation of the cartilage matrix could be caused by MMPs released from chondrocytes in the cartilage, infiltrating PMNs or from the fibroblastic granuloma. Studies suggest that the granuloma tissue plays the major role, similar to that of the fibroblastic/macrophage derived pannus in the arthritic joint.

The direct effect of drugs on the inflammatory response can be determined by measuring the granuloma weight and cellular infiltration. This model cannot be used to directly compare the efficacies of anti-inflammatory and anti-rheumatic drugs in the clinic, because cartilage matrix components of patients cannot be accurately measured. The comparison between the effects of drugs in this model and those developed by other researchers appears similar, with inhibition of collagen loss by MMP inhibitors [14], steroids [12] and no effect by slow acting anti-rheumatic drugs such as gold and D-penicillamine, although Bottomley et al. showed inhibitory effects in their mouse model [12]. Studies on this model have also shown the presence of a barrier to large proteins, which may be similar to that found in the synovial joint [18] and could reduce levels of the MMP inhibitor $\alpha_2$macroglobulin.

The disadvantage of this model is that evaluation of the effects of drugs on articular cartilage within an articulating, load-bearing synovial joint cannot be made.

Propionibacterium acnes monoarthritis in the rat

Introduction

The most widely used animal model of arthritis is adjuvant arthritis [20]. This model is routinely used in our laboratories, but there are limitations with this model. The Roche adjuvant arthritis studies are performed in AHH/r rats by injecting *Mtb* into the right hind paw on day 0. After 14–16 days these animals develop a severe arthritis and histological sectioning of injected hind paw has shown massive tissue invasion by PMNs, periosteal bone changes and cartilage/bone damage. However, the damage to cartilage is mainly due to mechanical fracturing of the small bones and not due to erosion by pannus-like tissue overlying the bones. Pannus formation by synoviocyte outgrowth is central to rheumatoid arthritis pathology, and it is this pannus that causes erosion of joint [21] probably by release of MMPs [22]. The description of pannus-related changes in studies on *Propionibacterium acnes* (*P. acnes*) monoarthritis in the rat by Trimble et al. (1987) led us to investigate this model [9].

Methods

The bacterium *P. acnes* was obtained from National Collection of Type Cultures (London) as freeze-dried cells (NCTC 737). The cells were initially stab-cultured into sterile 5 ml agar/broth tubes and incubated at 37° C for 7 days. The broth was prepared using 37 g/l of Brain Heart infusion (Oxoid, Unipath Ltd., Basingstoke) and mixed with 0.2% agar before autoclaving. The cells were then cultured in bottles containing 100 ml of sterile Brain Heart infusion and incubated at 37° C. After 3 days there was normally sufficient growth of the cells to harvest by centrifugation (1000 g for 10 min). The cells were then resuspended in 10% formalin in phosphate buffered saline for 1 h at room temperature. Two washes with phosphate buffered saline followed by centrifugation were performed to remove the formalin. The pelleted cells were weighed, then resuspended at a concentration of 40 mg/ml in phosphate buffered saline and stored at –80° C.

An aliquot of cells was thawed and mixed 1:1 with Freund's incomplete adjuvant (Sigma Chemical Co., Poole, Dorset) and sonicated for 1 min. 20 µl of the mixture was injected intra-articularly into the right hind knee joint of isoflurane anaesthetised rats. After 28 days the injection was repeated using the same concentration and formulation of *P. acnes* in Freund's incomplete adjuvant.

Changes in knee diameter were determined by measuring the left and right knee joints using a pair of electronic callipers (Mitutoyo Ltd., Tokyo) and by placing the jaws of the callipers on the lateral and medial aspects of the knee. Knee diameters were measured before the second injection of *P. acnes* into the knee, and on days 3, 7, and 14 after injection. The change in knee diameter was expressed as the difference between left (non-injected) and right knee diameters in the same animal.

On sacrifice, the right hind legs of the animals were dissected free of skin and the legs were removed at the hip joint. The majority of the muscle tissue except around the knee joint was removed by dissection. The legs were then placed on an aluminium holder (a 6 × 6 cm sheet with at 120° fold in the centre) so that all the knee joints were held at the same angle for histological sectioning. Knees were fixed by immersing in 10% buffered formalin solution for 2 to 3 days before transfer to decalcifying solution (10% EDTA) for a 3 week period. The legs were then trimmed to show the knee aspects and the tissues fixed and embedded in wax. Coronal sections were taken through the knee to visualise the patella, femoral and tibial condyles and also one or both meniscoids. Sections were stained using haematoxylin and eosin stain. In the drug studies, the microscopic field of view (final magnification = 145×) which showed the area from the synovial lining layer to the cruciate ligament of the lateral femoral condyle was used for all sections. Image analysis techniques were used to quantify these areas of articular cartilage and the overlying pannus. The camera image from the microscope was digitised (750 × 550 pixels) by a frame-grabbing board (Primagraphics, Virtuoso board) connected to a 486 computer running an image analysis computer programme (PC_image, Foster-Findlay Ltd., Newcastle-on-Tyne). The software provides both image analysis and image processing programmes for 8-bit images. The area of cartilage was defined by cell morphology whilst viewing the section through the microscope and the area determined by using the image analysis system to draw around the edge of the area. Pannus area was quantified in the same manner.

The effects of drugs on this model were determined by daily oral administration of the drugs (or subcutaneous for dexamethasone) from the day of challenge injection (day 0) to the termination of the experiment on day 14.

Results

Injection of *P. acnes* into groups of rats induced a significant increase in knee diameter of the injected right knee compared to the non-injected left knee (Fig. 2A). The maximal increase in knee diameter occurred on day 3, with significant increases on days 7 and 14.

Quantification of articular cartilage in the lateral femoral condyle compartment showed significant loss of cartilage on day 7 and 14. The volume of pannus directly overlying the cartilage was also quantified and cartilage area decreased in direct proportion to the increase in pannus (Fig. 2B).

In the time-course study, sections were taken through the knees of animals to show the lateral femoral condyle from the synovial lining layer to the cruciate ligament attachment site (Fig. 3A–D). Three days after challenge injection, extensive inflammatory cell infiltration and oedema inside the synovial capsule extended from the space between the patella and femur around the sides of the knee and into the femoral-tibial joint space (Fig. 3B). The inflammatory cells were loosely packed, mainly polymorphonuclear leukocytes (PMNs) and fibroblast-like cells with evidence of erythrocyte clumps possibly caused by haemorrhage. In most of the sections there was little damage to the cartilage or bone; in one section there was focal necrosis of the chondrocytes and early pannus-driven erosion of the cartilage. The fat pad layer between the femoral and tibial condyles was completely infiltrated

A 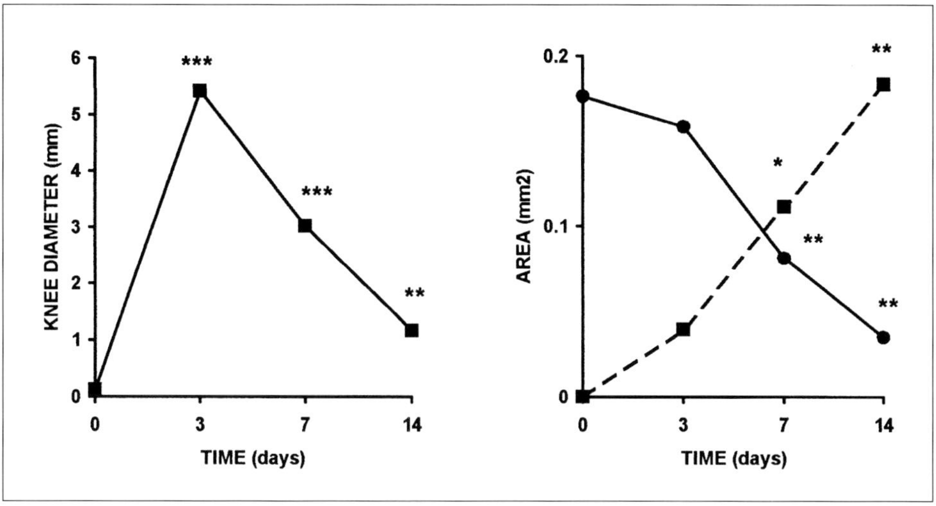 B

Figure 2
Time-course related changes in the P. acnes *model of monoarthritis.*
Rats were sensitised with an intra-articular injection of P. acnes *(40 µg in 1:1 saline:Freund's incomplete adjuvant). The injection was repeated 28 days later to induce the chronic inflammatory response. Changes in knee diameter (Fig. 2A) were determined by comparing left and right knee diameter values measured by electronic callipers. Cartilage (•) and pannus (▪) areas (Fig. 2B) were measured by image analysis of histological sections stained by haematoxylin and eosin. Measurements were taken of the lateral femoral condyles. Statistical significance is shown as* $^{*}p<0.05$, $^{**}p<0.01$ *and* $^{***}p<0.001$ *according to Student's* t-test *(unpaired, two-tailed).*

with inflammatory cells with the occasional large clear area possibly caused by the presence of oil from the injection. There was exudate within the joint space and presence of fibrin and PMNs.

By day 7 there was significant bone outgrowth to the sides of the femoral condyles due to periosteal cell hyperplasia (Fig. 3C). Synovitis was obvious, and the inflammatory exudate in this tissue seemed less extensive compared to day 3 sections. There were fewer PMNs but an increase in numbers of lymphocytes, macrophages and fibroblast-like cells. The margins of the fat pad layer showed densely packed synoviocyte hyperplasia. This cellular layer was also found overlying the articular cartilage on both the lateral and medial femoral condyles, but not on the patella or opposing femoral cartilage. The cartilage was mainly undamaged, but there were areas where chondrocytes were hyperplastic.

Sections taken from rats 14 days after challenge often showed extensive loss of articular cartilage by overgrowth of the inflammatory pannus layer (Fig. 3D). In

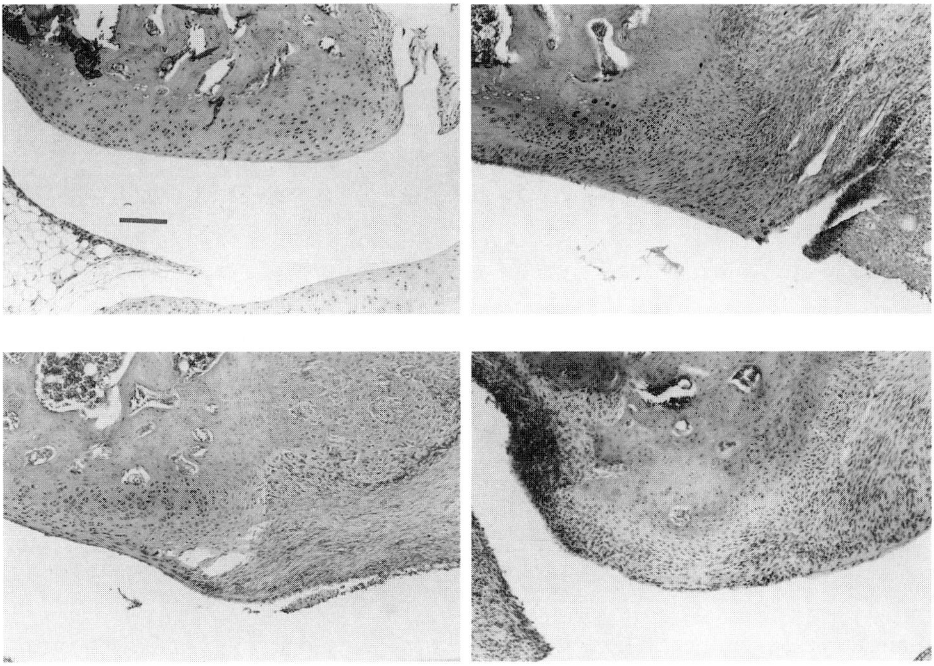

Figure 3

Histology photomicrographs of the lateral femoral condyle of normal rat and animals with P. acnes *induced arthritis.*

In the normal, non-arthritic knee the synovial lining layer is seen (RHS) connected to the articular cartilage of the condyle (Fig. 3A). The cartilage layer is differentiated from the bone layer by the presence of cartilage chondrocytes and haemopoetic (dark) areas in the bone. The joint space between the femoral and tibial condyles is partially filled by a fat cell layer (central and LHS). The black line denotes the scale of 100 µm. In the arthritic animals there is a pronounced thickening of the synovial lining layer as seen in the day 3 photomicrograph (Fig. 3B). In this section, there is massive infiltration by inflammatory cells (RHS) with the presence of oedema in the tissues up to the edge of the cartilage. there is also some overgrowth of the cartilage by inflammatory cells. By day 7 (Fig. 3C) the inflammatory tissue is less oedematous and has formed a thickened layer with overgrowth of the cartilage. There is also bone growth evident to the side of the condyle (top, RHS). The formation of the pannus layer is obvious by day 14 (Fig. 3D) as a thick dense layer of cells overlying the articular cartilage and invading the cartilage and bone (LHS). New bone formation on the side of the femoral condyles (RHS) has formed haemopoetic centres. There is also infiltration/proliferation of the fat pad layer by synovial lining layer cells or by macrophage/fibroblastic cells (bottom LHS).

Table 2 - Effect of drugs on the P. acnes *model of monoarthritis*

Drug	Dose mg/kg	% Inhibition		
		Knee swelling	Cartilage loss	Pannus formation
Ro32-3555	50	–4	49*	–15
Auranofin	10	0	–40	–23
D-Penicillamine	150	16	–32	–5
Indomethacin	2	56**	–2	33
Dexamethasone	0.1	29**	34	74*

The drugs were administered daily throughout the 14 days of the experiments. Dexamethasone administered subcutaneously, Ro32-3555, indomethacin and D-penicillamine orally (once per day). Statistical significance of the measured parameters was determined by comparison to vehicle treated animals. Student's t*-test was used for this analysis and significant effects shown as* *$p<0.05$, **$p<0.01$ *and* ***$p<0.001$. *Group size was 12 animals per dose.*

some areas the pannus layer had destroyed the cartilage and was actively invading the underlying bone structure. Osteoclasts invading bone at pannus/bone junctions were seen in some sections. In areas unaffected with pannus the cartilage chondrocytes had a normal morphological appearance. The pattern of pannus-related damage with the most consistent area was the lateral femoral condyles and at the insertion point of the cruciate ligament; pannus was occasionally found on the medial condyle and meniscoids but seldom at the intracondyle groove. The inflammatory cells in the joint were predominately monocytic or fibroblast-like, very few PMNs were visible. Lymphocytes were present in the inflammatory layer, occasionally grouped together but not as an obvious germinal centre or lymphocyte follicle. The new bone formed at the edge of the condyles showed pronounced haemopoetic cell development and looked similar to the existing bone in appearance.

This generalised bone growth (periostosis) is not an osteoarthritic change but a response to inflammation seen in rats, possibly because the growth plates do not close with age as seen in man. Periostosis also occurs in other rat arthritis models such as adjuvant and Type II collagen-induced arthritis and excessive new bone growth means that radiological assessment of bone changes in these models (or rats per se) is very difficult to accurately quantify.

Analysis of the whole of the joint on day 14 showed that the area of cartilage most consistently affected by pannus was the lateral femoral condyle, and this area was used to quantify the effects of drugs on cartilage and pannus. There was a degree of variability in cartilage degradation between animals, which is probably similar to that seen in RA patients. To obtain statistically significant effects any drug must inhibit degradation by about 50%. The degree of variation in animals can be reduced by having larger sized groups (12–15 animals), and rejecting animals that do not show pronounced knee swelling on day 3 after sensitisation.

Drugs administered to the animals showed a variety of effects. Ro32-3555 (50 mg/kg po) had no effect on knee swelling, but did significantly protect the articular cartilage without reducing the pannus tissue (Tab. 2). Auranofin and D-penicillamine dosed animals had smaller areas of cartilage than the control (vehicle-dosed) group, although this was not statistically significant. Indomethacin and dexamethasone both significantly reduced knee swelling and appear to have some effect on pannus formation (the indomethacin effect was not statistically significant) but did not protect cartilage from erosion. Only Ro32-3555, the collagenase specific inhibitor, protects cartilage in this model which shows great potential for these types of modulators in the clinic.

Conclusions

The injection of *P. acnes* into the knee of sensitised rats causes a short-term inflammatory swelling lasting 14 days. Investigation of the knee pathology using histological methods revealed that in this model there is a chronic inflammatory response characterised by synovitis, a lack of extensive infiltration by PMNs, pannus formation by fibroblast/macrophage-like cells resulting in cartilage degradation and the erosion of bone. This pattern of chronic inflammation induced damage is seen on femoral and tibial condyles as well as the patella-femoral junction. These changes are very similar to descriptions of rheumatoid arthritis pathology [23, 24].

The model has been used only for tertiary drug screen analysis because of the time-consuming histology and analysis required. The use of image analysis and strict positioning of the knee for sectioning has allowed us to accurately quantify cartilage degradation and pannus formation. Although variability limits the usefulness of the *P. acnes* model for general drug screening purposes, the similarities in pathology to (human) arthritis make this model extremely valuable for the evaluation of disease-modifying anti-rheumatic drugs.

References

1 Resnick D, Niwayama G (1988) Degenerative diseases of extraspinal locations. In: D Resnick, G Niwayama (eds): *Diagnosis of bone and joint disorders*. 2nd edition. Saunders, Philadelphia, 1365–1479

2 Buckland-Wright JC, Carmichael I, Walker SR (1986) Quantitative microfocal radiography accurately detects joint changes in rheumatoid arthritis. *Annals Rheum Dis* 45: 379–383

3 Carpenter TA, Everett JR, Hall LD, Harper GP, Hodgson RJ, James MF, Watson PJ (1994) High resolution magnetic resonance imaging of arthritic pathology in the rat knee. *Skeletal Radol* 23: 429–437

4 Heingrad D, Saxne T (1991) Molecular markers of processes in cartilage in joint disease. *Br J Rheumatol* 30 (Suppl 1): 21–24

5 Walton M: (1977) Degenerative joint disease in the mouse knee: radiological and morphological observations. *J Path* 123: 97–107

6 Meachim G (1963) The effect of scarification on articular cartilage in the rabbit. *J Bone Joint Surg* 58A: 230–242

7 Clay K, Seed MP, Clements-Jewery S (1989) Studies on intereleukin-1b induced glycosaminoglycan release from rat femoral head cartilage *in vitro*. *J Pharm Phamacol* 41: 503–504

8 Arner EC, Di Meo TM, Ruhl DM, Pratta MA (1989) *in vivo* studies on the effects of recombinant interleukin on articular cartilage. *Agents and Actions* 27: 254–257

9 Trimble BS, Evers CJ, Ballaron SA, Young JM (1987) Intraarticular injection of *Propionibacterium acnes* causes an erosive arthritis in rats. *Agents and Actions* 21: 281–283

10 Sin YA, Sedgwick AD, Willoughby DA (1984) Studies on the mechanism of cartilage degradation. *J Path* 142: 23–30

11 Chandler CL, Colville-Nash PR, Moore AR, Howat DW, Desa FM, Willoughby DA (1989) The effects of heparin and cortisone on an experimental model of pannus. *Int J Tiss Reac* 11: 113–116

12 Bottomley KMK, Griffiths RJ, Rising TJ, Steward A (1988) A modified mouse air pouch model for evaluating the effects of compounds on granuloma induced cartilage degradation. *Br J Pharmacol* 93: 627–635

13 DiPasquale G, Conaty J, Dea D, Perry K (1987) The potential use of implanted radiolabelled bovine nasal cartilage in dialysis tubing to evaluate agents affecting cartilage degradation. *Agents and Actions* 21: 331–333

14 Karran EH, Dodgson K, Harris SJ, Markwell RE, Harper GP (1995) A simple *in vivo* model of collagen degradation using collagen-gelled cotton buds: The effects of collagenase inhibitors and other agents. *Inflammation Res* 44: 36–46

15 Farndale RW, Sayer CA, Barrett AJ (1982) A direct spectrophotometric microassay for sulphated glycosaminoglycans in cartilage culture. *Connect Tissue Res* 9: 247–250

16 Bishop J, Greenham AK, Lewis EJ (1993) A novel *in vivo* model for the study of cartilage degradation. *J Pharmac and Toxicol Methods* 30: 19–25

17 Kwok-Chu H, Chi-Pui P (1989) Automated analysis of urinary hydroxyproline. *Clin Chem Acta* 185: 191–196

18 Levick JR (1981) Permeability of rheumatoid and normal human synovium to specific plasma proteins. *Arth Rheum* 24: 1550–1560

19 Lewis EJ, Bishop J, Bottomley KMK, Bradshaw D, Brewster M, Broadhurst MJ, Brown PA, Budd JM, Elliot L, Greenham AK et al (1997) Ro 32-3555, an orally active collagenase inhibitor, prevents cartilage breakdown *in vitro* and *in vivo*. *Br J Pharmacol* 121: 540–546

20 Billingham MEJ, Davies GE (1983) Experimental models of arthritis in animals as screening tests for drugs to treat arthritis in man. In: JR Vane, SH Ferreira (eds): *Handbook of experimental pharmacology*. Springer-Verlag, Berlin, 50/2: 108–144
21 Krane, SM (1974) Joint erosion in rheumatoid arthritis. *Arth Rheum* 1: 306–312
22 Krane SM, Amento EP, Goldring MB, Goldring SR, Stephenson ML (1988) Modulation of matrix synthesis and degradation in joint inflammation. In: Glavert AM (eds): *Research monographs in cell and tissue physiology*, 15. Elsevier, Amsterdam, 179–195
23 Gardner DL (1992) Rheumatoid arthritis: cell and tissue pathology. In: DL Gardiner (ed): *The pathological basis of the connective tissue diseases*. Edward Arnold, London, 444–526
24 Gardner DL (1992) Rheumatoid arthritis: pathogenesis, experimental studies and complications. In: DL Gardiner (ed): *The pathological basis of the connective tissue diseases*. Edward Arnold, London, 527–567

Angiogenesis

James D. Winkler, Jeffrey R. Jackson, Tai-Ping Fan[1] *and Michael P. Seed*[2]

Department of Immunopharmacology, SmithKline Beecham Pharmaceuticals, 709 Swedeland Road, King of Prussia, PA 19406 USA; [1]Department of Pharmacology, University of Cambridge, Cambridge, UK; and [2]Panceutics Ltd., P.O. Box 11358, Swindon, Wilts, SN3 4GP, UK

Introduction

Angiogenesis is defined as the production of new blood vessels from pre-existing blood vessels. It is a carefully controlled process that occurs in a healthy individual only under specific conditions at specific times, such as during embryogenesis, wound healing, ovulation and menses. At other times, the vasculature is extremely stable, with very low rates of endothelial cell turnover and little production of new vessels [1]. However, there are pathological conditions in which angiogenesis occurs, examples being cancer, arthritis, ocular diseases, psoriasis [2–4]. In fact, the hypothesis has been proposed that many of these diseases depend on the new vasculature to provide a basis for their proliferation [5–7]. Thus, inhibition of angiogenesis is proposed to be a potential therapeutic approach to chronic proliferative diseases.

It is important to stress in this book that there is an interdependence between angiogenesis and inflammation [4, 8]. Inflammation occurs at most sites of angiogenesis and inflammatory cells can and may provide most of the pro-inflammatory mediators that drive angiogenesis [9]. In turn, angiogenesis provides the support needed for chronic, proliferative inflammatory diseases, such as arthritis and psoriasis. Thus, it is appropriate to consider angiogenesis and its models in a book on in vivo models of inflammatory diseases.

General comments on angiogenesis measurement

By definition, the angiogenic response, the formation of new blood vessels, is a response that must be measured *in vivo*. Certain aspects of angiogenesis can be mimicked in culture systems, such as endothelial cell proliferation and migration [10], endothelial cell tube formation [11] or the production of pro-angiogenic growth factors in cell systems [12]. Although each of these aspects is important to study in isolation, they do not mimic the entire angiogenic process. Angiogenesis is

In Vivo Models of Inflammation, edited by D.W. Morgan and L.A. Marshall

a complex process, involving many cell types, in various states of development and activation, and a balance of angiogenic promoters and inhibitors. Thus angiogenesis models require the presence of several different cell types (endothelial, smooth muscle, pericyte, inflammatory), the right type of extracellular matrix and the right balance of angiogenic factors [13, 14]. The *in vitro* system that comes closest to these conditions is the aortic explant system [15, 16], making it an excellent transition system between *in vitro* and *in vivo* models. In this chapter, we will limit our discussion to *in vivo* angiogenesis models. The reader is referred to other reviews of *in vitro* and *in vivo* angiogenesis models, for completeness [1, 4, 10, 17–19].

The need to measure angiogenesis *in vivo* leads to the necessity of measuring angiogenesis in models that are more complex than *in vitro* systems. *in vivo* studies are encumbered with certain questions/issues, including: which angiogenesis model to employ; what stimulus to use to drive angiogenesis; what method to use to quantitate angiogenesis; what are the needs for throughput and quantitation; and what are the limitations of interpretation. This chapter will review these issues for a variety of angiogenesis models. Table 1 summarizes the key advantages and disadvantages of these models.

Chorioalantoic membrane

Brief overview

One often used *in vivo* model is the avian chorioalantoic membrane (CAM) model [20–22]. This model has been used for decades as a model of angiogenesis. One could argue about whether this is truly an *in vivo* model, but it does examine bonafide angiogenesis and its longevity and usefulness require its inclusion in this review.

Use, examples of data

Avian, often chicken, fertilized eggs are opened to expose the chorioalantoic membrane. Studies can be started at different times after fertilization, from 3 to 13 days, and run for several days. A disc allowing slow release of a known or presumed angiogenic factor, with or without inhibitor, is placed on the membrane. Angiogenesis is measured by visual inspection or by examining growth in the area using radiolabeled proline incorporation to quantitate collagen synthesis [23]. Examples of angiogenic factors studied with the technique include bFGF, VEGF and its homologues and thrombin [22, 24, 25]. Inhibitors studied using the CAM include metalloproteinase inhibitors, antagonists of $\alpha v\beta 3$, AGM-1470, and cAMP activators [26–29].

Table 1 – Summary of key advantages and disadvantages of in vivo *angiogenesis models*

Model	Description	Advantages	Disadvantages
Chorioalantoic membrane	A disc with an angiogenic factor, ± inhibitor, is placed on the egg chorioalantoic membrane. Angiogenesis measured by visual inspection or with radiolabeled proline.	• High through-put, • high local concentration of compounds	• Non-mammalian, • Difficult to quantitate
Matrix implants	Matrigel mixed with various angiogenic stimuli, implanted sc. After removal, angiogenesis assessed by morphometric analysis or by cellular infiltration.	• Ease of handling • Allows addition of specific mediators	• Presence of endogenous growth factors • Fixed matrix components
Granulomas	Granulomatous response initiated by pro-inflammatory stimuli; angiogenesis measured by dye injection, radioactive xenon gas or doppler blood flow.	• Complex, inflammatory stimuli • Quantitation straightforward	• Requires systemic drug exposure
Ocular	Pellets containing angiogenic factor or cellular extracts are implanted into the cornea.	• Allows *in vivo* measurement over time	• Difficult surgery • Difficult to quantitate
Wound healing	Surgical wounding of skin, blood flow measured by histological means, doppler flowmetry or xenon clearance.	• Manipulations easy • Topical treatment • Can use human skin	• Risk of infection • Drug stability problems
Window models	Windows placed in the skin or mesentery containing grids with factors to be studied or angiogenic fluids or cancer cells. Angiogenesis quantitated by counting number of vessels or by image analysis.	• Can study individual vessels • Small amount of compound needed	• Difficult surgery • Low throughput
Cancer	Injected cancer cells allowed to to grow over time. The resulting cancer growth is assessed by weight, number of foci, increased tissue burden or time to death.	• High clinical relevance • Can use human cancers	• Each cancer is different • Difficult to link results to angiogenesis

Advantages

The CAM may be thought of as an *in vitro/in vivo* transition. As it can look at angiogenesis stimulated by applied agents, it can test specific molecular targets. This model provides the potential for high throughput, to allow assessment of many samples. Because of the topical application of compounds, it can provide a high local concentration of agents for testing, using a low amount of material.

Limitations, comments

This system is of course non-mammalian, so there are limitations of interpreting results across species. It involved embryonic angiogenesis, which may have unknown differences from disease-driven angiogenesis. Many investigators have found this system to be too variable to provide dose-response data and methods to quantitate the results can be difficult. For these reasons, this model is often used for semi-quantitative results.

Matrix Implants

Brief overview

As mentioned in the introduction, one of the important factors for angiogenesis is the matrix in which the new blood vessel needs to grow. An importance advance in the field was made with the discovery of a matrix material, Matrigel, that could be made in large quantities and could be used to support angiogenesis studies [30].

Use, examples of data

Matrigel can be mixed with various angiogenic stimuli and implanted subcutaneously. After removal, the extent of angiogenesis can be assessed by morphometric analysis and the results expressed as a percentage of the matrigel that is vascularized. Or, assessment of cellular infiltration can be done using image analysis [31,32]. This method gives results that are consistent enough to assess inhibition of angiogenesis [33].

Advantages

One advantage is in handling of the material: it is a liquid at 4° C and a solid at body temperature. This allows for ease of mixing and injection, yet allows a solid material to be implanted. Thus, it does not require major surgical steps for implantation. Specific angiogenic factors can be added to the matrigel to promote angiogenesis [31, 33, 34]. In addition, authors have shown that more complex additions can be made, such as synovial fluid [35] and whole cells [36]. The solid implant can be removed at different times to examine the time-course of angiogenesis.

Limitations, comments

One limitation is that some preparations of matrigel contain endogenous growth factors, which must be removed to get a null background. The purity and constancy of the matrix preparation can be an issue as well. Matrigel is a mixture of matrix proteins; some researchers may want to study specific or different mixtures of matrix molecules and their effects on angiogenesis. To remove matrix material entirely from the model, researchers can use implants of inert material, such as tumor cells in alginate beads [37, 38]. As with the CAM assay, angiogenesis is driven by specific mediators and the link to disease-driven angiogenesis needs to be established.

Granuloma models

Brief overview

As mentioned in the introduction, chronic inflammatory diseases often have a strong angiogenic component, the prototypic example being rheumatoid arthritis. However, it is difficult to quantitate angiogenesis in models of arthritis, due to the small size of the pannus tissue and inaccessibility of the joint space. Models in which granulomatous tissue is allowed to grow in a subcutaneous air pouch provide more accessible means to study angiogenesis in inflammatory tissue [39–41].

Use, examples of data

A granulomatous response can be initiated by either injection of a pro-inflammatory stimuli or implantation of a foreign body, such as a piece of sponge or rat femur [40, 42, 43]. The granuloma is allowed to grow for various times and then angiogenesis can be measured by a number of means, including carmine dye injection, radioactive xenon gas and doppler blood flow measurements [18, 42]. These methods have allowed examination and quantitation of a number of angiogenesis inhibitors, including angiostatic steroids, lipid mediator inhibitors and cytokine inhibitors [43–46].

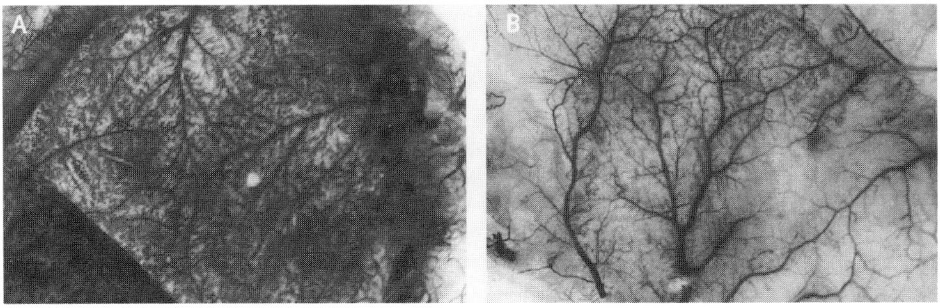

Figure 1
The effect of medroxyprogesterone on inflammatory angiogenesis. Cedarwood oil cleared 6-day granulomas at low magnification from mice treated with 30 mg/kg medroxyprogesterone b.i.d., i.p. (B) or vehicle only (A). Note the substantial reduction in the fine capillaries by the angiostatic steroid. (Reproduced from [8], with permission).

Advantages

This model is driven by complex stimuli, such as adjuvant with crotin oil, so in many ways it can mimic a complex inflammatory angiogenesis. The resulting granuloma has been shown to produce most of the known angiogenic factors [47]. In addition, the characteristics of the granuloma and its angiogenic response change over time after challenge, so that angiogenesis inhibitors can be assessed under different conditions. Quantitation of the vessel volume per gram tissue is straight forward. These models can also be used to study vascular regression. In addition, vascular casting can allow the entire granuloma to be viewed under low power, to get a complete view of the extend of drug effects on angiogenesis (Fig. 1).

Limitations, comments

The complexity of the stimuli that was mentioned as an advantage can also be a disadvantage if the researcher wants to study a specific angiogenic factor. Drug exposure must be systemic, as local injection in the air space can interfere with the granuloma formation. In the adjuvant with crotin oil model, the growth of the granuloma is not limited by angiogenesis [48], so that inhibition of angiogenesis is not reflected in the granuloma size. While a limitation, this observation can also be used to show that a given compound is not anti-inflammatory, as anti-inflammatory compounds inhibit granuloma growth. The model of cotton pellet implantation has granuloma growth more clsely linked to vascular development.

Ocular models

Brief overview

The mammalian cornea is a normally avascular space. That, together with it easy access and visibility in albino animals, makes it an attractive model in which to assess angiogenesis.

Use, examples of data

The first experiments used directly implanted tumor cells [49–51]. To give a sustained release, pellets, made of hydron or Matrigel, can be formed containing the desired angiogenic factor or cellular extracts and implanted into the stroma of the cornea of rabbits, rats or mice [52–54]. Strong vascular responses have been observed with various factors, including bFGF, VEGF and IL1 [53–55]. Inhibition of angiogenesis has been assessed for various compounds [52, 56].

Advantages

This model is often driven by implantation of single angiogenic factors, making it ideal for studying the effects of single factors in a complex system. As measurements can be made in living animals, angiogenesis can be assessed in the same animal over time. Treatments with angiogenesis inhibitors can be either systemic or within the implant.

Limitations, comments

The surgical technique required to perform ocular implants routinely and with minimal inflammation requires some training. To obtain a consistent response thus requires a certain level of experience. The quanitation of the response can be difficult, as various parameters can vary independently, such as the length of blood vessel growth, their density, the lateral extent of growth ("clock hours") and histological assessment.

Wound healing

Brief overview

Following injury to the skin, there is a complex process of healing that occurs. The injury induces an inflammatory response, proliferative and angiogenic mediators are produced, extracellular matrix molecules are made, and repair and differentiation occurs [57–59]. Angiogenesis is a major part of this process, accounting for up to 60% of the repair tissue [60]. As such, it is an *in vivo* model that can be used to study angiogenesis.

Use, examples of data

The model is initiated by surgical wounding of a portion of skin [60]. In addition, cryoinjury can be used in transplanted skin to delay angiogenesis, allowing easier assessment of angiogenic stimulants [61]. Blood flow in the area can be measured by histological means, by measuring Doppler flowmetry or by xenon clearance. Compounds can be added topically or given systemically. This model has been used to examine the expression of angiogenic factors and receptors during the healing process, a recent example being expression of the Tie2 receptor [62]. In addition, one can examine the effects of exogenously added factors on angiogenesis and wound healing [60, 63]

Advantages

One of the key advantages is access to a surface tissue, to allow manipulations and measurements. Treatments can be applied topically. Because of the decreased vascular supply that exists after the wounding, the tissue is hypoxic in the central portion, allowing this stimuli to be studied. In nude mice, human skin can be transplanted, to allow the study of angiogenesis driven by human-specific factors and inhibitors [64].

Limitations, comments

One concern is the possibility of infection in a surface wound, which could be detrimental to the animal as well as the study. The use of topical dressings helps decrease infection, but removes some of the advantages of easy access. Another limitation is a concern about factor or drug stability. Due to the tremendous remodeling of tissue and matrix that occurs in wounds, there are larger amounts of catalytic enzmyes present, which may hamper compound stability.

Window models

Brief overview

Another method used to get an assessable view of the angiogenic process are "window" models, in which blood vessels grow in a small space in tissue in which a window is placed. Thus, these models involve angiogenesis that occurs in a carefully controlled, visual environment. Such windows can include both cutaneous and mesenteric windows.

Use, examples of data

Cutaneous windows are often placed in the dorsal skin. Within the window, researchers can place grids containing factors to be studied or angiogenic fluids or cancer cells [65, 66]. Angiogenesis can be quantitated by counting the number of vessels growing in the grids or by image analysis. Tumor volume and vascularity can also be measured. A recent example of the use of this model is the report of inhibition of tumor growth and angiogenesis by soluble Tie2 receptor [67]. The more sophisticated use of this model is to measure vessel function in the live animal, looking at RBC velocity, blood flow and/or leukocyte adhesion and rolling [65]. The mesenteric window model can be used to study the angiogenic process induced to occur in the rat mesenteric tissue [68]. In this model, increases in angiogenesis have been observed with factors such as VEGF, IL8 and IL1 [69–71].

Advantages

This is an excellent model to study blood vessel function in individual vessels. Due to the small window size, studies can be done with small amounts of compounds or factors, which can be given locally.

Limitations, comments

These models require specific surgical techniques to be mastered and can have low throughput. Because of the difficulty of these models, relatively fewer studies have been done, so there is less literature on the effects of various agents in these models that would help guide future studies.

Cancer models

Brief overview

One of the driving forces in angiogenesis research is the hypothesis that inhibition of angiogenesis will be beneficial in cancer, as cancers will not be able to grow beyond a minimal size without a corresponding increase in vascular support [5,7].

Thus, it is reasonable that many investigators utilize cancer growth as a model for assessing inhibitors of angiogenesis.

Use, examples of data

The models generally involve injection of an inoculum of cancer cells into mice, which are allowed to grow over time. The resulting cancer growth is assessed by weight, number of foci, increased tissue burden or time to death. In addition, vascular density within the tumor can be measured, by vascular casting, image analysis or histological means [72, 73]. Models exist for assessing either primary tumor growth or growth of tumor metastases. Animal are usually treated systemically with potential angiogenesis inhibitors. Examples of studies done with inhibitors of angiogenesis include examinations of AGM-1470, angiostatin, endostatin, lincomide, VEGF antibody and batimastat [74–79].

Advantages

One advantage of this type of model is that it closely mimics the clinical disease and thus compounds that are active in such models may rapidly move into clinical development. It also allows one to study human cancers, which can be implanted in nude mice. Angiogenesis is thought to proceed rapid tumor growth, so that inhibition of angiogenesis can be read as decreases in tumor size. Animals can be treated in therapeutic regimes that mimic the clinical situation, to determine if resistance develops [80].

Limitations, comments

Each tumor model is unique and thus may have its own properties; some tumors may be more angiogenesis-dependent than others and each may produce different angiogenic factors. It may be necessary for each researcher to examine angiogenesis inhibitor "standards" in each tumor model, to determine their sensitivity. Treatments must work against the host vasculature, so that therapeutics, such as human-specific antibodies that do not react with rodent proteins, cannot be studied. It can be difficult to interpret the mechanism of inhibition of tumor growth, as both antiangiogenic and direct antitumor effects can have the same endpoint, inhibition of tumor growth.

Other models

In addition to the above mentioned models, researchers have used other models to measure angiogenesis. Not all *in vivo* models can be covered in this chapter and we have focused on those most widely used. Examples of other models include measurement of angiogenesis in the ovarian follicle during ovulation [81], in avascular retroperitoneal adipose tissue [82] and in areas made ischemic after ligation [83].

Evaluation of models

As can be seen from this chapter, there are a number of *in vivo* animal models that can be used to assess angiogenesis. As is true with all models, the decision of which animal model to use should be based on a number of variables, including the research question to be answered, the type of angiogenic stimuli to be studied, the need for quantitative data, the need for high throughput, the ability of the reagents to cross to different species, the availability of limiting compounds or factors and the expertise available.

One key question is whether the different models are representative of angiogenesis. Here the answer is incomplete. We can say that, where studied, the angiogenic process appears morphologically similar across species and across physiological and pathological conditions. That is, angiogenesis appears to occur in a similar manner in different species and under different conditions. In keeping with this observation, most angiogenic factors have been found to work in most models of angiogenesis, as mentioned throughout this chapter. However, this raises an important point. In the more complex models, such as tumor and granuloma angiogenesis, there can be a number of angiogenic factors at work and angiogenesis may be driven by different growth factors in different models. Thus some inhibitors would be expected to work in some models, but not others, depending on their molecular target.

The next question is, does inhibition of angiogenesis in these models translate into efficacy in animal models of disease or into clinical efficacy? Compounds that have been effective in angiogenesis models, such as CAM, ocular and granuloma models, have also shown efficacy in angiogenesis-dependent disease models, such as animal models of arthritis [84, 85]. As described herein, the cancer and wound healing models of angiogenesis are themselves models of disease, so compounds that inhibit in these models are presumed to be anti-cancer and anti-angiogenic at the same time. However, this assumption will not always hold true. Certainly, many compounds are anti-cancer without being anti-angiogenic. Even compounds that have been shown to be anti-angiogenic in some models must be examined critically in tumor models, as they may not have anti-cancer effects in these models by that mechanism. A good example is that of AGM-1470. This compound has been shown to be anti-angiogenic in a number of models and to have anti-cancer properties. However, recent evidence indicates that the mechanism by which this compound has anti-cancer activity is by inhibition of methioine aminopeptidase and may block cancer growth by mechanisms in addition to inhibition of angiogenesis [86]. The ultimate test of the utility of animal models of angiogenesis will be clinical trials of efficacy with specific angiogenesis inhibitors, a test for which we are anxiously waiting results.

Summary

This chapter reviews *in vivo* models of angiogenesis: chorioalantoic membrane, matrix implants, ocular, granuloma, vascular windows, wound healing and cancer. For each of these models, we review their uses, limitations and advantages. The decision of which animal model to use should be based on a number of variables, including the research question, angiogenic stimuli, the need for high throughput and quantitation, species and the availability of limiting compounds or angiogenic factors. While there are advantages to each model, the ultimate correlation with clinical efficacy remains to be proven.

References

1 Fan TP, Jaggar R, Bicknell R (1995) Controlling the vasculature: angiogenesis, anti-angiogenesis and vascular targeting of gene therapy. *Trends Pharmacol Sci* 16: 57–66

2 Polverini PJ (1995) The pathophysiology of angiogenesis. *Crit Rev Oral Biol Med* 6: 230–247

3 Chung SK, Ng A, Min HY, Strattonthomas J, Rosenberg S, Shuman M, Hwang DG (1995) Inhibition of basic fibroblast growth factor-induced corneal angiogenesis by a urokinase plasminogen-activator receptor antagonist. *Invest Ophthal Visual Sci* 36: S30

4 Folkman J, Brem H (1992) Angiogenesis and inflammation. In: JI Gallin, IM Goldstein, R Snyderman (eds): *Inflammation: Basic principles and clinical correlates*, Second edition. Raven Press Ltd, New York, 821–839

5 Folkman J (1972) Anti-angiogenesis: new concept for therapy of solid tumors. *Ann Surg* 175: 409–416

6 Folkman J, Shing Y (1992) Angiogenesis. *J Biol Chem* 267 (16): 10931–10934

7 Folkman J (1992) The role of angiogenesis in tumor growth. *Semin Cancer Biol* 3 (2): 65–71

8 Jackson JR, Seed MP, Kircher CH, Willoughby DA, Winkler JD (1997) The co-dependence of angiogenesis and chronic inflammation. *FASEB J* 11: 457–465

9 Polverini PJ (1997) Role of the macrophage in angiogenesis-dependent diseases. *EXS* 79: 11–28

10 Auerbach W, Auerbach R (1994) Angiogenesis inhibition: A review. *Pharmacol Ther* 63: 265–311

11 Okamura K, Morimoto A, Hamanaka R, Ono M, Kohno K, Uchida Y, Kuwano M (1992) A model system for tumor angiogenesis: involvement of transforming growth factor-alpha in tube formation of human microvascular endothelial cells induced by esophageal cancer cells. *Biochem Biophys Res Comm* 186: 1471–1479
12 Colville Nash PR, Willoughby DA (1997) Growth factors in angiogenesis: current interest and therapeutic potential. *Mol Med Today* 3: 14–23
13 Folkman J (1997) Angiogenesis and angiogenesis inhibition: an overview. EXS 79: 1–8
14 Iruela-Arispe ML, Dvorak HF (1997) Angiogenesis: a dynamic balance of stimulators and inhibitors. *Thromb Haemost* 78: 672–677
15 Villaschi S, Nicosia RF (1993) Angiogenic role of endogenous basic fibroblast growth factor released by rat aorta after injury. *Am J Pathol* 143: 181–190
16 Nicosia RF, Villaschi S (1995) Rat aortic smooth-muscle cells become pericyte-like cells during angiogenesis *in vitro*. *Faseb J* 9: A 587
17 Cockerill GW, Gamble JR, Vadas MA (1995) Angiogenesis: Models and modulators. *Int Rev Cytology* 159: 113–160
18 Colville-Nash PR, Seed MP (1993) The current state of angiostatic therapy, with special reference to rheumatoid arthritis. *Curr Opin Invest Drugs* 2: 763–813
19 Auerbach R, Auerbach W, Polakowski I (1991) Assays for angiogenesis: a review. *Pharmacol Ther* 51: 1–11
20 Folkman J (1974) Proceedings: Tumor angiogenesis factor. *Cancer Res* 34: 2109–2113
21 Nguyen M, Shing Y, Folkman J (1994) Quantitation of angiogenesis and antiangiogenesis in the chick embryo chorioallantoic membrane. *Microvasc Res* 47: 31–40
22 Wilting J, Christ B, Bokeloh M (1991) A modified chorioallantoic membrane (CAM) assay for qualitative and quantitative study of growth factors. Studies on the effects of carriers, PBS, angiogenin, and bFGF. *Anat Embryol Berl* 183: 259–271
23 Maragoudakis ME, Haralabopoulos GC, Tsopanoglou NE, Pipili Synetos E (1995) Validation of collagenous protein synthesis as an index for angiogenesis with the use of morphological methods. *Microvasc Res* 50: 215–222
24 Oh SJ, Jeltsch MM, Birkenhager R, McCarthy JE, Weich HA, Christ B, Alitalo K, Wilting J (1997) VEGF and VEGF-C: specific induction of angiogenesis and lymphangiogenesis in the differentiated avian chorioallantoic membrane. *Dev Biol* 188: 96–109
25 Tsopanoglou NE, Pipili Synetos E, Maragoudakis ME (1993) Thrombin promotes angiogenesis by a mechanism independent of fibrin formation. *Am J Physiol* 264: C1302–C1307
26 Anand Apte B, Pepper MS, Voest E, Montesano R, Olsen B, Murphy G, Apte SS, Zetter B (1997) Inhibition of angiogenesis by tissue inhibitor of metalloproteinase-3. *Invest Ophthalmol Vis Sci* 38: 817–823
27 Brooks PC, Montgomery AMP, Rosenfeld M, Reisfeld RA, Hu T, Klier G, Cheresh DA (1994) Integrin avb3 antagonists promote tumor regression by inducing apoptosis of angiogenic blood vessels. *Cell* 79: 1157–1164
28 Kusaka M, Sudo K, Fujita T, Marui S, Itoh F, Ingber D, Folkman J (1991) Potent anti-

angiogenic action of AGM-1470: comparison to the fumagilliam parent. *Biochem Biophys Res Commun* 174: 1070–1076
29 Tsopanoglou NE, Haralabopoulos GC, Maragoudakis ME (1994) Opposing effects on modulation of angiogenesis by protein kinase C and cAMP-mediated pathways. *J Vasc Res* 31: 195–204
30 Passaniti A, Taylor RM, Pili R, Guo Y, Long PV, Haney JA, Pauly RR, Grant DS, Martin GR (1992) A simple, quantitative method for assessing angiogenesis and antiangiogenic agents using reconstituted basement membrane, heparin, and fibroblast growth factor. *Lab Invest* 67: 519–528
31 Haralabopoulos GC, Grant DS, Kleinman HK, Maragoudakis ME (1997) Thrombin promotes endothelial cell alignment in Matrigel *in vitro* and angiogenesis *in vivo*. *Am J Physiol* 273: C239–C245
32 Angiolillo AL, Kanegane H, Sgadari C, Reaman GH, Tosato G (1997) Interleukin-15 promotes angiogenesis *in vivo*. *Biochem Biophys Res Commun* 233: 231–237
33 Montrucchio G, Lupia E, De Martino A, Battaglia E, Arese M, Tizzani A, Bussolino F, Camussi G (1997) Nitric oxide mediates angiogenesis induced *in vivo* by platelet-activating factor and tumor necrosis factor-alpha. *Am J Pathol* 151: 557–563
34 Joseph IB, Vukanovic J, Isaacs JT (1996) Antiangiogenic treatment with linomide as chemoprevention for prostate, seminal vesicle, and breast carcinogenesis in rodents. *Cancer Res* 56: 3404–3408
35 Lupia E, Montrucchio G, Battaglia E, Modena V, Camussi G (1996) Role of tumor necrosis factor-alpha and platelet-activating factor in neoangiogenesis induced by synovial fluids of patients with rheumatoid arthritis. *Eur J Immunol* 26: 1690–1694
36 Ito Y, Iwamoto Y, Tanaka K, Okuyama K, Sugioka Y (1996) A quantitative assay using basement membrane extracts to study tumor angiogenesis *in vivo*. *Int J Cancer* 67: 148–152
37 Plunkett ML, Hailey JA (1990) An *in vivo* quantitative angiogenesis model using tumor cells entrapped in alginate. *Lab Invest* 62: 510–517
38 Hoffmann J, Schirner M, Menrad A, Schneider MR (1997) A highly sensitive model for quantification of *in vivo* tumor angiogenesis induced by alginate-encapsulated tumor cells. *Cancer Res* 57: 3847–3851
39 Kimura M, Amemiya K, Yamada T, Suzuki J (1986) Quantitative method for measuring adjuvant-induced granuloma angiogenesis in insulin-treated diabetic mice. *J Pharmacobio-Dyn* 9: 442–446
40 Colville-Nash PR, Alam CAS, Appleton I, Browne JR, Seed MP, Willoughby DA (1995) The pharmacological modulation of angiogenesis in chronic granulomatous inflammation. *J Pharmacol Exp Ther* 274: 1463–1472
41 Andrade SP, Fan TP, Lewis GP (1987) Quantitative *in vivo* studies on angiogenesis in a rat sponge model. *Br J Exp Pathol* 68: 755–766
42 Hu DE, Hiley CR, Smither RL, Gresham GA, Fan TP (1995) Correlation of 133Xe clearance, blood flow and histology in the rat sponge model for angiogenesis. Further studies with angiogenic modifiers. *Lab Invest* 72: 601–610

43 Colville-Nash PR, El-Ghazaly M, Willoughby DA (1993) The use of angiostatic steroids to inhibit cartilidge destruction in an *in vivo* model of granuloma mediated cartilidge destruction. *Agents and Actions* 38: 126–134

44 Hori Y, Hu DE, Yasui K, Smither RL, Gresham GA, Fan TP (1996) Differential effects of angiostatic steroids and dexamethasone on angiogenesis and cytokine levels in rat sponge implants. *Br J Pharmacol* 118: 1584–1591

45 Jackson JR, Bolognese B, Hillegass L, Kassis S, Adams J, Griswold DE, Winkler JD (1998) Pharmacological effects of SB 220025, a selective inhibitor of p38 mitogen-activated protein kinase, in angiogenesis and chronic inflammatory disease models. *J Pharmacol Exp Ther* 284: 687–692

46 Jackson JR, Bolognese B, Hubbard WC, Marshall LA, Winkler JD (1998) Platelet-activating factor derived from 14 kDa phospholipase A2 contributes to inflammatory angiogenesis. *Biochem Biophys Acta* 1392: 145–152

47 Appleton I, Tomlinson A, Colville-Nash PR, Willoughby DA (1993) Temporal and spatial immunolocalization of cytokines in murine chronic granulomatous tissue. *Lab Invest* 69: 405–414

48 Kimura M, Suzuki J, Amemiya K (1985) Mouse granuloma pouch induced by Freund's complete adjuvant with croton oil. *J Pharmacobio-Dyn* 8: 393–400

49 Gimbrone MA,Jr., Cotran RS, Leapman SB, Folkman J (1974) Tumor growth and neovascularization: an experimental model using the rabbit cornea. *J Natl Cancer Inst* 52: 413–427

50 Fournier GA, Lutty GA, Watt S, Fenselau A, Patz A (1981) A corneal micropocket assay for angiogenesis in the rat eye. *Invest Ophthalmol Vis Sci* 21: 351–354

51 Muthukkaruppan V, Auerbach R (1979) Angiogenesis in the mouse cornea. *Science* 205: 1416–1418

52 Galardy RE, Grobelny D, Foellmer HG, Fernandez LA (1994) Inhibition of angiogenesis by the matrix metalloprotease inhibitor *N*-[2R-2-(hydroxamidocarbonymethyl)-4-methylpentanoyl)]-L-trypto phan methylamide. *Cancer Res* 54: 4715–4718

53 BenEzra D, Griffin BW, Maftzir G, Aharonov O (1993) Thrombospondin and *in vivo* angiogenesis induced by basic fibroblast growth factor or lipopolysaccharide. *Invest Ophthalmol Vis Sci* 34: 3601–3608

54 Kenyon BM, Voest EE, Chen CC, Flynn E, Folkman J, D'Amato RJ (1996) A model of angiogenesis in the mouse cornea. *Invest Ophthalmol Vis Sci* 37: 1625–1632

55 BenEzra D, Maftzir G (1996) Antibodies to IL-1 and TNF-alpha but not to bFGF or VEGF inhibit angiogenesis. *Invest Ophthalmol Vis Sci* 37: 4664

56 Friedlander M, Brooks PC, Shaffer RW, Kincaid CM, Varner JA, Cheresh DA (1995) Definition of 2 angiogenic pathways by distinct alpha(v) integrins. *Science* 270: 1500–1502

57 Pettet G, Chaplain MA, McElwain DL, Byrne HM (1996) On the role of angiogenesis in wound healing. *Proc R Soc Lond B Biol Sci* 263: 1487–1493

58 Sephel GC, Kennedy R, Kudravi S (1996) Expression of capillary basement membrane components during sequential phases of wound angiogenesis. *Matrix Biol* 15: 263–279

59 Takenaka H, Kishimoto S, Tooyama I, Kimura H, Yasuno H (1997) Protein expression of fibroblast growth factor receptor-1 in keratinocytes during wound healing in rat skin. *J Invest Dermatol* 109: 108–112

60 Arnold F, West DC (1991) Angiogenesis in wound healing. *Pharmacol Ther* 52 (3): 407–422

61 Lees VC, Fan TP (1994) A freeze-injured skin graft model for the quantitative study of basic fibroblast growth factor and other promoters of angiogenesis in wound healing. *Br J Plast Surg* 47: 349–359

62 Wong AL, Haroon ZA, Werner S, Dewhirst MW, Greenberg CS, Peters KG (1997) Tie2 expression and phosphorylation in angiogenic and quiescent adult tissues. *Circ Res* 81: 567–574

63 Okumura M, Okuda T, Okamoto T, Nakamura T, Yajima M (1996) Enhanced angiogenesis and granulation tissue formation by basic fibroblast growth factor in healing-impaired animals. *Arzneimittelforschung* 46: 1021–1026

64 Christofidou-Solomidou M, Bridges M, Murphy GF, Albelda SM, DeLisser HM (1997) Expression and function of endothelial cell alpha v integrin receptors in wound-induced human angiogenesis in human skin/SCID mice chimeras. *Am J Pathol* 151: 975–983

65 Dellian M, Witwer BP, Salehi HA, Yuan F, Jain RK (1996) Quantitation and physiological characterization of angiogenic vessels in mice: effect of basic fibroblast growth factor, vascular endothelial growth factor/vascular permeability factor, and host microenvironment. *Am J Pathol* 149: 59–71

66 Torres Filho IP, Hartley Asp B, Borgstrom P (1995) Quantitative angiogenesis in a syngeneic tumor spheroid model. *Microvasc Res* 49: 212–226

67 Lin P, Polverini P, Dewhirst M, Shan S, Rao PS, Peters K (1997) Inhibition of tumor angiogenesis using a soluble receptor established a role for Tie2 in pathologic vascular growth. *J Clin Invest* 100: 2072–2078

68 Norrby K (1992) On the quantitative rat mesenteric-window angiogenesis assay. *EXS* 61: 282–286

69 Norrby K (1996) Vascular endothelial growth factor and *de novo* mammalian angiogenesis. *Microvasc Res* 51: 153–163

70 Norrby K (1996) Interleukin-8 and *de novo* mammalian angiogenesis. *Cell Prolif* 29: 315–323

71 Norrby K (1997) Interleukin-1-alpha and *de novo* mammalian angiogenesis. *Microvasc Res* 54: 58–64

72 Inoue K, Ozeki Y, Suganuma T, Sugiura Y, Tanaka S (1997) Vascular endothelial growth factor expression in primary esophageal squamous cell carcinoma. Association with angiogenesis and tumor progression. *Cancer* 79: 206–213

73 Seed MP, Brown JR, Freemantle CN, Papworth JL, Colville Nash PR, Willis D, Somerville KW, Asculai S, Willoughby DA (1997) The inhibition of colon-26 adenocarcinoma development and angiogenesis by topical diclofenac in 2.5% hyaluronan. *Cancer Res* 57: 1625–1629

74 Ingber D, Fujita T, Kishimoto S, Sudo K, Kanamaru T, Brem H, Folkman J (1990) Syn-

thetic analogues of fumagillin that inhibit angiogenesis and suppress tumour growth. *Nature* 348: 555–557

75 O'Reilly MS, Boehm T, Shing Y, Fukai N, Vasios G, Lane WS, Flynn E, Birkhead JR, Olsen BR, Folkman J (1997) Endostatin: an endogenous inhibitor of angiogenesis and tumor growth. Cell 88: 277–285

76 Borgstrom P, Torres Filho IP, Hartley Asp B (1995) Inhibition of angiogenesis and metastases of the Lewis-lung cell carcinoma by the quinoline-3-carboxamide, Linomide. *Anticancer Res* 15: 719–728

77 O'Reilly MS, Holmgren L, Shing Y, Chen C, Rosenthal RA, Moses M, Lane WS, Cao Y, Sage EH, Folkman J (1994) Angiostatin: A novel angiogenesis inhibitor that mediates the suppression of metastases by a Lewis lung carcinoma. *Cell* 79: 315–328

78 Taraboletti G, Garofalo A, Belotti D, Drudis T, Borsotti P, Scanziani E, Brown PD, Giavazzi R (1995) Inhibition of angiogenesis and murine hemangioma growth by Batimastat, a synthetic inhibitor of matric metalloproteinases. *J Natl Cancer Inst* 87: 293–298

79 Kim KJ, Li B, Winer J, Armanini M, Gillett N, Phillips HS, Ferrara N (1993) Inhibition of vascular endothelial growth factor-induced angiogenesis suppresses tumour growth *in vivo*. *Nature* 362: 841–844

80 Boehm T, Folkman J, Browder T, O'Reilly MS (1997) Antiangiogenic therapy of experimental cancer does not induce acquired drug resistance. *Nature* 390: 404–407

81 Neeman M, Abramovitch R, Schiffenbauer YS, Tempel C (1997) Regulation of angiogenesis by hypoxic stress: from solid tumours to the ovarian follicle. *Int J Exp Pathol* 78: 57–70

82 Magovern CJ, Mack CA, Zhang J, Rosengart TK, Isom OW, Crystal RG (1997) Regional angiogenesis induced in nonischemic tissue by an adenoviral vector expressing vascular endothelial growth factor. *Hum Gene Ther* 8: 215–227

83 Bauters C, Asahara T, Zheng LP, Takeshita S, Bunting S, Ferrara N, Symes JF, Isner JM (1995) Site-specific therapeutic angiogenesis after systemic administration of vascular endothelial growth factor. *J Vasc Surg* 21: 314–325

84 Oliver SJ, Banquerigo ML, Brahn E (1994) Suppression of collagen-induced arthritis using an angiogenesis inhibitor, AGM- 1470, and a microtubule stabilizer, taxol. *Cell Immunol* 157: 291–299

85 Peacock DJ, Banquerigo ML, Brahn E (1995) A novel angiogenesis inhibitor suppresses rat adjuvant arthritis. *Cell Immunol* 160: 178–184

86 Griffith EC, Su Z, Turk BE, Chen S, Chang YH, Wu Z, Biemann K, Liu JO (1997) Methionine aminopeptidase (type 2) is the common target for angiogenesis inhibitors AGM-1470 and ovalicin. *Chem Biol* 4: 461–471

Asthma

William M. Selig[1] *and Richard W. Chapman*[2]

[1]Hoechst Marion Roussel, Inc., Route 202-206, P.O. Box 6800, Bridgewater, NJ 08807-0800, USA; [2]Schering-Plough Research Institute, 2015 Galloping Hill Road, Kenilworth, NJ 07033-0539, USA

Introduction

Asthma is a highly regulated and complex disease process involving interactions between lung immunology, neural control, and inflammation that result in reversible early and/or late phase airflow obstruction, airway hyperreactivity, cellular infiltration, and histopathologic changes (e.g. mucus and bronchial smooth muscle cell hypertrophy). Much of our current understanding of this disease process, albeit incomplete, has been obtained through preclinical studies using in vivo models developed in a variety of species to target cellular, biochemical, and molecular mechanisms. Some of the more commonly used species that will be discussed in this review include mouse, guinea pig, rat, rabbit, sheep, dog, and primate. Excluded from this list due to practical considerations is the horse which commonly exhibits an allergic response ("heaves") to hay mold spores [1]. Due to limitations of space, we have concentrated on the most recent applications, newer technologies, and drug therapies in these animal models whenever possible. Remember that no single animal model precisely mimics the human condition; however, each model helps advance the understanding of this complex disease process and establishes new, cost-effective therapeutic approaches.

Murine models

The popularity of murine models of asthmatic inflammation stems from the availability of numerous molecular and immunological probes developed for this species and the use of these models to study genetic aspects of asthma. A unique model exploiting these attractive features is discussed in detail in the following chapter by Wills-Karp et al. In addition, acute bronchoconstriction and airway hyperreactivity are also noted after antigen bronchoprovocation in mice [2–5]. Airway hyperreactivity following antigen provocation is more pronounced in high IgE producing mice [4] and is genetically inherent in A/J strain mice [6, 7]. Unfor-

In Vivo Models of Inflammation, edited by D.W. Morgan and L.A. Marshall

tunately, late phase airway obstruction does not occur in mice challenged with antigen [3].

Both allergen-challenged mice and human asthmatics exhibit evidence of pulmonary inflammation characterized by the influx of eosinophils and T cells [2, 3, 8, 9], airway epithelium denudation [8, 10, 11], interstitial airway edema [8, 10], thickening of the basement membrane [12], goblet cell hyperplasia, and mucus hypersecretion [8, 12, 13]. Repeated antigen challenge of mice results in airway goblet cell hyperplasia and lung fibrosis, a condition that duplicates the major histopathological features of chronic asthma [12].

Mast cells and their released products are believed to be critical in the development of various allergic respiratory disorders including asthma. Mast cell deficient mice have been used advantageously to investigate the role of mast cells involvement in antigen-induced pulmonary inflammation and airway hyperreactivity [14, 15]. Antigen challenge of sensitized, mast cell deficient mice resulted in diminished lung eosinophilia compared to controls [14] which could be restored by adoptive transfer of cultured bone-marrow derived mast cells. In contrast, mast cells appear to play less of a role in eosinophil recruitment into the lungs of allergic mice following repeated antigen challenge (i.e. chronic inflammatory conditions) [16].

Although IgE is the major homocytotropic antibody in both murine and human allergy [17], it should be noted that airway hyperresponsiveness, as well as pulmonary eosinophilia, in the allergic mouse may not depend upon traditional IgE/mast cell interactions. Antigen-challenged IgE deficient mice with a mutation of the C epsilon locus develop subsequent airway hyperresponsiveness and pulmonary eosinophilia [18]. This finding has been substantiated in studies demonstrating the presence of pulmonary eosinophilia in B cell deficient mice that have no capability to generate antigen-specific immunoglobulins [19].

Another advantage of the murine asthma model surrounds the fact that T cells infiltrate the lungs of mice following antigen provocation [2, 9, 20] and, similar to human asthma, are believed to be instrumental in driving a portion of the antigen-induced inflammatory response [17]. Much of the original work surrounding the concept of Th_1-versus Th_2-dependent cytokines is based upon work using murine T cells or experiments performed in mice [21]. Immunohistochemical, reverse transcriptase polymerase chain reaction, and enzyme-linked immunosorbent assay techniques have been developed to identify cytokines in this species and applied to characterize murine models of asthma [2, 9, 10, 13, 22–24]. Monoclonal antibodies directed towards specific T cell markers (e.g. CD_4, CD_8, CD_{44}) have been used in murine systems to facilitate the phenotyping of T cell subsets using flow cytometry [2, 9, 20, 24]. Collectively these methods have identified that antigen-induced pulmonary inflammation in mice, like the human, is driven by Th_2-dependent cytokines.

A major area of research is currently focused on the identification of bone marrow progenitors involved with the differentiation and release of leukocytes from the bone marrow following antigen challenge of allergic mice [8, 25]. The mouse offers distinct advantages for this type of experimentation. Many immunological probes required for this evaluation are commercially available. Likewise, the temporal profile of inflammatory cell egress from the bone marrow into the blood and lung tissue after antigen challenge has already been well documented in the mouse.

The primary disadvantage of murine models of asthma is the fact that the airway smooth muscle is poorly developed [26] and large doses of bronchoconstrictor agonists are required to elicit changes in pulmonary mechanics relative to assessing airway hyperreactivity [27]. Newer, improved methodologies for assessing lung mechanics in mice, including the use of forced oscillation techniques and measurement of lung function in conscious, free-roaming mice [3, 28], may enhance the study of antigen-induced hyperreactivity in this species.

Another disadvantage of murine asthma models is the fact that eosinophil degranulation has been difficult to demonstrate in mice following antigen challenge [29]. Detection using monoclonal antibodies directed to the various specific eosinophil granule proteins suggests that these proteins appear to be contained within the intact eosinophil following antigen challenge [29]. In addition, eosinophils are rarely found in the airway epithelium following antigen challenge of the mouse [4, 8, 19].

Cytokines, which regulate growth, differentiation, and effector functions in the lungs, are currently receiving a great deal of attention as therapeutic areas in inflammatory diseases such as asthma. Interleukin-4 (IL-4) and IL-5, expressed in the lungs of both antigen challenged mice [2, 5, 8–10, 13, 26, 30] and human asthmatics [17], are essential cytokines involved in eosinophil function. Pretreating allergic mice with monoclonal antibodies to murine IL-4 or IL-5 before challenge inhibits the development of antigen-induced pulmonary eosinophilia and/or airway hyperreactivity [2, 5, 11, 30–32]. Confirmation of this important relationship has been made in mice that have been genetically engineered to have either an IL-5 deficiency (IL-5 knockouts) or constitutively overexpress IL-5 in the lung (IL-5 transgenics). In allergic IL-5 knockout mice, pulmonary eosinophilia and airway hyperreactivity did not develop in response to antigen challenge, but was noted when the IL-5 gene was reintroduced to these mice [33]. Likewise, IL-5 transgenic mice exhibit pathological changes in the lungs consistent with chronic human asthma including goblet cell hyperplasia, epithelial hypertrophy, and focal collagen deposition [34]. Such studies in mice have defined an important role for IL-5 in the pulmonary pathophysiology of antigen challenge and it is likely that this cytokine will be important in human asthma as well.

Guinea pig models

The guinea pig has been widely utilized as a species to study various components of the classical asthmatic response. Antigen challenge of allergic guinea pigs produces an acute, histaminergic bronchoconstriction, transient airway hyperreactivity that is agonist-dependent, late phase responses, and sustained lung eosinophilia [35–38]. Clinically relevant compounds, β_2 receptor agonists and corticosteroids, can prevent some or all components of the allergic response in guinea pigs [38–40]. There is controversy regarding the development of the late phase response in this species and its preclinical utility relative to compound evaluation based primarily on methodological concerns.

The obvious advantage of the guinea pig allergic model, nonetheless, surrounds the ease of sensitization, the immunologic robustness of the model, and the fact that the allergic response is mechanistically multifactorial in that it is linked to a plethora of peptides, eicosanoids, cytokines, and chemokines. The major drawbacks of using this "allergic" species include: (a) the predominance of an IgG-dependent immunologic mechanism which may not relate to the atopic asthma condition [41, 42], (b) the prevalence toward hypereosinophilia under basal conditions [43], (c) the apparent temporal dissociation between airway hyperreactivity and lung eosinophilia [44, 45], and (d) the inherent biological (e.g. seasonal) variability associated with the model. Despite these limitations, the quinea pig model can be effectively utilized to delineate putative mechanisms involved in the pathogenesis of asthma.

The allergic guinea pig model has not been substantially altered since its inception some 90 years ago. The most notable modifications have surrounded sensitization procedures utilized to impart antigenicity. Sensitization is normally performed by intraperitoneal, subcutaneous, and/or inhalation routes of administration over a 2-week period using antigens in combination with adjuvants. Underwood et al. [46] previously demonstrated that guinea pigs could be readily sensitized to varying degrees with a host of antigens including ovalbumin, *Ascaris suum* extract, and house dust extract. Hsiue et al. [41] substituted crude mite extract (100 µg) from *Dermatophagoides farinae* (a common aeroallergen in humans) plus $Al(OH)_3$ (4 mg) for intraperitoneal ovalbumin injections to sensitize guinea pigs. This sensitization produced profound antigen-induced lung inflammation compared to ovalbumin challenge, although animals did not exhibit sensitization to aerosolized extract (whole body exposure to 10 mg/ml for 30 min). Reidel et al. [42] used experimental viral infection via intranasal Parainfluenza-3 virus (PI-3, 200 µl per nostril) in combination with aerosolized ovalbumin sensitization (0.5% for 3 min 4 and 16 days following inoculation) to augment antigen-induced airway reactivity through an increase in respiratory epithelial permeability. The simple modification of combining silica (50 mg/kg; 0.007 µm) and *Bordetella pertussis* (0.1 ml/kg) with ovalbumin (10–20 µg/kg given twice intraperitoneally over a 2-week period) appears to increase the occurrence of the allergic late phase response to approximately 60% through a postulated silica-induced cytokine-dependent mechanism [47]. Thus, modifications of sensitizing procedures in this model can be used to tailor the system to explore specific mechanisms associated with the pathogenesis of asthma.

Allergic guinea pigs have been routinely pretreated with histamine-1 receptor antagonists to prevent fatal antigen-induced bronchoconstriction. While this technique is widely accepted in this model, this type of pharmacological intervention may in fact diminish or abolish secondary release of mediators that propagate allergen-induced lung inflammation through divergent mechanisms. An alternative is the use of low dose aerosol ovalbumin challenge ($\leq 0.1\%$) for set periods (≥ 30 min) in the absence of antihistamine protective cover [35, 44].

Although not widely utilized, there have been recent attempts to modify the technology for monitoring both lung function and inflammation in the guinea pig model. Invasive techniques to look at lung mechanics in animals following antigen challenge are being replaced by both restrained and unrestrained plethysmographic methodologies which monitor respiratory patterns and/or lung mechanics for extended periods of time [26, 48, 49]. Unfortunately, these systems are generally technically limited in their usefulness when compared to systems now in use to monitor antigen-induced lung inflammation in this species.

Improvements in methodologies for monitoring allergen-induced lung inflammation in the guinea pig include the adaptation of a noninvasive, transoral bronchoalveolar lavage (2×2 ml sterile saline) in the lightly anesthetized animal [47]. This transoral delivery and retrieval of fluid technically allows for the use of individual animals to serve as their own controls in cross-over studies. Use of molecular probes (discussed below) has also enabled researchers to explore mechanisms and future targets more effectively and examine the role of various inflammatory mediators in the allergic guinea pig model.

A significant amount of work has recently focused on the efficacy of phosphodiesterase (PDE) inhibitors, specifically PDE IV, in the allergic guinea pig (Tab. 1) [50–56]. The anti-spasmogenic and anti-eosinophilic activity of PDE IV inhibitors effectively abolishes the key pathophysiologic features of airway hyperreactivity and lung inflammation in this model through inhibition of cyclic AMP catalysis via PDE (i.e. enhanced cyclic AMP-mediated airway smooth muscle relaxation and inhibition of cellular migration and activation). In a recent study, however, Underwood et al. [55] demonstrated that PDE IV inhibitors may act synergistically with endogenous catecholamines to suppress mast cell activation or degranulation. Inhibition of PDE IV isoenzyme, identified in various inflammatory cells associated with the allergic guinea pig model (i.e. eosinophils, neutrophils, lymphocytes, and macrophages), results in inflammatory cell suppression [52].

The allergic guinea pig is currently utilized as a sensitive and reproducible model of cytokine- and chemokine-induced eosinophilic inflammation. Intraperitoneally administered anti-murine interleukin-5 (anti-IL-5, 10–30 mg/kg) monoclonal antibodies consistently reduced antigen-induced peripheral blood, bronchoalveolar lavage fluid, and tissue eosinophilia in the guinea pig [57, 58]. Anti-IL-5 also produced dose-dependent inhibition of antigen-induced airway hyperreactivity in the aforementioned studies. Extensions of this type of study examining eosinophilic

Table 1 - The effect of phosphodiesterase inhibitors on antigen-induced responses in the guinea pig

Compound	**Route**	**Inhibition**					**Ref.**
		EPR	**LPR**	**AH**	**PMN**	**EOS**	
RP 73401 (PDE IV)	i.t.	Yes	–	–	–	Yes	50
Rolipram (PDE IV)	i.v.	Yes	–	–	–	Yes	55
ORG 20241 (PDE III/IV)	i.p.	No	No	Yes	No	Yes	51
Ro 20-1724 (PDE IV)	i.p.	No	No	Yes	Yes	Yes	56
CDP840 (PDE IV)	i.p.	Yes	–	–	No	Yes	53
CP-80,633 (PDE IV)	p.o.	Yes	–	–	–	–	54

i.t., intratracheal; i.v., intravenous; i.p., intraperitoneal; p.o., oral; EPR, early phase response; LPR, late phase response; AH, airway hyperreactivity; PMN, pulmonary neutrophilia; EOS, pulmonary eosinophilia; –, not tested or not shown

inflammation in the allergic guinea pig model have progressed into the arena of chemokine identification utilizing molecular tools to probe mRNA levels of chemoattractants of interest. Using Northern blot analysis of eotaxin mRNA expression, Rothenberg et al. [43] demonstrated a six-fold increase in lung eotaxin mRNA at 3 h following antigen challenge. More importantly, a relatively high constitutive level of lung eotaxin mRNA expression may be suggestive of an ongoing recruitment of eosinophils into guinea pig lung under basal conditions as suggested above. Humbles et al. [59] developed a radioimmunoassay using ^{125}I-labeled eotaxin to demonstrate that eotaxin levels peaked prior to eosinophil accumulation in the lung tissue and lavage fluid (i.e. 6 h as opposed to 12–24 h) following antigen challenge of guinea pigs. Despite differences in sensitization and challenge, this correlates with the early elevations in eotaxin mRNA noted in the previous allergic guinea pig study of Rothenberg et al. [43]. In combination, these studies help explain the kinetics of eotaxin generation in the allergic guinea pig and the contribution of this chemokine to eosinophil pathobiology in allergic disease.

Rat models

The primary strain of rat used to model allergic lung responses appears to be the highly inbred Brown Norway (BN) although Wistar, Lewis, Fisher, Sprague-Dawley, and Donryu strains have also been used on a limited basis [61–64]. Rats are generally sensitized by the intraperitoneal or subcutaneous route using various combinations of ovalbumin (0.1–5.0 mg), aluminum hydroxide (100-200 mg), and *Bordetella pertussis* vaccine (1 ml of 109 heat-killed bacilli). Animals are challenged with high concentrations of antigen (1–10%) for greater periods of time compared to the allergic guinea pig [61–64]. In contrast to allergic guinea pigs that exhibit a mast cell-dependent histaminergic bronchoconstriction, allergic rats appear to be serotonergic bronchoconstrictors [26]. Commonalities between the allergic BN rat model and the human condition include: (a) an allergic response immunologically driven by IgE [61, 63, 64], (b) the more consistent presence of both early and late phase airway responses in a greater subpopulation of animals compared to the guinea pig [63, 65, 66], (c) evidence of airway hyperreactivity to aerosolized methacholine, acetylcholine or serotonin albeit at relatively high concentrations [61, 67, 68], and (d) tissue and bronchoalveolar lavage fluid accumulation of neutrophils, eosinophils, and lymphocytes [61, 64–66]. Both β_2-receptor agonists and steroids act as anti-inflammatory agents in the antigen-challenged allergic BN rat [69, 70].

The main advantage of utilizing the allergic rat model to establish preclinical drug efficacy surrounds the fact that this is also the primary rodent species used for toxicologic and pharmacokinetic studies and, thus, allows for the establishment of a therapeutic window (i.e. efficacy versus toxicity). The presence of an impressive inflammatory response makes the allergic rat a reliable model for examining antigen-driven leukocytic responses at both the cellular and tissue level (see below). In contrast, antigen-induced airway hyperreactivity to various agonists is not generally as remarkable in the allergic rat as that noted in the allergic guinea pig model. (Keep in mind that the BN rat is a weak bronchoconstrictor compared to the guinea pig on an agonist to agonist basis.)

There have been no substantial changes in technology in the allergic BN rat model relative to sensitization, challenge, and monitoring lung function since its original development. As is the case with the guinea pig model, the most notable advances have evolved in the area of monitoring/estimating antigen-induced lung inflammation. Use of molecular probes has extended our mechanistic understanding of the pathophysiology surrounding asthma relative to this model. Recent examples of this have centered on defining the role of cytokines in antigen-induced lung inflammation in this species. Renzi et al. [63] demonstrated that allergic BN rats, which are high IgE and late phase airway responders, and Sprague-Dawley rats, which produce minimal IgE and minor antigen-induced airway responses, express increased mRNA for Th-2 cytokines (IL-4 and IL-5) and Th-1 cytokines (IL-1 and interferon gamma), respectively. In fact, Haczku et al. [68] corroborated and extended these results by demonstrating that interferon-γ (IFNγ) mRNA was actually reduced in antigen-challenged BN rats. Taken together, these data mechanistically support the role of antigen-stimulated Th-2 cell induction and the resultant production of pro-eosinophilic IL-4 and IL-5 in the allergic BN rat model. Furthermore, the results establish a strong link between antigen-induced lymphocyte accumulation and resulting airway changes, which include eosinophilia, in this model. In a subsequent study, Schneider et al. [64] used histology and bronchoalveolar lavage in combination with more sensitive myeloperoxidase and eosinophil peroxidase enzyme assays to characterize the "kinetic" accumulation of macrophages, neutrophils, eosinophils, and lymphocytes in the lungs of allergic BN, Fischer, and Lewis rats. The later two strains exhibited little to no antigen-induced inflammation compared to the intense pulmonary inflammation (eosinophils > neutrophils > macrophages > lymphocytes) noted in allergic BN rats. This dissimilarity was credited to the lack of IgE production in Fischer and Lewis rats but could also be attributed to the lack of a Th-2 dependent response in this species and easily studied as outlined above.

Due to the intensity and duration of the cellular inflammatory response following antigen challenge in the allergic BN rat, this model has been used extensively to explore the putative role of adhesion pathways in the pathogenesis of asthma. In recent studies [66, 67, 71–73], monoclonal antibodies against various adhesion molecules including very late antigen-4 (VLA-4; expressed on eosinophils and lymphocytes), lymphocyte function-associated antigen-1 (LFA-1; expressed on most leukocytes), and intercellular adhesion molecule-1 (ICAM-1; the ligand for LFA-1) were intravenously or intraperitoneally administered (generally 1–6 mg/kg) to animals prior to antigen challenge. Depending on the experimental paradigm (i.e. whether antibodies were administered intravenously or intraperitoneally), these monoclonal antibodies alone or in combination consistently inhibited antigen-induced early and late phase airway responses as well as airway leukocyte accumulation (primarily eosinophils and lymphocytes) (Tab. 2). The results were time- and dose-dependent. Despite the fact that these studies were conducted under different conditions, the results support the use of the allergic BN rat as a suitable animal model for studying adhesion phenomena as it relates to pulmonary disease.

Rabbit models

The allergic rabbit demonstrates many of the pathophysiological features of human asthma including acute bronchoconstriction, late phase airway obstruction, and airway hyperresponsiveness [74–76]. In contrast to the aforementioned models, the paradigm used to sensitize rabbits plays a critical role in the consistent development of these pulmonary responses. Sensitizing antigens include *Alternaria tenius* [74, 75], ragweed or dust-mite extract [76, 77] that is admixed to an adjuvant (aluminum hydroxide gel) and injected intraperitoneally within 24 h of birth to ensure the predominant production of IgE antibodies [74]. If delayed, sensitization pro-

Table 2 - The effect of adhesion blocking agents on antigen-induced responses in the Brown Norway rat

Compound	Route	Inhibition					Ref.
		EPR	LPR	AH	PMN	EOS	
VLA-4 mAB	i.v.	Yes	Yes	–	No	No	66
	i.v.	–	–	Yes	No	No	67
	i.p.	–	–	–	–	Yes	72
	i.p.	–	–	–	–	Yes	73
LFA-1 mAB	i.v.	–	–	Yes	No	Yes	67
ICAM-1 mAB	i.p.	–	–	–	–	Yes	73
VLA-4 mAB + LFA-1 mAB	i.v.	–	–	Yes	No	Yes	67
ICAM-1 mAB + LFA-1 mAB	i.v.	Yes	Yes	–	Yes	No	71
LFA-1 mAB + Mac-1 mAB	i.v.	Yes	Yes	–	Yes	No	66

mAB, monoclonol antibody; i.p., intraperitioneal; i.v., intravenous; EPR, early phase response; LPR, late phase response; AH, airway hyperreactivity; PMN, pulmonary neutrophilia; EOS, pulmonary eosinophilia; –, not tested or not shown

duces both IgE and IgG antibodies and a subsequent diminished pulmonary response to antigen provocation [74]. Weekly booster injections of the antigen/alum mixture out to 4 months are necessary to ensure full sensitization [71–74]. Extreme care must be taken when sensitizing pups because the rabbit doe may reject the offspring under stressful conditions.

Antigen challenge of sensitized rabbits also produces an inflammatory condition in the lungs characterized by airway edema and inflammatory cell influx [76, 78–81]. As with the allergic guinea pig, antigen-induced airway hyperreactivity does not necessarily correlate with the presence of inflammatory cells in the BAL fluid in the allergic rabbit [82]. From a practical standpoint, the identification of antigen-induced allergic rabbit bronchoalveolar lavage and tissue eosinophilia is difficult because of the presence of another type of granulocyte, the heterophil, that is functionally analogous to the neutrophil, but resembles the eosinophil both morphologically and histochemically [82]. Thus, specialized staining techniques are usually required to identify the rabbit eosinophil. These include the use of the peroxidase substrate 3,3'-diaminobenzidine together with nickel chloride that preferentially stains the cytoplasm of the rabbit eosinophil but not heterophil [83]. Another technique used to discriminate rabbit eosinophils from heterophils involves staining the BAL cells with hematoxylin and 1% chromotrope 2R [82].

The major advantage of the allergic rabbit is the fact that it is one of the few species that demonstrates a late phase airway obstruction [74-76] which, similar to humans, is attenuated by corticosteroids but not by β-adrenergic agonists [84]. Another distinct advantage of the rabbit over other species is the fact that adenosine is an important mediator of the allergic response [77, 85, 86]. Adenosine causes bronchoconstriction in allergic, but not in naïve rabbits [77]. This condition is also seen in human asthmatics but not in healthy individuals. It appears that the adenosine A_1-receptor is the important receptor subtype producing altered pulmonary pathology in the allergic rabbit [87].

Despite the cumbersome sensitization procedure outlined above, several groups have used this model to successfully evaluate various drug candidates [76–81, 85, 86]. In a recent study using a novel technology, Nyce and Metzger [85] demonstrated that aerosolized phosphorothioate antisense oligodeoxynucleotide pretreatment, which targeted the adenosine A_1-receptor, reduced bronchial hyperresponsiveness to histamine in dust-mite allergen challenged rabbits.

Sheep models

The naturally allergic, *Ascaris suum* sensitive sheep model has been routinely used to study early- and/or late-phase airway responses in single or dual responders [88]. Other key features of this antigen-induced model include: (a) heightened, transient airway reactivity to inhaled methacholine (0.1–10.0% for 1 min) [89] or carbachol (0.2–4.0% for 10 breaths) [90] and (b) lung inflammation characterized by bronchoalveolar lavage and/or tissue accumulation of macrophages, neutrophils, eosinophils, and lymphocytes [89, 90].

In this model, restrained, unanesthetized animals are instrumented through the nasal passages with an esophageal balloon and cuffed endotracheal tube containing a tracheal catheter to measure pleural and tracheal pressure, respectively. This, along with measurement of thoracic gas volume using whole body plethysmography, are the predominant features distinguishing this model from other preparations mentioned in this chapter. More germane to the characterization of antigen-induced airway inflammation has been introduction of a double-balloon nasotracheal tube into the allergic animal to create a tracheal chamber used to lavage a restricted portion of the upper airway [91]. This technology can be applied to the study of airway epithelial function (i.e. mucus secretion) in any of the large animal allergic models.

Despite the limited availability of this preparation (i.e. use by a small group of laboratories), the allergic sheep model has been extensively utilized to demonstrate intravenous, oral or inhaled efficacy of compounds from various therapeutic areas [88, 89, 92–94]. Some of the more recently tested compounds are summarized in Table 3. Notably some of these compounds have gone on to be tested or are being tested in man.

Table 3 - The effect of various compounds on antigen-induced responses in the allergic sheep

Compound	Route	Inhibition					Ref.
		EPR	LPR	AH	PMN	EOS	
α4 Integrin antibody MoAb HP1/2	aero. or i.v.	No	Yes	Yes	–	–	88
Endothelin antagonist BQ-123	aero. or i.v.	No	Yes	Yes	–	–	93
Tryptase inhibitor APC17731	aero.	Yes	Yes	Yes	–	Yes	92
Antiallergic/antiinflam-matory TBY-2285	p.o.	Yes	Yes	Yes	–	Yes	90
LTD_4 Antagonist Montelukast	i.v.	Yes	Yes	–	–	–	94
Non glucocorticoid 21-Aminosteroid U-74006F	i.v.	Yes	Yes	Yes	No	Yes	89

i.v., intravenous; p.o., oral; aero., aerosol; EPR, early phase response; LPR, late phase response; AH, airway hyperreactivity; PMN, pulmonary neutrophilia; EOS, pulmonary eosinophilia; –, not tested or not shown

Dog models

Dogs readily develop spontaneous allergic disorders; however, these responses are generally characterized as atopic dermatological as opposed to respiratory abnormalities [95]. Likewise, the respiratory physiology and anatomy of this species is characterized by wide airways and numerous collaterals such that commonly noted clinical respiratory pathophysiologies (e.g. hyperreactivity, late phase responses) are difficult to detect in the general canine population [96]. Despite these limitations, researchers have developed allergic dog models which exhibit antigen-induced acute and/or late phase responses [97–99], airway hyperreactivity [97], and pulmonary eosinophil and neutrophil accumulation [97, 98, 100, 101]. Interestingly, antigen-induced airway hyperreactivity in this species may persist for extended periods of time measured in months [99].

Relative to practical immunological approaches, these models range from natively allergic Basenji-greyhounds [102] or mongrel dogs which are naturally or actively sensitized with *Ascaris suum* larvae to inbred dogs which are sensitized from birth to an allergen such as ragweed [98, 99–101, 103]. Neonatal exposure to allergen (intraperitoneal injection of short ragweed extract in aluminum hydroxide within 24 h of birth followed by weekly or biweekly injections for 4 months [97, 101]) or postnatal sensitization (combinations of short ragweed, mixed pollen grasses, and live distemper virus injected within a month of birth and repeated bimonthly in the case of the grass extracts [99]) paradigms results in consistent, high IgE titers. Although widely accepted, use of these perinatal sensitization paradigms requires a great deal of effort, monitoring, and long term commitment.

The most obvious advantage of utilizing an allergic dog preparation based upon the ragweed neonatal sensitization paradigm is the ability to manipulate production of IgE titers in animals using selective breeding. Sensitive enzyme-linked immunoassays have recently been developed to monitor ragweed specific IgE levels and evaluate the role of IgE in the development of allergic dog models [96]. There is a growing regard for the use of genetic approaches to produce colonies of allergic animals with high IgE titers to study hereditary atopy. Interestingly, animals must contact allergen early (within the first week of birth) to develop an IgE-mediated response which is readily boostable upon repeat exposure [95]. This appears to be a dominant inherited genetic trait that may be similar to what is noted in human atopic families.

Improvements in methodologies to examine the pathophysiology of the asthmatic response in allergic dog models have been sparse and may reflect the difficulty in working with this species on a routine basis. Freed et al. [104] have developed a bronchoscopic procedure that allows for measurement of segmental or collateral system resistance. This method may be applicable for segmental antigen challenge and eliminate the problems associated with whole lung antigen exposure (i.e. fatal bronchoconstriction). An improved method for monitoring antigen-induced inflammation in allergic dogs involves the use of an *in situ* isolated tracheal preparation. This system has been used to study the kinetics of neutrophil elastase [105] and interleukin-8 release [106] following antigen superfusion of ragweed extract (42 protein nitrogen units/ml) over the trachea during an 8 h period.

The allergic dog has not been used routinely to test the therapeutic efficacy of anti-asthma compounds. Becker et al. [97] did examine the effect of a leukotriene synthesis inhibitor, MK-0591, in the allergic dog model and demonstrated that this compound reduced antigen-induced acute bronchoconstriction and airway hyperreactivity but not lung lavage eosinophilia in this model. Recent studies have centered on the use of this model to explore mechanisms surrounding antigen-induced stimulation of bone marrow granulocyte progenitor cells and corticosteroid-mediated suppression of this phenomenon [107, 108]. These findings have been partially corroborated in human asthmatics in which increases in neutrophil-macrophage and eosinophil-basophil progenitors have been noted as early as 6 h following antigen exposure [109]. In the dog, there is speculation that a serum-derived, undefined hematopoietic factor (which is blocked by steroids) is responsible for induction of antigen-induced increases in bone marrow progenitors and the resulting lung inflammation [110].

Primate models

Primate models of asthma have been used for more than 20 years to investigate pathophysiological and inflammatory responses in the lungs following antigen provocation. Rhesus [111, 112], cynomolgus [113–116] or squirrel [117, 118] monkeys can be used because they all possess a natural sensitivity to *Ascaris suum*, presumably due to the prior sensitization by nematode infestation. In our experience using cynomolgus monkeys, approximately 40% of the animals tested for pulmonary sensitivity to *Ascaris* show a positive acute bronchoconstrictor response to *Ascaris* inhalation challenge. Similar to studies in rhesus [119] and squirrel monkeys [118], we have found no correlation between cutaneous and pulmonary sensitivity to *Ascaris* in cynomolgus monkeys. Sensitization to *Ascaris* can also be performed in non-reactive monkeys by active [111, 112] or passive [111] sensitization to antigen. There are limitations to the utility of these sensitization paradigms due to a lack of antigen-induced airway hyperreactivity and refractory antigen responses upon multiple exposure, respectively.

The characteristic features of antigen-challenged monkeys with a natural sensitivity to the antigen include an acute bronchoconstriction [112, 114, 116–118], late phase airway obstruction [114, 118], inflammatory cell influx into the lungs [112, 113, 115, 116], and airway hyperresponsiveness [113, 115, 116]. In general, there is a wide range in sensitivity and reactivity to the antigen between monkeys, but responses within a particular monkey are very reproducible. The reaginic antibody resides in an immunoglobulin type analogous to immunoglobulin E of man [111, 112, 120]. This makes the allergic primate a very useful model for studying the immunological aspects of asthma [121, 122].

Bronchoconstriction occurs immediately after antigen challenge of sensitized monkeys [112, 114, 116–118]. Histamine and leukotrienes are the major mast cell mediators involved with the acute allergic bronchoconstrictor response [116, 123] and these mast cell mediators are found in the BAL fluid after antigen challenge [112–114]. The acute allergic bronchoconstrictor response is reversed by β-adrenergic bronchodilators [116]. Airway microvascular leakage also occurs soon after the antigen challenge resulting in an increase in protein concentration in the BAL fluid [124].

Approximately 6 h after the antigen challenge, a late phase airway obstruction occurs [114, 118]. This response is unaffected by bronchodilators but inhibited by corticosteroids [114, 118] suggesting that it is the result of an inflammatory response in the lungs rather than contraction of airway smooth muscle. There is also

Table 4 - The effect of various compounds on antigen-induced responses in the primate

Species	Compound	Route	Inhibition					Ref.
			EPR	LPR	AH	PMN	EOS	
Cynomolgus	LTB_4 Antagonist (CP105,696)	p.o.	–	–	Yes	Yes	No	124
	LTD_4 Antagonist (ICI198,615)	i.m.	No	–	Yes	No	Yes	128
	PDE IV Inhibitor (Rolipram)	s.c.	No	–	Yes	Yes	Yes	130
	PDE IV Inhibitor (CP-80,633)	s.c.	–	–	–	Yes	Yes	54
	IL-5 Antibody (TRFK-5)	i.v.	–	–	Yes	Yes	Yes	115
	ICAM-1 Antibody	i.v.	–	–	Yes	–	Yes	134, 135
	ELAM-1 Antibody	i.v.	No	Yes	–	Yes	–	135
Rhesus	Bradykinin antagonist (NPC17731)	aero.	Yes	–	–	–	–	133
	5-Lipoxygenase inhibitor	p.o.	Yes	Yes	–	Yes	–	125, 131
	LTD_4 Antagonist	aero.	Yes	–	–	–	–	129
	Lazaroid (U-83836E)	p.o.	–	Yes	Yes	–	Yes	132
Squirrel	Prostaglandin antagonist (L-640,035)	p.o.	Yes	–	–	–	–	117
	LTD_4 Antagonist (MK-0476)	p.o.	Yes	Yes	–	–	–	94

i.m., intramuscular; i.v., intravenous; p.o., oral; s.c., subcutaneous; aero., aerosol; EPR, early phase response; LPR, late phase response; AH, airway hyperreactivity; PMN, pulmonary neutrophilia; EOS, pulmonary eosinophilia; –, not tested or not shown

an influx of BAL neutrophils at 6 h after challenge [114, 118] and these inflammatory cells may contribute to antigen-induced airway obstruction [114, 125]. The mere presence of neutrophils in the BAL fluid does not necessarily cause airway obstruction [126], but activation and degranulation of the neutrophils would likely cause a profound inflammatory response resulting in mucosal congestion, airway edema, and airway narrowing.

Bronchial hyperresponsiveness is also seen in antigen-challenged allergic monkeys [113, 115, 116, 124] and can be measured 24 h after the antigen challenge [115]. At this time there is a large influx of eosinophils into the lungs [113, 115, 116] and the presence of major basic protein in the BAL fluid indicates that eosinophil activation and degranulation has occurred [113]. Our experience in cynomolgus monkeys suggests that both pulmonary eosinophilia and bronchial hyperresponsiveness occur 24 h after a single challenge with *Ascaris suum* extract [115]. Other paradigms have been used to produce antigen-driven bronchial hyperresponsiveness and they involve multiple antigen provocation over a few days [113, 116]. Although the multiple antigen challenge procedure may give a more pronounced degree of airway hyperresponsiveness [113], it is technically more demanding than the single antigen exposure and involves rather complex dosing schedules when evaluating the effects of drugs [116].

Primate models of asthma have been used to evaluate standard anti-allergy and anti-asthmatic compounds [116, 118, 127]. In addition, some of the newer therapies that may prove to be useful in human asthma have been tested in primates. These include leukotriene B_4 and D_4 receptor antagonists [94, 124, 128, 129], phosphodiesterase IV inhibitors [54, 130], 5-lipoxygenase inhibitors [125, 131], lazaroid U-83836E [132], a bradykinin receptor antagonist [133], a monoclonal antibody to IL-5 [115] and monoclonal antibodies to ICAM-1 and ELAM-1 adhesion molecules [134, 135] (Tab. 4).

In conclusion, the animal models of "asthma" discussed above provide researchers with useful in vivo systems to explore important components of the human asthmatic response. These types of studies using various molecular, immunological, and pharmacological approaches will continue to define both the key mechanisms contributing to the pathogenesis of asthma as well as novel therapeutics for the treatment of this disease.

References

1 Gray PR, Derksen FJ, Broadstone RV, Robinson NE, Johnson HG, Olson NC (1992) Increased pulmonary production of immunoreactive 15-hydroxyeicosatetraenoic acid in an animal model of asthma. *Am Rev Respir Dis* 145: 1092–1097

2 Garlisi CG, Falcone A, Hey JA, Paster TM, Fernandez X, Rizzo CA, Minnicozzi M, Jones H, Billah MM, Egan RW, Umland SP (1997) Airway eosinophils, T cells, Th2-type cytokine mRNA and hyperreactivity in response to aerosol challenge of allergic mice with previously established pulmonary inflammation. *Am J Respir Cell Mol Biol* 17: 642–651

3 Hessel EM, Van Oosterhout AJM, Hofstra CL, DeBie JJ, Garssen J, Van Loveren H, Verheyen AKCP, Saveloul HFJ, Nijkamp FP (1995) Bronchoconstriction and airway hyperresponsiveness after ovalbumin inhalation in sensitized mice. *Eur J Pharmacol* 293: 401–412

4 Eum S-Y, Hailé S, Lefort J, Huerre M, Vargaftig BB (1995) Eosinophil recruitment into the respiratory epithelium following antigenic challenge in hyper-IgE mice is accompanied by interleukin-5-dependent bronchial hyperresponsiveness. *Proc Natl Acad Sci* 92: 12290–12294

5 Hamelmann E, Oshiba A, Loader J, Larsen GL, Gleich G, Lee J, Gelfand EW (1997) Antiinterleukin-5 antibody prevents airway hyperresponsiveness in a murine model of airway sensitization. *Am J Respir Crit Care Med* 155: 819–825
6 Levitt RC, Mitzner W (1989) Autosomal recessive inhertiance of airway hyperreactivity to 5-hydroxytryptamine. *J Appl Physiol* 67: 1125–1132
7 Levitt RC, Mitzner W, Kleeberger SR (1990) A genetic approach to the study of lung physiology: understanding biological variability in airway responsiveness. *Am J Physiol* 258: L157–L164
8 Kung TT, Jones H, Adams III GK, Umland SP, Kreutner W, Egan RW, Chapman RW, Watnick AS (1994) Characterization of a murine model of allergic pulmonary inflammation. *Int Arch Allergy Immunol* 105: 83–90
9 Garlisi CG, Falcone A, Billah MM, Egan RW, Umland SP (1996) T cells are the predominant source of interleukin-5 but not interleukin-4 mRNA expression in the lungs of antigen-challenged allergic mice. *Am J Respir Cell Mol Biol* 15: 420–428
10 Kaminuma O, Mori A, Ogawa K, Nakata A, Kikkawa H, Naito K, Suko M, Okudaira H (1997) Successful transfer of late phase eosinophil infiltration in the lung by infusion of helper T cell clones. *Am J Respir Cell Mol Biol* 16: 448–454
11 Hogan SP, Koskinen A, Foster PS (1997) Interleukin-5 and eosinophils induce airway damage and bronchial hyperreactivity during allergic airway inflammation in BALB/c mice. *Immunol Cell Biol* 75: 284–288
12 Blyth DI, Pedrick MS, Savage TJ, Hessel EM, Fattah D (1996) Lung inflammation and epithelial changes in a murine model of atopic asthma. *Am J Respir Cell Mol Biol* 14: 425–438
13 Ohkawara Y, Lei XF, Stampfli MR, Marshall JS, Xing Z, Jordana M (1997) Cytokine and eosinophil responses in the lung, peripheral blood, and bone marrow compartments in a murine model of allergen-induced airways inflammation. *Am J Respir Cell Mol Biol* 16: 510–520
14 Kung TT, Stelts D, Zurcher JA, Jones H, Umland SP, Kreutner W, Egan RW, Chapman RW (1995) Mast cells modulate allergic pulmonary eosinophilia in mice. *Am J Respir Cell Mol Biol* 12: 404–409
15 Martin TR, Takeishi T, Katz HR, Austen KF, Drazen JM, Galli SJ (1993) Mast cell activation enhances airway responsiveness to methacholine in the mouse. *J Clin Invest* 91: 1176–1182
16 Brusselle GG, Kips JC, Tavernier JM, Van der Heyden JG, Cavelier CA, Pauwels RA, Bluethmann H (1994) Attenuation of allergic airway inflammation in IL-4 deficient mice. *Clin Exp Allergy* 24: 73–83
17 Corrigan CJ, Kay AB (1992) T cells and eosinophils in the pathogenesis of asthma. *Immunol Today* 13: 501–507
18 Mehlhop PD, Van de Rijnn M, Goldberg AB, Brewer JP, Kurup VP, Martin TR, Oettgen HC (1997) Allergen-induced hyperreactivity and eosinophilic inflammation occur in the absence of IgE in a mouse model of asthma. *Proc Natl Acad Sci* 94: 1344–1349
19 Korsgren M, Erjefält JS, Korsgren O, Sundler F, Persson CGA (1997) Allergic

eosinophil-rich inflammation develops in lungs and airways of B cell-deficient mice. *J Exp Med* 185: 885–892
20 Van Oosterhout AJ, Hofstra CL, Shields R, Chan B, van Ark I, Jardieu PM, Nijkamp FP (1997) Murine CTLA4-IgG treatment inhibits airway eosinophilia and hyperresponsiveness and attenuates IgE upregulation in a murine model of allergic asthma. *Am J Respir Cell Mol Biol* 17: 386–392
21 Mosmann TR, Cherwinski H, Bond MW, Giedin MA, Coffman RL (1986) Two types of murine helper T cell clone I: Definition according to profiles of lymphokine activities and secreted proteins. *J Immunol* 136: 2348 –2357
22 Gonzalo J-A, Lloyd CM, Kremer L, Finger E, Martinez-A C, Siegelman MH, Cybulsky M, Gutierrez-Ramos J-C (1996) Eosinophil recruitment to the lung in a murine model of allergic inflammation. The role of T cells, chemokines and adhesion receptors. *J Clin Invest* 98: 2332–2345
23 Chu HW, Wang JM, Boutet M, Boulet LP, Laviolette M (1996) Immunohistochemical detection of GM-CSF, IL-4 and IL-5 in a murine model of allergic bronchopulmonary aspergillosis. *Clin Exp Allergy* 26: 461–468
24 Bell SJD, Metzger WJ, Welch CA, Gilmour MI (1996) A role for Th2 T memory cells in early airway obstruction. *Cell Immunol* 170: 185–194
25 Elsas MICG, Joseph D, Elsas PX, Vargaftig BB (1997) Rapid increase in bone-marrow eosinophil production and response to eosinopoietic interleukins triggered by intranasal allergen challenge. *Am J Respir Cell Mol Biol* 17: 404–413
26 Karol MH (1994) Animal models of occupational asthma. *Eur Respir J* 7: 555–568
27 Martin TR, Gerard NP, Galli SJ, Drazen JM (1988) Pulmonary responses to bronchoconstrictor agonists in the mouse. *J Appl Physiol* 64: 2318–2323
28 Hamelmann E, Schwarze J, Takeda K, Oshiba A, Larsen GL, Irvin CG, Gelfand EW (1997) Non invasive measurement of airway responsiveness in allergic mice using barometric plethysmography. *Am J Respir Crit Care Med* 156: 766–775
29 Stelts D, Falcone A, Egan RW, Falcone A, Garlisi CG, Gleich GJ, Kreutner W, Kung TT, Nahrebne DK, Chapman RW, Minnicozzi M (1998) Eosinophils retain their granule major basic protein in a murine model of allergic pulmonary inflammation. *Am J Respir Cell Mol Biol* 18: 463–470
30 Corry DB, Folkesson HG, Warnock ML, Erle DJ, Matthay MA, Wiener-Kronish JP, Locksley RM (1996) Interleukin 4, but not interleukin 5 or eosinophils is required in a murine model of acute airway hyperreactivity. *J Exp Med* 183: 109–117
31 Zhou CY, Crocker IC, Koenig G, Romero FA, Townley RG (1997) Anti-interleukin-4 inhibits immunoglobulin E production in a murine model of atopic asthma. *J Asthma* 34: 195–201
32 Kung TT, Stelts, DM, Zurcher JA, Adams III GL, Egan RW, Kreutner W, Watnick AS, Jones H, Chapman RW (1995) Involvement of IL-5 in a murine model of allergic pulmonary inflammation: prophylactic and therapeutic effect of an anti-IL-5 antibody. *Am J Respir Cell Mol Biol* 13: 360–365
33 Foster PS, Hogan SP, Ramsay AJ, Matthaei KI, Young IG (1996) Interleukin 5 deficien-

cy abolishes eosinophilia, airways hyperreactivity, and lung damage in a mouse asthma model. *J Exp Med* 183: 195–201
34 Lee JJ, McGarry MP, Farmer SC, Denzler KL, Larson KA, Carrigan PE, Brenneise IE, Horton MA, Haczku A, Gelfand EW et al (1997) Interleukin-5 expression in the lung epithelium of transgenic mice leads to pulmonary changes pathognomonic of asthma. *J Exp Med* 185: 2143–2156
35 O'Donnell M, Garippa J, Rinaldi N, Selig WM, Tocker JE, Tannu SA, Wasserman MA, Welton A, Bolin DR (1994) Ro 25-1553: A novel long-acting vasoactive intestinal peptide agonist. *J Pharmacol Exp Ther* 270: 1289–1294
36 Itoh K, Takahashi E, Mukaiyama O, Satoh Y, Yamaguchi T (1996) Relationship between airway eosinophilia and airway hyperresponsiveness in a late asthmatic model of guinea pigs. *Int Arch Allergy Immunol* 107: 86–94
37 Underwood S, Foster M, Raeburn D, Bottoms S, Karlsson J-A (1995) Time-course of antigen-induced airway inflammation in the guinea pig and its relationship to airway hyperresponsiveness. *Eur Respir J* 8: 2104–2113
38 Yamada N, Kadowaki S, Umezu K (1992) Development of an animal model of late asthmatic responses in guinea pigs and effects of anti-asthmatic drugs. *Prostaglandins* 43: 507–521
39 Matsumoto T, Ashida Y, Tsukuda R (1994) Pharmacological modulation of immediate and late airway response and leukocyte infiltration in the guinea pig. *J Pharmacol Exp* Ther 269: 1236–1244
40 Savoie C, Plant M, Zwikker M, van Staden CJ, Boulet L, Chan CC, Rodgers IW, Pon DJ (1995) Effect of dexamethasone on antigen-induced high molecular weight glycoconjugate secretion in allergic guinea pigs. *Am J Respir Cell Biol* 13: 133–143
41 Hsiue TR, Lei HY, Hsieh AL, Wang TY, Chang HY, Chen CR (1997) Mite-induced allergic airway inflammation in guinea pigs. *Int Arch Allergy Immunol* 112: 295–302
42 Riedel F, Krause A, Slenczka W, Rieger CHL (1996) Parainfluenza-3-virus infection enhances allergic sensitization in the guinea pig. *Clin Exp Allergy* 26: 603–609
43 Rothenberg ME, Luster AD, Lilly CM, Drazen JM, Leder P (1995) Constitutive and allergen-induced expression of eotaxin mRNA in the guinea pig lung. *J Exp Med* 181: 1121–1216
44 Banner KH, Paul W, Page CP (1996) Ovalbumin challenge following immunization elicits recruitment of eosinophils but not bronchial hyperresponsiveness in guinea pigs: Time course and relationship to eosinophil activation status. *Pulmonary Pharmacol* 9: 179–187
45 Chapman ID, Morley J (1996) Eosinophil accumulation and airway hyperreactivity. *Eur Respir J* 9: 1331–1333
46 Underwood DC, Osborn RR, Hand JM (1992) Lack of late-phase airway responses in conscious guinea pigs after a variety of antigen challenges. *Agents Actions* 37: 191–194
47 Heuer HO, Wenz B, Jennewein HM, Urich K (1996) Characterization of a novel airway late phase model in the sensitized guinea pig which uses silica and *Bordetella pertussis* as adjuvant for sensitization. *Eur J Pharmacol* 317: 361–369

48 Santing RE, Olymulder CG, Zaagsma J, Meurs H, (1994) Relationship between allergen-induced early- and late-phase obstructions, bronchial hyperreactivity and inflammation in conscious, unrestrained guinea pigs. *J Allergy Clin Immunol* 93: 1021–1030
49 Griffiths-Johnson D, Karol M (1991) Validation of a noninvasive technique to assess development of airway hyperreactivity in an animal model of immunological pulmonary hypersensitivity. *Toxicol* 65: 283–294
50 Raeburn D, Underwood SL, Lewis SA, Woodman VR, Battram CH, Tomkinson A, Sharma S, Jordan R, Souness JE, Webber SE et al (1994) Anti-inflammatory and bronchodilator properties of RP 73401, a novel and selective phosphodiesterase type IV inhibitor. *Br J Pharmacol* 113: 1423–1431
51 Santing RE, Olymulder CG, Van der Molen K, Meurs H, Zaagsma J (1995) Phosphodiesterase inhibitors reduce bronchial hyperreactivity and airway inflammation in unrestrained guinea pigs. *Eur J Pharmacol* 275: 75–82
52 Banner KH, Moriggi E, Da Ros B, Schioppacassi G, Semeraro C, Page CP (1996) The effect of selective phosphodiesterase 3 and 4 isoenzyme inhibitors and established anti-asthma drugs on inflammatory cell activation. *Br J Pharmacol* 119: 1255–1261
53 Hughes B, Howat D, Lisle H, Holbrook M, James T, Gozzard N, Blease K, Hughes P, Kingaby R, Warrellow G et al (1996) The inhibition of antigen-induced eosinophilia and bronchoconstriction by CDP 840, a novel stero-selective inhibitor of phosphodiesterase type 4. *Br J Pharmacol* 118: 1183–1191
54 Turner CR, Cohan VL, Cheng JB, Showell HJ, Pazoles CJ, Watson JW (1996) The in vivo pharmacology of CP-80,633, a selective inhibitor of phosphodiesterase 4. *J Pharmacol Exp Ther* 278: 1349–1355
55 Underwood DC, Matthews JK, Osborn RR, Bochnowicz S, Torphy TJ (1997) The influence of endogenous catecholamines on the inhibitory effects of rolipram against early- and late-phase response to antigen in the guinea pig. *J Pharmacol Exp Ther* 280: 210–219
56 Danahay H, Broadley KJ (1997) Effects of inhibitors of phosphodiesterase on antigen-induced bronchial hyperreactivity in conscious sensitized guinea pigs and airway leukocyte infiltration. *Br J Pharmacol* 120: 289–297
57 Mauser PJ, Pitman A, Witt A, Fernandez X, Zurcher J, Kung T, Jones H, Watnick AS, Egan RW, Kreutner W, Adams GK (1993) Inhibitory effect of the TRFK-5 anti IL-5 antibody in a guinea pig model of asthma. *Am Rev Respir Dis* 148: 1623–1627
58 Akutsu I, Kojima T, Kariyone A, Fukuda T, Makino S, Takatsu K (1995) Antibody against interleukin-5 prevents antigen-induced eosinophil infiltration and bronchial hyperreactivity in the guinea pig airways. *Immunol Lett* 45: 109–116
59 Humbles AA, Conroy DM, Marleau S, Rankin SM, Plaframan RT, Proudfoot AEI, Wells TNC, Li D, Jeffrey PK, Griffiths-Johnson DA et al (1997) Kinetics of eotaxin generation and its relationship to eosinophil accumulation in allergic airways: Analysis in a guinea pig model in vivo. *J Exp Med* 186: 601–612
60 Watanabe A, Hayashi H (1990) Allergen-induced biphasic bronchoconstriction in rats. *Int Arch Allergy Appl Immunol* 93: 26–34

61 Kips JC, Cuvelier CA, Pauwels RA (1992) Effect of acute and chronic antigen inhalation on airway morphology and responsiveness in actively sensitized rats. *Am Rev Respir Dis* 145: 1306–130

62 Misawa M, Chiba Y (1993) Repeated antigenic challenge-induced airway hyperresponsiveness and airway inflammation in actively sensitized rats. *Japan J Pharmacol* 61: 41–50

63 Renzi PM, Al Assaad AS, Yang J, Yasruel Z, Hamid Q (1996) Cytokine expression in the presence or absence of late airway responses after antigen challenge of sensitized rats. *Am J Respir Cell Mol Biol* 15: 367–373

64 Schneider T, van Velzen D, Moqbel R, Issekutz AC (1997) Kinetics and quantification of eosinophil and neutrophil recruitment to allergic lung inflammation in a Brown Norway rat model. *Am J Respir Cell Mol Biol* 17: 702–712

65 Renzi PM, Olivenstein R, Martin JG (1993) Inflamamtory cell populations in the airways and parenchyma after antigen challenge in the rat. *Am Rev Respir Dis* 147: 967–974

66 Rabb HA, Olivenstein R, Issekutz TB, Renzi PM, Martin JG (1994) The role of the leukocyte adhesion molecules VLA-4, LFA-1, and Mac-1 in allergic airway responses in the rat. *Am J Respir Crit Care Med* 149: 1186–1191

67 Laberge S, Rabb H, Issekutz TB, Martin JG (1995) Role of VLA-4 and LFA-1 in allergen-induced airway hyperresponsiveness and lung inflammation in the rat. *Am J Respir Crit Care Med* 151: 822–829

68 Haczku A, Macary P, Haddad EB, Huang TJ, Kemeny DM, Moqbel R, Chung KF (1996) Expression of Th-2 cytokines interleukin-4 and -5 and of Th-1 cytokine interferon-g in ovalbumin-exposed sensitized Brown Norway rats. *Immunol* 88: 247–251

69 Howell RE, Jenkins LP, Fielding LE, Grimes D (1995) Inhibition of antigen-induced pulmonary eosinophilia and neutrophilia by selective inhibitors of phosphodiesterase types 3 or 4 in Brown Norway rats. *Pulmonary Pharmacol* 8:83–89

70 Elwood W, Lotvall JO, Barnes PJ, Chung KF (1992) Effect of dexamethasone and cyclosporin A on allergen-induced airway hyperresponsiveness and inflammatory cell responses in sensitized Brown Norway rats. *Am Rev Respir Dis* 145: 1289–1294

71 Nagase T, Fukuchi Y, Matsuse T, Sudo E, Matsui, Orimo H (1995) Antagonism of ICAM-1 attenuates airway and tissue responses to antigen in sensitized rats. *Am J Respir Crit Care Med* 151: 1244–1249

72 Richards IN, Kolbasa KP, Hatfield CA, Winterrowd GE, Vonderfecht SL, Fidler SF, Griffin RL, Brashler JR, Krzesicki RF, Sly LM et al (1996) Role of very late activation antigen-4 on the antigen-induced accumulation of eosinophils and lymphocytes in the lungs and airway lumen of sensitized Brown Norway rats. *Am J Respir Cell Mol Biol* 15: 172–183

73 Taylor BM, Kolbasa KP, Chin JE, Richards IM, Fleming WE, Griffin RL, Fidler SF, Sun FF (1997) Roles of adhesion molecules ICAM-1 and a4 integrin in antigen-induced changes in microvascular permeability associated with lung inflammation in Brown Norway rats. *Am J Respir Cell Mol Biol* 17: 757–766

74 Shampain MP, Behrens BL, Larsen GL, Henson PM (1982) An animal model of late pulmonary responses to Alternaria challenge. *Am Rev Respir Dis* 126: 493–498
75 Minshall EM, Riccio MM, Herd CM, Douglas GJ, Seeds EAM, McKenniff MG, Sasaki M, Spina D, Page CP (1993) A novel animal model for investigating persistent airway hyperresponsiveness. *J Pharmacol Toxicol Methods* 30: 177–188
76 Metzger WJ (1990) Late phase asthmatic responses in the allergic rabbit. In: W Dorsch (ed): *Late phase allergic reactions*. CRC, Boca Raton, FL, 347–362
77 Ali S, Mustafa SJ, Metzger WJ (1994) Adenosine receptor-mediated bronchoconstriction and bronchial hyperresponsiveness in an allergic rabbit model. *Am J Physiol* 266: L271–L277
78 Behrens BL, Clark RAF, Presley DM, Graves JP, Feldsien DC, Larsen GL (1987) Comparison of the evolving histopathology of early and late cutaneous and asthmatic responses in rabbits after a single antigen challenge. *Lab Invest* 56: 101–113
79 Gozzard N, el-Hashim A, Herd CM, Blake SM, Holbrook M, Hughes B, Higgs GA, Page CP (1996) Effect of the glucocorticosteroid budesonide and a novel phosphodiesterase type 4 inhibitor CDP840 on antigen-induced airway responses in neonatally immunised rabbits. *Br J Pharmacol* 118: 1201–1208
80 el-Hashim AZ, Jacques CA, Herd CM, Lee TM, Page CP (1997) The effect of R 15.7/HO, an anti-CD18 antibody on the late airway response and airway hyperresponsiveness in an allergic rabbit model. *Br J Pharmacol* 121: 671–678
81 Gozzard N, Herd CM, Blake SM, Holbrook M, Hughes B, Higgs GA, Page CP (1996) Effects of theophylline and rolipram on antigen-induced airway responses in neonatally immunized rabbits. *Br J Pharmacol* 117: 1405–1412
82 Minshall E, Spina D, Page CP (1996) Effects of neonatal immunization and repreated allergen exposure on airway responsiveness in the rabbit. *J Appl Physiol* 80: 2108–2119
83 Song B-Z, Donoff RB, Tsuji T, Todd R, Gallagher GT, Wong DTW (1993) Identification of rabbit eosinophils and heterophils in cutaneous healing wounds. *Histochem J* 25: 762–771
84 Larsen GL, Shampain MP, Marsh WR, Behrens BL (1984) An animal model of the late asthmatic response to antigen challenge. In: AB Kay (ed): *Asthma – physiology, immunopharmacology and treatment*. Academic Press, London, UK, 245–262
85 Nyce JW, Metzger WJ (1997) DNA antisense therapy for asthma in an animal model. *Nature* 385: 721–725
86 Ali S, Mustafa SJ, Metzger WJ (1994) Adenosine-induced bronchoconstriction and contraction of airway smooth muscle from allergic rabbits with late-phase airway obstruction: Evidence for an inducible adenosine A1 receptor. *J Pharmacol Exp Ther* 268: 1328–1334
87 el-Hashim A, D'Agostino B, Matera MG, Page C (1996) Characterization of adeonsine receptors involved in adenosine-induced bronchoconstriction in allergic rabbits. *Br J Pharmacol* 119: 1262–1268
88 Lobb RR, Repinsky B, Leone DR, Abraham WM (1996) The role of $\alpha 4$ integrins in lung pathophsyiology. *Eur Respir J* 22: 104s–108s

89 Fujimoto K, Kubo K, Okada K, Kobayashi T, Sekiguchi M, Sakai A (1996) Effect of the 21-aminosteroid U-740067 on antigen-induced bronchoconstriction and bronchoalveolar eosinophilia in allergic sheep. *Eur Respir J* 9: 2044–2049
90 Abraham WM, Ahmed A, Cortes A, Sielczak M, Wantanabe A (1996) Effect of TYB-2285 on antigen-induced airway responses in sheep. *Pulmonary Pharmacol* 9:49–58
91 Mariassy AT, Abraham WM, Wanner A (1994) Effect of antigen on the glyconjugate profile of tracheal secretions and the epithelial glycocalyx in allergic sheep. *J Allergy Clin Immunol* 93: 585–593
92 Clark JM, Abraham WM, Fishman CE, Forteza R, Ahmed A, Corts A, Warne RL, Moore WR, Tanaka RD (1995) Tryptase inhibitors block allergen-induced airway and inflammatory responses. *Am J Respir Crit Care Med* 152: 2076–2083
93 Noguchi K, Ishikawa K, Yano M, Ahmed A, Cortes A, Abraham WM (1995) Endothelin-1 contributes to antigen-induced airway hyperresponsiveness. *J Appl Physiol* 79: 700–705
94 Jones TR, Labelle M, Belley M, Champion E, Charette L, Evans J, Ford-Hutchinson AW, Gauthier JY, Masson P et al (1995) Pharmacology of montelukast sodium (Singulair™), a potent and selective leukotriene D_4 receptor antagonist. *Can J Physiol Pharmacol* 73: 191–201
95 de Weck AL, Mayer P, Stumper B, Schiessel B, Pickat L (1997) Dog allergy, a model for allergy genetics. *Int Arch Allergy Immunol* 113: 55–57
96 Collie DDS, DeBoer DJ, Muggenburg BA, Bice DE (1997) Evaluation of association of blood and bronchoalveolar eosinophil number and serum total immunoglobulin. *Am J Vet Res* 58: 34–39
97 Becker AB, Black C, Lilley MK, Bajawa K, Ford-Hutchinson AW, Simons FER, Tagari P (1995) Antiasthmatic effects of a leukotriene biosynthesis inhibitor (MK-0591) in allergic dogs. *J Appl Physiol* 78: 615–622
98 Itabashi S, Ohrui T, Sekizawa, Aikawa T, Nakazawa H, Sasaki H (1993) Late phase asthmatic resposnes cause peripheral airway hyperresponsiveness in dogs treated with metopirone. *Int Arch Allergy Immunol* 101: 215–220
99 Chung KF, Becker AB, Lazarus SC, Frick OL, Nadel JA, Gold WM (1985) Antigen-induced airway hyperresponsiveness and pulmonary inflammation in allergic dogs. *J Appl Physiol* 58: 1347–1353
100 Woolley MJ, Lane CG, Ellis R, Stevens WH, Woolley KL, O'Byrne PM (1995) Role of airway eosinophils in the development of allergen-induced airway hyperresponsiveness in dogs. *Am J Respir Crit Care Med* 152: 1508–1512
101 Baldwin F, Becker AB (1993) Bronchoalveolar eosinophilic cells in a canine model of asthma: Two distinct populations. *Vet Pathol* 30: 97–103
102 Emala C, Hirshman C (1996) Animal models of bronchial hyperreactivity. In: IP Hall (ed): *Genetics of asthma and atopy*. Monogr Allergy. Karger, Basel, 33: 35–52
103 Richards IM, Griffin RL, Oostveen JA, Elfring G, Conder GA (1988) Role of cyclooxygenase products of arachidonic acid metabolism in *Ascaris* antigen-induced bronchoconstriction in sensitized dogs. *J Pharmacol Exp Ther* 245: 735–741

104 Freed AN, Omori C, Hubbard WC, Adkinson NF (1994) Dry air- and hypertonic aerosol-induced bronchoconstriction and cellular responses in canine lung periphery. *Eur Respir J* 7: 1308–1316
105 Tabachnik E, Schuster A, Gold WM, Nadel JA (1992) Role of neutrophil elastase in allergen-induced lysozyme secretion in the dog trachea. *J Appl Physiol* 73: 695–700
106 Kaneko T, Massion PR, Hara M, Nadel JA (1996) Ragweed antigen causes interleukin-8 production in sensitized dog trachea. *Am J Respir Crit Care Med* 153: 136–140
107 Woolley MJ, Denburg JA, Ellis R, Dahlback M, O'Byrne PM (1994) Allergen-induced changes in bone marrow progenitors and airway responsiveness in dogs and the effect of inhaled budesonide on these parameters. *Am J Respir Cell Mol Biol* 11: 600–606
108 Inman MD, Denburg JA, Ellis R, Dahlback M, O'Byrne PM (1997) The effect of treatment with budesonide or PGE2 in vitro on allergen-induced increases in canine bone marrow progenitors. *Am J Respir Cell Mol Biol* 17: 634–641
109 Denburg JA, Inman MD, Wood L, Ellis R, Sehmi R, Dahlback M, O'Byrne PM (1997) Bone marrow progenitors in allergic airways disease: Studies in canine and human models. *Int Arch Allergy Immunol* 113; 181–183
110 Inman MD, Denburg JA, Ellis R, Dahlback M, O'Byrne PM (1996) Allergen-induced increases in bone marrow progenitors in airway hyperresponsiveness in dogs: Regulation by a serum hemopoietic factor. *Am J Respir Cell Mol Biol* 15: 305–311
111 Patterson R, Kelly JF (1974). Animal models of the asthmatic state. Ann Rev Med 25: 53–68
112 Johnson HG, Stout BK (1989) *Ascaris suum* ova-induced bronchoconstriction, eosinophilia, and IgE antibody responses in experimentally infected primates did not lead to histamine hyperreactivity. *Am Rev Respir Dis* 139: 710–714
113 Gundel RH, Gerritsen ME, Gleich GJ, Wegner CD (1990) Repeated antigen inhalation results in a prolonged airway eosinophilia and airway hyperresponsiveness in primates. *J Appl Physiol* 68: 779–786
114 Gundel RH, Wegner CD, Letts LG (1992) Antigen-induced acute and late-phase responses in primates. *Am Rev Respir Dis* 146: 369–373
115 Mauser PJ, Pitman AM, Fernandez X, Foran SK, Adams III GK, Kreutner W, Egan RW, Chapman RW (1995) Effect of an antibody to interleukin-5 in a monkey model of asthma. *Am J Respir Crit Care Med* 152: 467–472
116 Turner CR, Anderson CJ, Smith WK, Watson JW (1996) Characterization of a primate model of asthma using anti-allergy/anti-asthma agents. *Inflamm Res* 45: 239–245
117 McFarlane CS, Piechuta H, Hall RA, Ford-Hutchinson AW (1984) Effects of a contractile prostaglandin antagonist (L-640,035) upon allergen-induced bronchoconstriction in hyperreactive rats and conscious squirrel monkeys. *Prostaglandins* 28: 173–182
118 Hamel R, McFarlane CS, Ford-Hutchinson AW (1986) Late pulmonary responses induced by Ascaris allergen in conscious squirrel monkeys. *J Appl Physiol* 61: 2081–2087
119 Patterson R, Harris KE (1992) IgE-mediated rhesus monkey asthma: natural history and individual animal variation. *Int Arch Allergy Immunol* 97: 154–159

120 Ferreira FD, Mayer P, Sperr WH, Valent P, Seiberler S, Ebner C, Liehl E,Scheiner O, Kraft D, Valenta R (1996) Induction of IgE antibodies with predefined specificity in rhesus monkeys with recombinant birch pollen allergens, Bet v 1 and Bet v 2. *J Allergy Clin Immunol* 97: 95–103
121 Fox JA, Hotaling TE, Struble C, Ruppel J, Bates DJ, Schoenhoff MB (1996) Tissue distribution and complex formation with IgE of an anti-IgE antibody after intravenous administration in cynomolgus monkeys. *J Pharmacol Exp Ther* 279: 1000–1008
122 Meng YG, Singh N, Wong WL (1996) Binding of cynomolgus monkey IgE to a humanized anti-human IgE antibody and human high affinity IgE receptor. *Mol Immunol* 33: 635–642
123 Osborn RR, Hay DWP, Wasserman MA, Torphy TJ (1992) SK&F 104353, a selective leukotriene receptor antagonist, inhibits leukotriene D4-and antigen-induced bronchoconstriction in cynomolgus monkeys. *Pulmon Pharmacol* 5: 153–157
124 Turner CR, Breslow R, Conklyn MJ, Andresen CJ, Patterson DK, Lopez-Anaya A, Owens B, Lee P, Watson JW, Showell HJ (1996) *In vitro* and *in vivo* effects on leukotriene B_4 antagonism in a primate model of asthma. *J Clin Invest* 97: 381–387
125 Gundel RH, Torcellini CA, Clarke CC, DeSai S, Lazer ES, Wegner CD (1990) The effects of a 5-lipoxygenase inhibitor on antigen-induced mediator release, late phase bronchoconstriction and cellular infiltrates in primates. *Adv Prostaglandin Thromb Leuk Res* 21: 457–460
126 Allen DL, Herman DR, Williams GD, Spaethe SM, Dorato MA, Wolff RK (1995) Effect of an LTB_4 aerosol exposure on pulmonary function, cell populations, and mediators in the lungs of rhesus monkeys. *Inhalation Toxicol* 7: 1141–1152
127 Gundel RH, Kikade P, Torcellini CA, Clark CC, Watrous J, Desai S, Homon CA, Farina PR, Wegner CD (1991) Antigen-induced mediator release in primates. *Am Rev Respir Dis* 144: 76–82
128 Turner CR, Smith WB, Andresen CJ, Swindell AC, Watson JW (1994) Leukotriene D_4 receptor antagonism reduces airway hyperresponsiveness in monkeys. *Pulm Pharmacol* 7: 49–58
129 Patterson R, Harris KE, Krell RD (1988) Effect of a leukotriene D4 (LTD4) antagonist on LTD_4 and *Ascaris* antigen-induced airway responses in rhesus monkeys. *Int Arch Allergy Appl Immunol* 86: 440–445
130 Turner CR, Andresen CJ, Smith WB, Watson JW (1994) Effect of rolipram on responses to acute and chronic antigen exposure in monkeys. *Am J Respir Crit Care Med* 149: 1153–1159
131 Katayama S, Sakuma Y, Abe S, Tsunoda H, Yamatsu I, Katayama K (1993) Inhibition of IgE-mediated leukotriene generation and bronchoconstriction in primates with a new 5-lipoxygenase inhibitor, E6080. *Int Arch Allergy Immunol* 100: 178–184
132 Johnson HG, Stout BK (1993) Late phase bronchoconstriction and eosinophilia as well as methacholine hyperresponsiveness in Ascaris-sensitive rhesus monkeys were reversed by oral administration of U-83836E. *Int Arch Allergy Immunol* 100: 362–366
133 Hogan MB, Harris KE, Protter AA, Patterson R (1996) A bradykinin antagonist inhibits

both bradykinin and the allergen-induced airway response in primates. *Proc Assoc Am Phys* 109: 269–274
134 Wegner CD, Gundel RH, Reilly P, Haynes N, Letts LG, Rothlein R (1990) Intercellular adhesion molecule-1 (ICAM-1) in the pathogenesis of asthma. *Science* 247: 456–459
135 Gundel RH, Wegner CD, Torcellini CA, Clarke CC, Haynes N, Rothlein R, Smith CW, Letts LG (1991) Endothelial leukocyte adhesion molecule-1 mediates antigen-induced acute airway inflammation and late-phase airway obstruction in monkeys. *J Clin Invest* 88: 1407–1411

Allergen-induced airway inflammation and airway hyperreactivity in mice

Marsha A. Wills-Karp[1], Andrea Keane-Myers[2], Stephen H. Gavett[3], and Douglas Kuperman[1]

[1]Johns Hopkins School of Hygiene and Public Health, 615 N. Wolfe St., Baltimore, MD 21205, USA; [2]Laboratory of Molecular Immunology, Schepens Eye Research Institute, Harvard Medical School, Boston, MA 02114, USA; [3]Pulmonary Toxicology Branch, USEPA, Health Effects Research Laboratory, Research Triangle Park, NC 27711, USA

Introduction

Allergic asthma is a chronic, often debilitating disease which has been increasing in both prevalence and mortality worldwide, despite increased use of currently available medications. Although extensive research has been devoted to the study of this disease, the fundamental mechanisms which underlie the development and perturbation of the asthmatic state remain elusive.

Asthma is characterized structurally by the presence of chronic inflammation of the airways, with intense infiltration of the bronchial mucosa by lymphocytes, eosinophils and mast cells, along with epithelial desquamation [1, 2]. Functionally asthma is characterized by abnormal bronchoconstriction, with hyperresponsiveness of the airways in response to challenge to a variety of stimuli. These stimuli include cold air, exercise, diverse pharmacological agents, and allergen challenge [1, 3]. Structure and function interact at the level of airway geometry in that bronchoconstriction takes place in airways already narrowed by inflammation.

A large percentage of asthma is allergic in nature and studies demonstrate that sensitivity to common household allergens is significantly associated with the development of asthma and airway hyperresponsiveness [4–6]. In these individuals challenge with allergen leads to the rapid onset of mucosal edema, increases in airway smooth muscle tone, airway narrowing and an increase in neutrophil, mast cells and eosinophil numbers in bronchoalveolar lavage fluids [3]. This immediate response is thought to be the result of the action of mediators such as histamine released by mast cells and/or basophils. Some allergic asthmatics also develop late phase responses beginning 3 to 6 h after antigen challenge and lasting up to 48 h in the absence of therapy. In these responses, airway narrowing is associated with BAL fluid which is characterized primarily by increased numbers of eosinophils [7–9]. The degree of eosinophilia in both the blood and BAL fluid of asthmatics correlates with the clinical severity of disease as well as the degree of baseline bronchial hyperresponsiveness [7, 10, 11]. Although the exact mechanisms by which eosinophils exert pathophysiological effects on airways remains unclear,

In Vivo Models of Inflammation, edited by D.W. Morgan and L.A. Marshall

eosinophils are known to secrete a number of lipid mediators which have profound effects on a range of airway cell types [10, 12]. Allergic asthma is further characterized by the production of abnormal amounts of IgE in response to inhalation of aeroallergens [4, 5, 13]. Serum levels of IgE correlate with severity of disease in these patients.

As therapeutic interventions that reduce airway inflammation ameliorate disease symptoms, inflammation is hypothesized to play a key role in the development of allergic asthma. As the primary regulator of the inflammatory cascade, much interest has recently been focussed on the role of the T lymphocyte in asthma [14–17]. In particular, a subset (Th2) of $CD4^+$ T helper cells which has been distinguished functionally by its pattern of cytokine secretion is thought to play a key role [18, 19]. Th2 cells are thought to promote pulmonary allergic responses through their secretion of the cytokines IL-4 and IL-5, which promote IgE production and eosinophilia, respectively [20, 21]. Despite the inferential evidence for this hypothesis, studies are needed in order to determine the mechanisms by which these cytokines elicit the pathological features of asthma. As $CD4^+$ T cells are critical components of the host immune response, ethical concerns preclude study of the pathogenic role of these cells in human asthmatics. Thus, there is a clear need for predictive animal models to determine the mechanisms responsible for the pathological changes in structure and function in asthmatic airways with the hope that more specific immunotherapeutics can be developed to treat bronchial asthma.

Animal models of allergic inflammation

As discussed in the previous chapter by Selig and Chapman, few non-human species spontaneously develop allergic inflammation and/or airway hyperresponsiveness; therefore, most animal models have been based on elicitation of allergic airway responses via systemic sensitization and local lung challenge with soluble protein antigens. Several well-established animal models of allergic inflammation currently exist, namely the ovalbumin-sensitized and -challenged guinea pig model [22, 23], the *Ascaris*-exposed dog model [24], and the Basenji-Greyhound dog model [25]. Unfortunately, tools to examine specific components of the immune system are limited in these models. On the other hand a wealth of immunological reagents are available for the study of immune responses in the mouse, making it a particularly valuable model. In addition, one of the attractive features of utilizing mice to study the pathogenesis of disease is the availability of over 200 well characterized inbred strains and the ability to delete or over-express specific genes through knockout and transgenic technologies. In this chapter, we describe a mouse model of antigen-induced airway hyperresponsiveness and summarize the knowledge gained to date from study of this model.

Establishment of a murine model of allergen-induced airway hyperresponsiveness

Choice of murine strains

As murine strains are known to exhibit different immune responses to various immunologic stimuli, it is necessary to consider carefully the choice of strains when establishing a model of allergic asthma. We initially based our selection of murine strains on a previous report from our group on the airway responsiveness of various murine strains to i.v. delivery of bronchoconstrictor agents. Specifically, Levitt and Mitzner [26] had reported that there was a wide variation in airway reactivity induced by intravenous acetylcholine challenge between different inbred strains of mice. They identified the most responsive strain as the A/J and the least responsive as C3H/HeJ mice. They went on to show that inheritance of this trait was an autosomal recessive trait which they have subsequently mapped to a region of mouse chromosome 6 [27]. Based on the magnitude of the inherent airway contractile responses to acetylcholine stimulation in A/J mice under naïve (noninflammatory) conditions, we selected this strain for further study under conditions of allergen challenge.

Antigen sensitization and challenge

When we began this effort, the only antigen which had been reported to induce inflammation in the murine lung was sheep red blood cells (SRBC). Curtis and colleagues [28] had shown that i.p. sensitization with SRBC and intratracheal challenge with SRBC-induced an inflammatory response in the murine lung. Thus we initially examined the responses of A/J mice to SRBC exposure using a similar approach to that reported by Curtis and his colleagues. Specifically, mice were sensitized by intraperitoneal injection with SRBC (1×10^8) followed by an intratracheal instillation of SRBC (6×10^8 in 50 µl volume) 2 weeks after the primary sensitization (Fig. 1). Airway reactivity was assessed by the method reported by Levitt and Mitzner [26] in their initial survey of murine strains. Briefly, airway pressure changes in response to i.v. acetylcholine were examined in tracheotomized, paralyzed, mechanically ventilated mice. Other analyses performed included cytology of bronchoalveolar lavage (BAL) fluid, serum antibody measurements, lung histology, cytokine bioassays of BAL fluid, and cytokine mRNA analysis of lung tissues.

Kinetics of allergen-induced AHR and airway inflammation

Using the exposure regime described above, airway hyperresponsiveness was seen as early as 24 h after challenge in A/J mice (Fig. 1). The peak of airway hyperrespon-

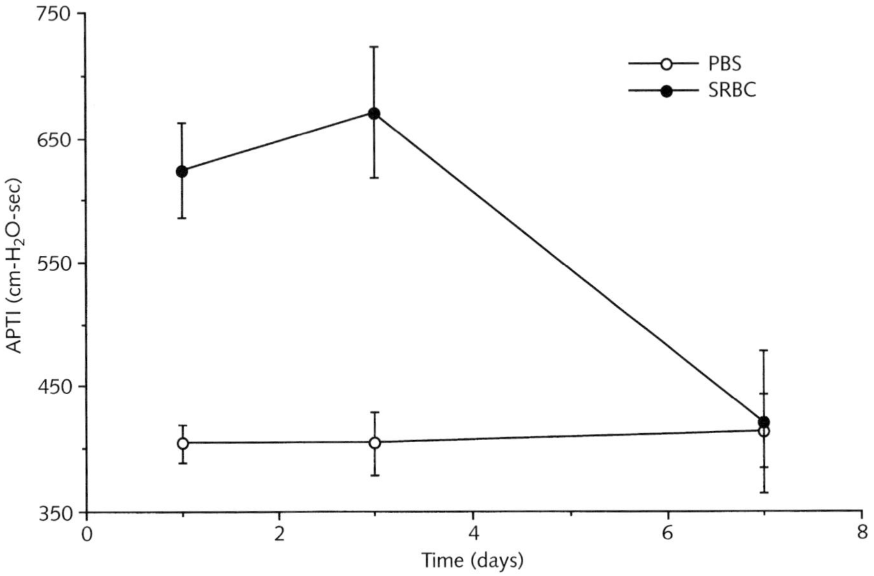

Figure 1
Kinetics of antigen-induced airway hyperresponsiveness in A/J mice. Mice were sensitized by an i.p. injection of 1×10^8 *SRBC (0.5 ml) or PBS (controls) 2 weeks prior to an intratracheal challenge with either SRBC (6×10*^{8} *cells) or PBS alone. Changes in airway pressure following i.v. administration of acetylcholine (25 mg/kg) were measured at different times after the intratracheal challenge. APTI, airway pressure-time index. Data are expressed as means and SEM of 4–6 mice/group.*

siveness occurred 3 days after exposure and was returned to control levels 7 days after the single antigen challenge. Along with this exaggerated airway contractile response, antigen treatment also induced airway inflammation in A/J mice characterized by increased numbers of neutrophils, lymphocytes and eosinophils in the BAL fluid. Initially, increases in neutrophils were observed in the lavage fluid which returned to control levels by 48 h after challenge. On the other hand, eosinophil numbers in the BAL progressively increased between 24–72 h and continued to rise 7 days after challenge. As both AHR and eosinophilia were observed at 72 h, this time period was chosen as the optimal time for subsequent analyses. Interestingly,

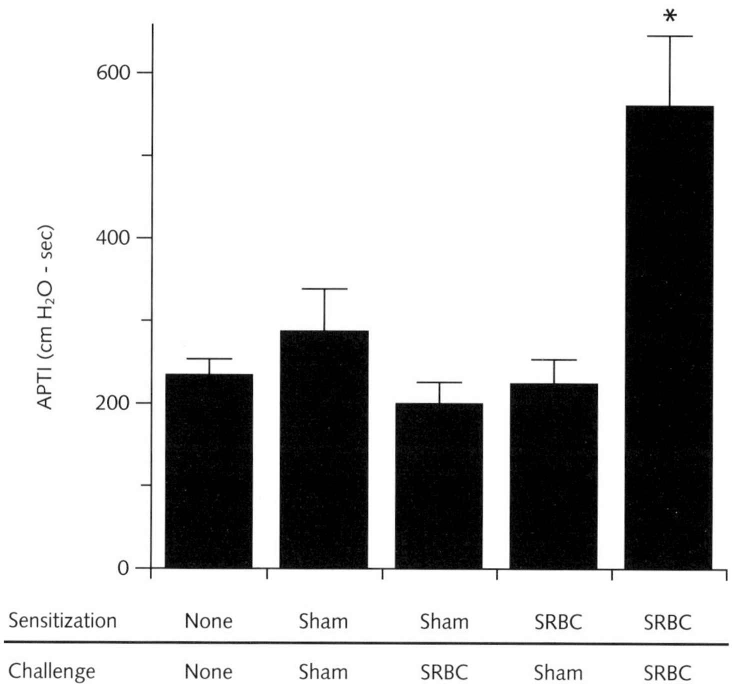

Figure 2
*Effect of various sensitization and challenge protocols on airway reactivity to intravenously administered acetylcholine in A/J mice. Mice were untreated or sensitized and challenged with SRBC or PBS vehicle (sham) as indicated. Acetylcholine induced changes in airway pressure were determined 3 days after challenge. Data are the means and SEM of 5–19 mice per group. *$p < 0.05$ compared with all other groups.*

the time-course of pulmonary eosinophilia does not coincide with that of airway hyperresponsiveness, leading us to speculate that eosinophils are not required for the development of airway hyperresponsiveness in this model. However, the time lag between entry of cells into the airways from the vasculature and their recovery in lavage fluids may complicate our interpretation of the data.

Examination of a combination of sensitization and challenge regimes, revealed that both systemic sensitization and local lung challenge were necessary to elicit inflammatory changes and airway hyperresponsiveness in A/J mice [29] (Fig. 2). Systemic challenge alone does not result in any change in inflammatory cell content or airway reactivity changes, nor does a single intratracheal challenge alone. Interestingly, Gelfand and colleagues [30] have shown that prolonged (two 20-day consecutive aerosol exposures) aerosol exposure can substitute for systemic sensitization. However as a practical consideration, 40 consecutive day exposures greatly increases the effort and costs of this type of experiment.

Methods of lung antigen challenge

Although aerosol challenge more closely reflects the natural exposure to environmental allergens, intratracheal instillation was intially chosen over aerosol delivery for our studies for a number of reasons. First, when comparing airway responses in a number of different strains it has been shown that breathing frequency and tidal volumes differ among inbred murine strains which would likely result in differential deposition of antigen between murine strains [31]. Delivery of the antigen directly to the airways allows us to be reasonably sure that the delivered dose of antigen is equivalent between strains. In addition, unless mice are exposed by nose-only aerosol challenge, they often groom themselves and run the risk of ingesting the antigen which may lead to development of oral tolerance.

As the intratracheal methods of antigen delivery involved a surgical technique which was not performed uniformly by individuals in my laboratory, we sought to develop a less invasive technique which would still allow delivery of equivalent amounts of antigen to the lungs of various murine strains. We have adopted a non-invasive method first communicated to us by Dr. Ian Gilmour (personal communication), which we have coined aspiration challenge. Briefly, mice are anesthesized by i.p. injection of ketamine and xylazine (45 and 8 mg/kg, respectively) mixture and placed in a supine position on a board which is placed at a 45° angle [32]. The animal's tongue is extended with a pair of lined forceps and antigen is placed on the back of the tongue. Extension of the tongue prevents swallowing and the animal is forced to inspire the antigen. Examination of this technique using dye has shown that the antigen is evenly distributed throughout the lung and that no antigen is observed in the esophagus or stomach. This technique is quick and efficient, taking roughly 1–2 min after achieving a suitable level of anesthesia.

Subsequently, we have examined the airway responses of A/J to several other antigens such as HDM, ovalbumin and *Schistosome mansoni* eggs (Tab. 1). In each case, a similar increase in airway hyperresponsiveness and eosinophilic inflammatory response was noted suggesting that the genetic predisposition of this strain is an important factor in their susceptibility to allergic asthmatic symptoms. In this regard, we have examined the effects of the antigen challenge protocol described above (a single sensitization followed by a single lung challenge) on airway reactivity in other murine strains using ovalbumin as the sensitizing antigen. As shown in Table 2, there is a distribution of responses to antigen challenge amongst murine

Table 1 - Comparison of the effects of various antigen challenges on AHR, airway inflammation and IgE levels in A/J mice.

Group	AHR (cm-$H_2$0-sec)	EOS (10^6)	Total IgE (ng/ml)
Control	1133 + 190	0	2.0
OVA	2300 + 40	1.0	6.6+
HDM	2465 + 452	0.93 + 0.22	47 + 8
S. mansoni eggs	1600 + 90	8.7 + 1.5	N.D.

Ova-challenged mice were injected i.p. with 10 µg ovalbumin and challenged with 50 µl of a 1.5% solution of ovalbumin by aspiration 2 weeks later. House dust mite treated animals received 10 µg house dust mite (Greer Laboratories, N.C.) i.p. followed by aspiration challenge with 100 µg HDM 2 weeks later. Schistosome mansoni *eggs (5000 eggs) were administered i.p. 2 months prior to an i.v. injection with 6000 parasite eggs. Data are expressed as means + SEM.*

Table 2 - Comparison of immune responses to allergen challenge in murine strains.

Strain	Inherent AR	Ag-Induced AHR	EOS	IgE
A/J	++++	++++	++++	+++
AKR	+++	+++	+	+
Balb/c	++	++	++	–
C57BL/6	++	–	++	–
C3H/HeJ	++	–	+++	–

Mice were sensitized with an i.p. injection of 10 µg ovalbumin 2 weeks prior to an aspiration challenge with 50 µl of a 1.5% solution of ovalbumin. 72 h after aspiration challenge, airway reactivity to i.v. acetylcholine challenge, BAL cellularity and serum antibody levels were examined. Inherent airway reactivity is defined as the response of naïve animals to i.v. acetylcholine challenge (50 µg/kg).

strains, with the A/J strain being the most responsive and the C3H/HeJ strain being the least. Although the inherent responsiveness to cholinergic challenge of the high and low responders is generally predictive of their response to antigen challenge, this phenomenon does not hold up for the intermediate responders. For instance, naïve Balb/c and C57BL/6 are equally reactive to acetylcholine, but only Balb/c show additional increases in airway reactivity to Ach after antigen challenge. Interestingly, the presence of eosinophils in the BAL does not correlate with the degree of AHR across murine strains. Although there are clear differences in the sensitivity of murine strains to antigen challenge using the protocol described here, allergic responses can be elicited in most murine strains with more aggressive sensitization and challenge protocols (i.e. multiple sensitizations and challenges with the use of adjuvants) [33–35]. Some of the confusion in the literature regarding the role of specific mediators and cells in allergy is derived from comparison of responses between murine strains. Although many investigators view this as a weakness of murine models, it can also be viewed as an opportunity to tease out the multiple pathways which may contribute to the development of allergic asthmatic symptoms. Clearly human asthma is not due to a defect in a single gene and is composed of many subphenotypes (i.e. allergic vs. non-allergic; exercise-induced, aspirin-sensitive, mild versus severe disease). In this regard, we are currently exploiting the differences in antigen-induced airway hyperresponsiveness between A/J and C3H mice to determine the genes important in the susceptibility of A/J mice to allergen-induced airway hyperresponsiveness.

Serum antibody detection

As elevated IgE levels are a characteristic of allergic asthma, we have examined both total and antigen-specific IgE levels in the serum of A/J mice using monoclonal antibodies for ELISA detection of IgE. As shown in Table 1, when A/J mice are sensitized and challenged with either ovalbumin or house dust mite, significant elevations in serum levels of total IgE are observed. Similar results are obtained when antigen-specific IgE levels are assessed. These results demonstrate that allergen-induced AHR in this model is associated with elevations in serum levels of IgE as in allergic asthmatics.

Mucus cell hyperplasia

Early during our characterization of this model, we noticed physical differences in the appearance of the epithelial cell layer in allergen-challenged animals. To determine whether these apparent structural changes were due to increased mucus content, we stained sections from control and challenged mice with periodic acid schiff reagent (PAS) to detect acid mucoglycoproteins. We found striking increases in the number of mucus containing cells in the epithelial layer of allergic mice [36]. This phenomenon has recently been observed in several murine strains [37, 38]. As mucus hypersecretion is present in a number of pulmonary disorders, this model may provide a useful tool for development of reagents designed to reduce mucus secretion.

Pulmonary cytokine measurements

As studies in human asthmatics have clearly shown that the disease is associated with a Th2 cytokine pattern in the lung, we routinely assess pulmonary cytokine production. Two approaches of cytokine determination have been used in this model. The first is to measure cytokine messenger RNA levels in the whole lung of sensitized and challenged animals [39]. One of the major advantages of this technique is that you can examine levels of many cytokines on the same sample as large quantities of RNA are recovered from the whole lung. Two disadvantages of this technique are: (1) it is laborious and expensive, and (2) message levels are not always reflective of protein levels in tissues. Given these two caveats, we have shown that the message levels are a reasonable measure of cytokine protein levels in lavage fluid.

Another technique we have used is to examine cytokine levels in the lavage fluid by ELISA [39]. This method is easy and less time consuming, however the number of cytokines which can be evaluated in a single BAL sample is limited. Another limitation of this method is the relative instability of certain cytokines such as IL-4 in lavage fluids.

Utlizing the above approaches to cytokine measurement, we have consistently shown that antigen challenge of A/J mice results in significant elevations in BAL IL-4 and IL-5 protein levels, with no increases in the Th1 cytokine, IFNγ (Fig. 3). The effects of antigen challenge on lung cytokine message levels were similar to those on protein levels [39]. These results suggest that as in human asthma, the hyperresponsiveness of A/J mice following antigen challenge is associated with the differentiation of CD4 T cells into Th2-cytokine producing cells.

Steroid dependence of allergic responses in A/J mice

As steroid treatment is a main-stay of therapy for asthma, we have examined the ability of the steroid dexamethasone to inhibit antigen-induced airway hyperresponsiveness in A/J mice. Treatment of ova-sensitized A/J mice with 1 mg/kg of dexamathesone 1 day before antigen challenge and daily thereafter for 3 days (Fig. 4), signficantly suppressed antigen-induced AHR. The suppression of AHR was concomitant with a significant reduction in BAL eosinophils. Thus inflammatory responses in this model are ablated by steroid treatment as they are in human asthmatics.

In this model, we were able to simulate in mice several of the hallmarks of human asthma including AHR, pulmonary inflammation characterized by eosinophilia, elevated serum IgE levels and goblet cell hyperplasia. These results suggest that this model provides an excellent tool with which to begin determining the mechanisms by which certain individuals mount Th2-mediated immune responses to ubiquitious allergens and subsequently develop airway obstruction.

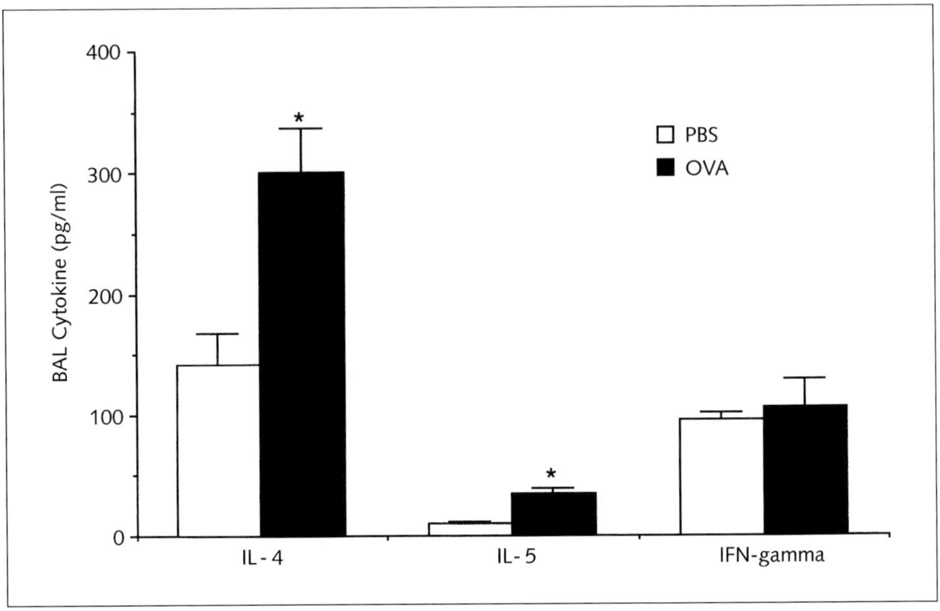

Figure 3
*Measurement of IL-4, IL-5 and IFNγ levels in bronchoalveolar lavage fluids of mice. Cytokine levels (IL-4, IL-5, IFNγ) were measured in unconcentrated lavage fluid by ELISA from mice sensitized and challenged with either OVA or PBS (control) as described in the legend of Figure 4. Data are means and SEM, N = 6–8. *denotes values significantly different from PBS group, $p < 0.05$.*

Role of T cells and T cell-derived cytokines in this model

As stated above, one of the major advantages of a murine model is the availability of a plethora of reagents to probe specific components of the immune response. Thus as a substantial amount of inferential evidence suggested that T cells may be pivotal in the development of allergic response in humans, we conducted a study to directly examine the role of $CD4^+$ T cells in the allergic inflammatory response in A/J mice. The dependence of antigen-induced inflammation and airway reactivity to cholinergic challenge on $CD4^+$ T cell function was determined using the rat anti-mouse CD4 (L3T4) monoclonal antibody GK1.5. Briefly, mice were sensitized with SRBC in this case and given 0.5 mg GK1.5 2 days prior to the intratracheal instillation of SRBC. The efficacy of GK1.5 depletion of $CD4^+$ T cells in the lung was verified by flow cytometric analysis of enzymatically digested lung cells. Depletion of $CD4^+$ T cells significantly inhibited the increased airway responsiveness induced

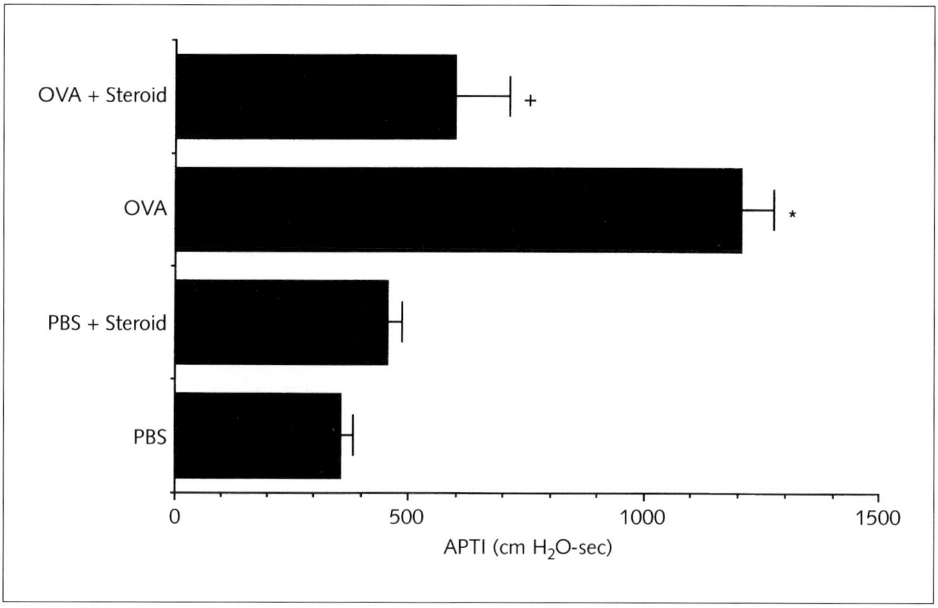

Figure 4
*Steroid inhibition of antigen-induced airway hyperresponsiveness in A/J mice. Mice were i.p. sensitized with either 50 μl of PBS (designated PBS) or ovalbumin (10 μg) 2 weeks prior to an aspiration challenge with either 50 μl of a 1.5% ovalbumin solution or PBS alone. Mice were treated with dexamethasone (1 mg/kg) 1 day before antigen challenge and daily thereafter for three days. Data are the means and SEM of 6–8 mice/group. *denotes values significantly different from PBS; + denotes those values significantly different from OVA group.*

by antigen treatment in the mice not receiving GK1.5. A significant reduction in the numbers of antigen-induced BAL fluid macrophages, neutrophils, lymphocytes, and esosinophils was noted in the mice with CD4 depletion. The ablation of antigen-induced AHR and cell infiltration by T helper cell depletion provides conclusive evidence that $CD4^+$ lymphocytes mediate antigen-induced AHR and pulmonary cellular infiltration. Subsequently these studies have been confirmed by other groups of investigators [40, 41].

Interestingly, we had previously shown that the inherent responsiveness of A/J mice to acetylcholine challenge described by Levitt and Mitzner [26] could be inhibited by administration of the T-cell suppressant agent, cyclosporin A (CsA) [42]. This was an intriguing observation since the animals exhibited no overt inflammation. Following our demonstration that CD4 depletion eliminated the antigen induced response in these mice, we also examined the effect of CD4 and CD8 depletion on the inherent response to acetylcholine. We were disappointed to find that neither depletion of CD4 or CD8 cells reproduced the effect of cyclosporin on reactivity in these mice. Thus, the exact pathways by which CsA affected inherent airway reactivity remain unknown.

Th2 cytokine involvement in airway hyperresponsiveness

The immunopathogenic role of CD4 T cells is thought to be conferred by the elaboration of the Th2 cytokines, IL-4 and IL-5, however the exact mechanisms by which these cytokines induce the symptoms and the pathology associated with asthma are currently not well understood. To gain a better understanding of the role of these cytokines we conducted experiments to assess their *in vivo* role in the development of allergic airway responses in A/J mice.

Interleukin-4

Of the Th2 cytokines, IL-4, is a pleiotropic cytokine which is critical in Th2 cell differentiation [43] as well as in other processes which are potentially important in allergic responses such as: (1) upregulation of VCAM-1 expression which preferentially leads to eosinophil recruitment into tissues [44], (2) induction of B cell class switching to IgE production [45] , and (3) inducing mast cell growth in conjunction with IL-3 [45]. To examine the role of IL-4 in the asthmatic response in A/J mice we treated animals with a monoclonal antibody directed against the IL-4 receptor [36]. Specifically, SRBC-sensitized mice were given 3 mg of rat anti-IL-4 receptor mAb (M1, IgG2a) 3 days prior to local lung challenge mice with SRBC (6×10^8). 3 days after challenge, airway reactivity and BAL were measured. As previously mentioned SRBC sensitization and challenge resulted in significant increases in airway reactivity to acetylcholine concomitant with elevations in the number of recovered eosinophils in the BAL in A/J mice (Tab. 3). Blockade of the IL-4R significantly reduced antigen-induced AHR and virtually ablated eosinophils in BAL fluids, while it had no inhibitory effect on antigen-induced increases in IL-4 and IL-5 protein levels in BAL fluids. These results suggested that the type-2 cytokine pattern was established during the sensitization period and that blockade of the IL-4 receptor at the time of challenge was not able to reverse this pattern. Therefore, the effectiveness of IL-4R blockade on AHR and eosinophilia was likely due to blockade of IL-4 receptor ligation on cells other than T cells such as mast cells, B cells or eosinophils. At the time we published this study other investigators had shown that neutralization of the IL-4 protein prior to challenge was not effective at suppressing antigen-induced AHR [46]. At the time the discrepancies in the results were difficult to

Table 3 - The Role of IL-4 in airway inflammation and AHR in A/J mice

Group	AHR (cm-$H_2$0-sec)	EOS ($\times 10^5$)	IL-4 (pg/ml)	IL-5 (pg/ml)	IFNγ (pg/ml)
Control+Isotype	492+63	0	5+2	17+3	135+13
Control + Anti-IL-4R	434+63	0	1+0.7	12+3	53+19
SRBC+Isotype	1134+177*	7.7+3.0*	45+14*	53+26*	212+37*
SRBC+Anti-IL-4R	636+68+	0.32+0.09	256+56+	63+17*	220+70*

Mice were sensitized to SRBC by i.p. injection of SRBC (1 $\times 10^8$) and intratracheal challenge with SRBC (6 $\times 10^8$) 2 weeks later. Mice were treated with 3 mg of either rat anti-mouse IL-4R (M1, IgG2a) or the isotype control mAb (GL117, IgG2a) 3 days prior to antigen challenge. Numbers of BAL eosinophils, BAL cytokine levels and in vivo *airway reactivity to i.v. acetylcholine challenge were assessed 3 days after the intratracheal antigen challenge. *denotes values significantly different from control groups, $p < 0.05$; + denotes values significantly different from SRBC+ isotype mAb group.*

resolve, however recent evidence that IL-13 and IL-4 share the IL-4 receptor alpha chain suggests that the effects of the M1 antibody may have been to block IL-13 responses [47]. Preliminary studies conducted in our laboratory suggest that indeed this interpretation may be correct.

Suprisingly, blockade of the IL-4 receptor significantly reduced the number of mucus-containing cells in lung sections from antigen-exposed animals, suggesting that the increase was IL-4 dependent. Whether the effects of IL-4 were direct or indirect is unknown. In this study, the decrease in mucus-containing cells after IL-4 blockade was concordant with the decrease in eosinophils in both the BAL and lung tissue, suggesting an association between eosinophil and goblet cell numbers. Several studies have supported the finding that IL-4 is important in antigen induced mucus production [38, 48]. For instance, IL-4 has recently been shown to induce the expression of a specific mucin gene called MUC5 in airway epithelial cells [48]. Consistent with this finding, we have recently shown that mice deficient in Stat6, the molecule which mediates IL-4 signaling, do not display increases in mucus containing cells after antigen challenge [49]. However, Cohn et al. [37] have recently demonstrated that antigen-induced mucus production is not mediated via IL-4 in the mouse lung. Again, since IL-4 and IL-13 both signal through the Stat6 pathway, one interpretation of these studies is that IL-13 mediates the mucus hyperplasia previously thought to be IL-4 dependent. Recent studies in our laboratory confirm a role for IL-13 in antigen-induced mucus hypersecretion.

Although eosinophils were suppressed by IL-4 receptor blockade in this study, IL-5 levels were still elevated, suggesting that the suppression of eosinophils was a result of an IL-4 receptor mediated process. Although these studies do not rule out a role for IL-5 as it is likely important early in the immune response to mobilize eosinophils from the bone marrow, they do, however, suggest that recruitment of eosinophils to the lungs following antigen challenge involves an IL-4 receptor mediated process.

One of the potential mechanism(s) by which IL-4 may induce AHR is its ability to upregulate endothelial VCAM-1 expression on endothelium. Interaction of VCAM-1 with the very late activation antigen-4 (VLA-4) promotes eosinophil recruitment [44]. Therefore, we examined the role of VCAM-1 in allergen-induced eosinophilia in A/J mice by treating antigen-sensitized mice with a combination of monoclonal antibodies directed against murine VLA-4 and VCAM-1 one day prior to antigen challenge. Surprisingly, antigen-induced AHR was not reduced by treatment with this combination of antibodies. However, there was a 60% reduction in BAL eosinophils. These results suggest that VCAM-1 is only partially involved in eosinophil recruitment into the murine lung following antigen provocation, and that other IL-4 mediated processes are also important for tissue eosinophilia. In support of this concept recent reports suggest that IL-4 induces eotaxin production in fibroblasts [50].

Taken together the studies in this model demonstrate the importance of IL-4 and/or IL-13 in mediating inflammation and constriction of murine airways in response to antigen exposures. Due to the pleiotrophic effects of these cytokines on specific immune responses to antigen exposure, the exact mechanisms of their actions remain to be defined. However these studies do highlight the possibility that genes important in the regulation of, production of, or response to IL-4 or IL-13 may underlie allergic asthma and that antagonism of either of these cytokines or their receptors may limit the symptoms of asthma. In support of this conclusion, recent studies have shown that a polymorphism in the alpha chain of the IL-4 receptor is associated with allergic asthma [51].

Interleukin-5

Both allergic and nonallergic asthmatics have significant increases in eosinophils in both lavage fluids and bronchial biopsies [7, 9–11]. In this regard, one of the Th2 cytokines, IL-5, has been shown to be the primary determinant of eosinophil differentiation, activation and survival in tissues [52, 53]. As such, there has been considerable interest in the role of IL-5 in the allergic diathesis. To examine the potential role of IL-5 in A/J mice, we treated SRBC-sensitized mice with rat anti-mouse IL-5 (TRFK5) 2 days prior to intratracheal challenge. In our model, neutralization of IL-5 *in vivo* resulted in significant suppression of eosinophils, but did not, however influence airway reactivity. Our studies are consistent with the findings of Corry et al. [33], who demonstrated that anti-IL-5 reduced eosinophilia, but did not reduce AHR in Balb/c mice. In contrast, other investigators have shown that IL-5 deficient animals do not develop AHR [34]. A potential explanation for the discrepancies in the literature may be provided by a report by Kranveld et al. [54] in which they demonstrate that IL-5 may be important in neurokinin production as well as in eosinophilia.

Administration of the Th1-inducing cytokine interleukin-12 ablates allergic airway responses

To further test the hypothesis that Th2 cytokines are important in the pathogenesis of asthma, we examined the ability of IL-12, a cytokine known to induce Th1 differentiation, on the development of allergic responses in A/J mice. IL-12 is a heterodimeric cytokine that is primarily produced by monocytes and macrophages in response to infections [55]. IL-12 induces cell-mediated immune functions, upregulates Th1 cytokines, especially IFNγ and inhibits or downregulates Th2 cytokines, IL-12 has also been shown to inhibit the development of IL-4-producing Th2 cells in response to antigens such as *Dermatophagoides pteronyssinus* [56] and to inhibit IL-4 production in bulk cultures of peripheral blood leukocytes from allergic patients [57]. Thus, we hypothesized that IL-12 treatment of A/J mice should abrogate the allergic response if it is indeed a result of Th2 cell differentiation. We demonstrated that administration of IL-12 at either 0.1 or 1 µg for 5 days surrounding the SRBC instillation challenge completely prevented the antigen-induced airway hyperresponsiveness and BAL eosinophilia [39]. This inhibitory effect of IL-12 on AHR was associated with a reduction in antigen-induced IL-4 and IL-5 levels and with a significant increase in IFNγ levels. To determine whether the inhibitory effects of IL-12 were mediated via IFNγ, we treated sensitized and challenged mice with IL-12 concurrently with a monoclonal antibody to IFNγ. We demonstrated that the inhibitory effects of IL-12 on AHR were dependent on IFNγ. In contrast, IL-12's effects on eosinophils were only partially dependent on IFNγ production. As asthma is generally established prior to diagnosis, intervention strategies which only address the initial sensitization will not be useful in therapy for asthma. Thus based on the efficacy of IL-12 in preventing AHR, we sought to determine whether IL-12 treatment would reverse already established pulmonary inflammation in these mice. To do so, sensitized mice were given two intratracheal challenges with SRBC (6×10^8 SRBC each) separated by a 7-day period. IL-12 was given before the second challenge by the same treatment schedule mentioned above. Surprisingly, IL-12 very effectively reversed both established AHR and eosinophilia. These findings have been verified by several other groups using different strains of mice and different treatment schedules [58, 59]. These results strongly support the role of Th2 cytokines in the pathogeneis of asthma and also suggest that local administration of IL-12 or agents that induce production of this cytokine may be useful in the treatment of chronic allergic conditions such as atopic asthma.

These findings are consistent with recent reports that the number of IL-12 producing cells in the lungs of asthmatics are markedly reduced compared to those in normal individuals [60]. Furthermore, successful steroid treatment is accompanied by significant increases in IL-12 producing cells in the lungs of asthmatic individuals following allergen exposure. These studies lend further support for the hypothesis that aberrant Th2 cytokine production mediates the pathology of asthma and suggest that dysregulation of IL-12 production may contribute to this effect.

Importance of T cell costimulatory pathways in allergic asthma

As CD4 T cell activation following allergen exposure is clearly an important initial step in differentiation of T cells into Th2 or Th1 cells, we were interested in determining whether factors important in T cell activation were important in Th2 cytokine differentiation in A/J mice. It has been shown that CD4 T cell activation requires two distinct signals from antigen presenting cells [61]. The first signal, which confers specificity, is provided by the interaction of the TCR with MHCII complexes on antigen presenting cells. A second costimulatory signal can be provided by APC-borne ligands for the CD28 and CTLA-4 receptors on T cells, TCR ligation in the absence of costimulation induces Ag-specific T cell anergy [62, 63]. The ligands for CD28 and CTLA-4 are B7-1 and B7-2. Blockade of the B7/CD28 pathways with CTLA4-IG a soluble fusion protein has been shown to effectively inhibit T cell activation *in vitro* and *in vivo*. Recently, some studies have suggested that B7/CD28 interactions may not only be important in T cell activation, but may also play a role in T cell differentiation with B7-1 favoring development of Th1 cells and B7-2 favoring Th2 cells [64, 65]. On the other hand, other studies have suggested that B7-1 and B7-2 molecules can substitute for each other during Th2 differentiation [66, 67]. Thus, we examined the relative contribution of B7-1 and B7-2 to the induction of Th2-mediated allergic airway responses in A/J mice. Administration of anti-B7-2 mAb to ova-sensitized and challenged mice abolished allergen-induced airway hyperresponsiveness, pulmonary eosinophilia, and elevations in serum IgG1 and IgE levels [68]. Blockade of B7-2 interactions also reduced both lung IL-4 and IL-5 mRNA and BAL IL-4 and IL-5 protein levels with no significant changes in IFNγ message or protein levels. In contrast, treatment with anti-B7-1 mAbs had no effect on allergen-induced AHR, IgE production, or cytokine production, however, it significantly suppressed pulmonary eosinophilia. These studies lend support for the concept that eosinophils do not necessarily mediate AHR. These results suggest that B7/CD28-CTLA-4 costimulation is required for the development of many of the physiologic features of allergic asthma in this model, possibly by promoting a pathologic type 2 associated immune response. As expression of these costimulatory molecules on APCs is known to vary, differential expression of these molecules may lead to either development of tolerance or progression of pathologic immune responses in response to allergen provocation. Perhaps blockade of B7-2/CD28 interactions in the lungs of asthmatics may provide a novel therapeutic approach to the treatment of allergic airway disorders.

Conclusions

In summary, we have described a murine model of asthma in which allergen provocation of A/J mice results in a phenotype which closely resembles human asthma. Specifically, allergen sensitization and challenge of these mice induces airway hyperresponsiveness, eosinophilic inflammation, mucus hyperplasia and elevated serum IgE levels. The development of allergen-induced AHR and inflammation in this strain is CD4 T cell dependent and associated with elevations in lung levels of Th2 cytokines. The preponderance of evidence in this model supports the critical role of the Th2 cytokines, IL-4 and/or IL-13 in mediating inflammation and constriction of murine airways in response to antigen exposures. Due to the pleiotrophic effects of these cytokines on specific immune responses to antigen exposure, the exact mechanisms of their actions remain to be defined. In contrast, the data generated in this model do not support a requirement for the Th2-cytokine, IL-5 or eosinophils for the development of AHR. Collectively, these studies highlight the possibility that genes important in the regulation of, production of, or response to IL-4 or IL-13 may underlie allergic asthma and that antagonism of either of these cytokines or their receptors may limit the symptoms of asthma.

Clearly, in a relatively short time period great strides have been made in our understanding of the immunopathogenesis of asthma through the study of murine models of asthma. However, a model as the name implies, is only an approximation of human asthma, and as such, information gained using these models awaits verification in humans.

Acknowledgements
This work was supported in part by grants from the National Institute of Health (R01HL58527) and from the Center for Indoor Air Research.

References

1 McFadden ER, Gilbert IA (1992) Asthma. *N Engl J Med* 327: 1928–1937
2 Beasley R, Roche WR, Roberts JA, Holgate ST (1989) Cellular events in the bronchi in mild asthma and after bronchial provocation. *Am Rev Respir Dis* 139: 806–817

3 Pauwels R (1989) The relationship between airway inflammation and bronchial hyperresponsiveness. *Clin Exp Allergy* 19: 395–398

4 Sears MR, Burrows B, Flannery EM, Herbison G, Hewitt CJ, Hodaway MD (1991) Relation between airway responsiveness and serum IgE in children with asthma and in apparently normal children. *N Eng J Med* 325: 1067–1071

5 Sears MR, Herbison GP, Holdaway MD, Hewitt CJ, Flannery EM, Silva PA (1989) The relative risks of sensitivity to grass pollen, house dust mite and cat dander in the development of childhood asthma. *Clin Exp Allergy* 19: 419–424.

6 Carter M, Perzanowski M, Raymond A, Platts-Mills TAE (1992) Risk factors for asthma in inner city children. *J Peds* 121: 862–866

7 Bousquet J, Chanez P, Lacoste JY, Barneon G, Ghavanian N, Enander I, Verge P, Ahlstedt S, Simony-La fontaine J, Godard P, Michel FB (1990) Eosinophilic inflammation in asthma. *N Eng J Med* 323: 1033–1039

8 Cockcroft DW (1989) Airway hyperresponsiveness and late asthmatic responses. *Chest* 94: 178–80

9 De Monchy, JGR, Kauffman HF, Venge P (1985) Bronchoalveolar eosinophilia during allergen-induced late asthmatic reactions. *Am Rev Respir Dis* 131: 373–6

10 Gleich GJ (1990) The eosinophil and bronchial asthma: current understanding. *J Allergy Clin Immunol* 85: 422–436

11 Wardlaw AJ, Dunnette S, Gleich GJ, Collins JV, Kay AB (1988) Eosinophils and mast cells in bronchoalveolar lavage in subjects with asthma: relationship to bronchial hyperreactivity. *Am Rev Respir Dis* 137: 62–69

12 Gundel, RH, Letts LG, Gleich GJ (1991) Human eosinophil major basic protein induces airway constriction and airway hyperresponsiveness in primates. *J Clin Invest* 87: 1470–1473

13 Stenius B, Wide L, Seymour WM, Holford-Strevens V, Pepys J (1971) Clinical significance of specific IgE to common allergens. *Clin Allergy* 1: 37–55

14 Gerblich A A, Salik H, Schuyler MR (1991) Dynamic T-cell changes in peripheral blood and bronchoalveolar lavage after antigen bronchoprovocation in asthmatics. Am. Rev. Respir.Dis. 143: 533–537.

15 Corrigan CJ, Kay AB (1990) $CD4^+$ T-lymphocyte activation in acute severe asthma: relationship to disease severity and atopic status. *Am Rev Respir Dis* 141: 970–977

16 Robinson, DS, Hamid Q, Ying S, Tsicopoulos A, Barkans J, Bentley AM, Corrigan C, Durham SR, Kay AB (1992) Predominant TH2-like bronchoalveolar T-lymphocyte population in atopic asthma. *N Engl J Med* 326: 298–304

17 Walker C, Bode E, Boer L, Hansel TT, Blaser K, Virchow JC Jr (1992) Allergic and nonallergic asthmatics have distinct patterns of T-cell activation and cytokine production in peripheral blood and bronchoalveolar lavage. *Am Rev Respir Dis* 146: 109–115

18 Mosmann TR, Cherwinski H, Bond MW, Gieldin MA, Coffman RL (1986) Two types of murine helper T cell clone. I. Definition according to profiles of lymphokine activities and secreted proteins. *J Immunol* 136: 2348–2357

19 Street NE, Mosmann TR (1991) Functional diversity of T-lymphocytes due to secretion of different cytokine patterns. *FASEB* 5: 171–177

20 Campbell HD, Tucker WQ, Hort Y, Martinson ME, Mayo G, Clutterbuck EJ, Sanderson CJ, Young IG (1987) Molecular cloning, nucleotide sequence, and expression of the gene encoding human eosinophil differentiation factor (interleukin 5). *Proc Natl Acad Sci USA* 84: 6629–6633

21 Finkelman FD, Katona IM, Urban JF, Holmes J, Ohara J, Tung AS, Sample JV, Paul WE (1988) IL-4 is required to generate and sustain *in vivo* IgE responses. *J Immunol* 141: 2335–2341

22 Iijima H, Ishii M, Yamauchi K, Chao CL, Kimura K, Shimura S, Shindoh Y, Inoue H, Mue S, Takishima T (1987) Bronchoalveolar lavage and histologic characterization of late asthmatic response in guinea pigs. *Am Rev Respir Dis* 136: 922–929

23 Ishida K, Kelly LJ, Thompson RJ, Beattie LL, Schellenberg RR (1989) Repeated antigen challenge induces airway hyperresponsiveness with tissue eosinophilia in guinea pigs. *J Appl Physiol* 67: 1133–1139

24 Pritchard DI, Eady RP, Harper ST, Jackson DM, Orr TS, Richards IM, Trigg S, Wells E (1983) Laboratory infection of primates with *Acaris suum* to provide allergic bronchoconstriction. *Clin Exp Immunol* 54: 469–476

25 Hirshman CA (1985) Basenji-greyhound models of asthma. *Chest* 87: 172–178

26 Levitt RC, Mitzner W (1988) Expression of airway hyperreactivity to acetylcholine as a simple autosomal recessive trait in mice. *FASEB J* 2: 2605–2608

27 Ewart SL, Mitzner W, DiSilvestre DA, Myers DA, Levitt RC (1996) Airway hyperresponsiveness to acetylcholine: Segregation analysis and evidence for linkage to murine chromosome 6. *Am J Respir Cell Mol Biol* 14: 487–495

28 Curtis JL, Kaltreider HB (1989) Characterization of bronchoalveolar lymphocytes during a specific antibody-forming cell response in the lungs of mice. *Am Rev Respir Dis* 139: 393–400

29 Gavett SH, Chen X, Finkelman FD, Wills-Karp M (1994) Depletion of murine CD4+ T-lymphocytes prevents antigen-induced airway hyperreactivity and pulmonary eosinophilia. *Am J Cell Mol Biol* 10: 587–593

30 Renz H, Smith HR, Henson JE, Ray BS, Irvin CG, Gelfand EW (1992) Aerosolized antigen exposure without adjuvant causes increased IgE production and increased airway responsiveness in the mouse. *J Allergy Clin Immunol* 89: 1127–1138

31 Tankersley CG, Fitzgerald RS, Kleeberger SR. (1994) Differential control of ventilation among inbred strains of mice. *Am J Physiol* 267: R1371–R1377

32 Keane-Myers AM, Gause WC, Finkelman FD, Xhou X-d, Wills-Karp M (1998) Development of murine allergic asthma is dependent upon B7-2 costimulation. *J Immunol* 160: 1036–1043

33 Corry DB, Folkesson HG, Warnock ML, Erle DJ, Matthay MD, Wiener-Kronish WP, Locksley RM (1996) Interleukin 4, but not interleukin 5 or eosinophils, is required in a murine model of acute airway hyperreactivity. *J Exp Med* 183: 109–117

34 Foster PS, Hogan SP, Ramsay AJ, Matthaei KI, Young IG(1996) Interleukin 5 deficien-

cy abolishes eosinophilia, airways hyperreactivity, and lung damage in a mouse asthma model. *J Exp Med* 183: 195–201

35 Lefort J, Bachelet C, Leduc D, Vargaftig BB (1996) Effect of antigen provocation of IL-5 transgenic mice on eosinophil mobilization and bronchial hyperresponsiveness. *J Allergy Clin Immunol* 97: 788–799

36 Gavett SH, O'Hearn DJ, Karp CL, Patel EA, Schofield BH, Finkelman FD, Wills-Karp M (1997) Interleukin-4 receptor blockade prevents airway responses induced by antigen challenge in mice. *Am J Physiol* 272 (*Lung Cell Mol Physiol 16*): L253–L261

37 Cohn L, Homer RJ, Marinov A, Rankin J, Bottomly K (1997) Induction of airway mucus production by T helper 2 (Th2) cells: a critical role for interleukin 4 in cell recruitment but not mucus production. *J Exp Med* 186: 1737–1747

38 Rankin J A, Picarella DE, Geba GP, Temann U, Prasad B, DiCosmo B, Tarallo A, Stripp B, Whitsett J, Flavell RA (1996) Phenotypic and physiologic characterization of transgenic mice expressing interleukin 4 in the lung: lymphocytic and eosinophilic inflammation without airway hyperreactivity. *Proc Natl Acad Sci USA* 93: 7821–7825

39 Gavett SH, O'Hearn D, Li X, Huang S, Finkelman FD, Wills-Karp M (1995) Interleukin 12 inhibits antigen-induced airway hyperresponsiveness, inflammation and Th2 cytokine expression in mice. *J Exp Med* 182: 1–10

40 Garlisi CG, Falcone A, Kung TT, Stelts D, Pennline KJ, Beavis AJ, Smith SR, Egan RW, Umland SP (1995) T cells are necessary for Th2 cytokine production and eosinophil accumulation in airways of antigen-challenged allergic mice. *Clin Immunol Immunopathol* 75: 75–83

41 Nakajima H, Iwamoto I, Tomoe S, Matsumaria R, Tomioka H, Takatsu K, Yoshida S (1992) $CD4^+$ T lymphocytes and interleukin-5 mediate antigen-induced eosinophil infiltration into mouse trachea. *Am Rev Respir Dis* 146: 374–377

42 Ewart S, Gavett S, Margolick J, Wills-Karp M (1996) Cyclosporin-A attenuates airway hyperresponsiveness in mice but not through inhibition of $CD4^+$ or $CD8^+$ T cells. *Am J Respir Cell Mol Biol* 14: 627–634

43 Swain SL, Weinberg AD, English M, Hutson G (1990) IL-4 directs the development of Th2-like helper effectors. *J Immunol* 145: 3796–3806

44 Schleimer RP, Sterbinsky S, Kaiswer S (1992) IL-4 induced adherence of human eosinophils and basophils but not neutrophils to endothelium. Association with expression of VCAM-1. *J Immunol* 148: 1086–1092

45 Madden KB, Urban Jr. JF, Ziltener HJ, Schrader JW, Finkelman FD Katona IM (1991) Antibodies to IL-3 and IL-4 suppress helminth-induced intestinal mastocytosis. *J Immunol* 147: 1387–1391

46 Coyle AJ, Le Gros G, Bertrand C, Tsuyuki S, Heusser CH, Kopf M, Anderson GP (1995) Interleukin-4 is required for the induction of lung Th2 mucosal immunity. *Am J Respir Cell Mol Biol* 13: 54–59

47 Lin J-X, Migone T-S, Friedman M, Weatherbee JA, Zhou L, Yamauchi A, Bloom ET, Mietz J, John S, Leonard WJ (1995) The role of shared receptor motifs and common

Stat proteins in the generation of cytokine pleiotrophy and redundancy by IL-2, IL-4, IL-7, IL-13, and IL-15. *Immunity* 2: 331–339
48 Temann U-APrasad B, Gallup MW, Basbaum C, Ho SB, Flavell RA, Rankin JA (1997) A novel role for murine IL-4 *in vivo*: induction of MUC5AC gene expression and mucin hypersecretion. *Am J Respir Cell Mol Biol* 16: 471–478
49 Kuperman D, Schofield BH, Wills-Karp M, Grusby MJ (1998) Signal transducer and activator of transcription factor 6 (Stat6)-deficient mice are protected from antigen-induced airway hyperresponsiveness and mucus production. *J Exp Med* 187: 1–10
50 Mochizuki M, Bartels J, Mallet AI, Christophers E, Schroder J-M (1998) IL-4 induces eotaxin: a possible mechanism of selective eosinophil recruitment in helminth infection and atopy. *J Immunol* 160: 60–68
51 Hershey GKK, Friedrich MF, Esswein LA, Thomas ML, Chatila TA (1997) The association of atopy with a gain-of-function mutation in the subunit of the interleukin-4 receptor. *N Engl J Med* 337: 1720–5
52 Lopez AF, Sanderson CJ, Gamble JR, Campbell HR, Young IG, Vadas, MA (1988) Recombinant human interleukin-5 is a selective activator of human eosinophil function. *J Exp Med* 167: 219–224
53 Wang JM, Rambaldi A, Biondi A, Chen ZG, Sanderson CJ, Mantovani A (1989) Recombinant human interleukin-5 is a selective eosinophil chemoattractant. *Eur J Immunol* 19: 701–705
54 Kraneveld A D, Nijkamp FP, Van Oosterhout AJM (1997) Role of neurokinin-2 receptor in interleukin-5-induced airway hyperresponsiveness but not eosinophilia in guinea pigs. *Am J Respir Crit Care Med* 156: 367–374
55 Trinchieri G (1994) Interleukin-12: A cytokine produced by antigen-presenting cells with immunoregulatory functions in the generation of T-helper cells type 1 and cytotoxic lymphocytes. *Blood* 84: 4008–27
56 Manetti R, Parronchi P, Giudizi MG, Piccinni M-P, Maggi E, Trinchieri G, Romagnani S (1993) Natural killer cell stimulatory factor (interleukin-12 [IL-12] induces T helper type 1 (Th1)-specific immune responses and inhibits the development of IL-4-producing Th cells. *J Exp Med* 177: 1199–1204
57 Kiniwa M, Gately M, Gubler U, Chizzonite R, Fargeas C, Delespesse G (1992) Recombinant interleukin-12 suppresses the synthesis of IgE by interleukin-4 stimulated human lymphocytes. *J Clin Invest* 90: 262–266
58 Kum TS, DeKruyff RH, Rupper R, Maecker HT, Levy S, Umetsu DT (1997) An ovalbumin-Il-12 fusion protein is more effective than ovalbumin plus free recombinant IL-12 in inducing a T helper cell type 1-dominated immune response and inhibiting antigen-specific IgE production. *J Immunol* 158: 4137–4144
59 Kips JC, Brusselle GG, Joos GF, Peleman RA, Devos RR, Tavernier JH, Pauwels RA (1995) Importance of interleukin-4 and interleukin-12 in allergen-induced airway changes in mice. *Int Arch Allergy Immunol* 107: 115–118
60 Naseer T, Minshall EM, Leung DYM, Laberge S, Ernst P, Martin RJ, Hamid Q (1997)

Expression of IL-12 and IL-13 mRNA in asthma and their modulation in response to steroid therapy. *Am J Respir Crit Care Med* 155: 845–851
61 Schwartz RH (1992) Costimulation of T lymphocytes: the role of CD28, CTLA-4, and B7/BB1 in interleukin-2 production and immunotherapy. *Cell* 71: 1065–1068
62 Bluestone JA (1995) New perspectives of CD28-B7-mediated T cell costimulation. *Immunity* 2: 555–559
63 June CH, Bluestone JA, Nadler LM, Thompson CB (1994) The B7 and CD28 receptor families. *Immunol Today* 15: 321–331
64 Kuchroo VK, Das MP, Brown JA, Ranger AM, Zamvil SC, Sobel RA, Weiner HL, Nabavi N, Glimcher LH (1995) B7-1 and B7-2 costimulatory molecules activate differentially the Th1/Th2 developmental pathways: application to autoimmune disease therapy. *Cell* 80: 707–718
65 Freeman GJ, Boussiotis VA, Anumanthan A, Bernstein GM, Ke X, Rennert PD, Gray GS, Gribben JG, Nadler LM (1995) B7-1 and B7-2 do not deliver identical costimulatory signals, since B7-2 but not B7-1 preferentially costimulates the initial production of IL-4. *Immunity* 2: 523–532
66 Natesan M, Razi-Wolf Z, Reiser H (1996) Costimulation of IL-4 production by murine B71 and B7-2 molecules. *J Immunol* 156: 2783–2791
67 Levine BL, Ueda Y, Craighead N, Huang ML, June CH (1995) CD28 ligands CD80 (B7-1) and CD86 (B7-2) induced long-term autocrine growth of $CD4^+$ T cell and induce similar pattern of cytokine secretion *in vitro*. *Int Immunol* 7: 891–904
68 Keane-Myers AM, Gause WC, Finkelman FD, Xhou X-D, Wills-Karp M (1998) Development of murine allergic asthma is dependent upon B7-2 costimulation. *J Immunol* 160: 1036–1043

Chronic obstructive pulmonary disease

David C. Underwood

Department of Pulmonary Pharmacology, SmithKline Beecham Pharmaceuticals, 709 Swedeland Road, King of Prussia, PA 19406, USA

Introduction

This chapter is designed as a reference for researchers in the field of *in vivo* lung research, particularly lung injury and COPD, and to provide a description of the diverse manifestations and potential models to mirror them. With respect to COPD, each particular subclass may exhibit all or only a few of these manifestations. Therefore, no single animal model of COPD exists. Further, chronic pathological findings in the lung may result from a continuous form of injury, but more than likely are the culmination of a series of acute, repetitive insults. As such, we must describe the acute aspects, the degree of resolution of each insult, and the product of a number of insults to assess the effect that resistance, accumulation of damage, or potentiation with repeated challenges impact the eventual outcome. The diverse nature of acute lung injury and COPD requires us to focus on a limited number of descriptors provided by animal models of the diseases.

In this chapter several animal models are discussed in which particular quantifiable parameters reflective of COPD are described. The models may be categorized under a clinical subclass of COPD (e.g. bronchitis, emphysema) or as a particular manifestation or complication associated with the pathology or symptomatology of the disease (e.g. fibrosis, inflammatory cell trafficking, cough, pulmonary hypertension). Inflammatory cell localization and activity are important in a number of other pulmonary dysfunctional states such as asthma and ARDS. Therefore, some of this material may be applicable to those diseases as well. Because of diversity of species employed and specific quantitative methods in these models, the reader is encouraged to seek details from original manuscripts.

Chronic obstructive pulmonary disease

Chronic obstructive pulmonary disease (COPD) is not one disease, but a disease spectrum encompassing complications from reductions in airflow due to architec-

In Vivo Models of Inflammation, edited by D.W. Morgan and L.A. Marshall

tural changes and inflammation, located primarily in the airways, but including some manifestations in the pulmonary vasculature. The clinical state is generally characterized as chronic bronchitis, fibrosis or emphysema usually associated with airflow obstruction, with significant overlap among the subclassifications [1]. It is prevalent in about 20 million men and women in the United States, mostly over 40 years of age. The mortality is about 20/100 000, making it the fourth leading cause of death overall, which rose 33% from 1979 to 1991 [2]. The primary causes are generally recognized as cigarette smoking > occupational and environmental exposure > idiopathic or congenital. Like asthma, the airflow obstruction may be accompanied by airway hyperresponsiveness, cough and enhanced mucus production. However, the disease-related remodeling of the airways and pulmonary vasculature in COPD is more demonstrable, and the airflow reductions, generally assessed by measurement of upper and central airway flow, are only partially reversible. The extent and consequences of COPD range from chronic bronchitis, symptomatically characterized as increased cough and sputum production for 3 months in 2 successive years, to actively atrophic states such as pathologically characterized fibrosis and emphysema [1]. Treatment of COPD has suffered from early understandable, but oversimplified classifications based on a poor understanding of the etiology of the disease. These included: (1) bronchitis, generally associated with hypoxemic cor pulmonale ("blue bloater" or type B COPD), and (2) emphysema with relatively normal blood gases and pulmonary circulation, and evidence of gross overinflation ("pink puffer" or type A disease). We now know that COPD is much more complex, and that there are multiple lesions which span classification subtypes [3]. Because of the general irreversibility of airflow obstruction with present therapy, especially in severe fibrosis or emphysema, treatment approaches have been directed towards optimizing the ventilation/perfusion relationship (oxygen therapy and surgical lung reduction) and reducing progress of the disease (cessation of exposure). Thus, quality of life assessment is a crucial marker, and currently considered unapproachable through animal modeling. Much of our laboratory knowledge of COPD-related airways disease is the result of chemical industry- and government-supported studies of environmentally-related pulmonary disease. A recent review concerning the effects of air pollutants provides substantial insight for selection of animal models for COPD [4].

Experimental chronic bronchitis

Chronic bronchitis is characterized by cough and mucus hypersecretion. In an early extensive review, Reid and Jones [5] thoroughly describe structural and histochemical end points neccesary for an experimental animal model of chronic bronchitis: (1) increased size of submucosal glands; (2) increased number of secretory cells in surface epithelium, especially at peripheral levels where they are normally absent;

and (3) alteration in the forms of glycoprotein (from neutral to acid) recognized within secretory cells due to increased sulfation or sialylation [6]. Because these changes represent exaggerations of normal structure rather than specific lesions, quantitation of temporal- and exposure-dependent manifestations is important for proper assessment. Models to demonstrate airway functional changes in COPD are less obvious in the literature, but may represent a particularly relevant approach towards early diagnosis of the disease. Airway obstruction and increased airway responsiveness to inhaled spasmogens (histamine, methacholine) are common features [7, 8], but many inhaled stimuli employed in assessing asthma (cold air, distilled H_2O and propranolol) are ineffective in demonstrating hyperresponsiveness in COPD [9]. The airway obstruction that occurs probably results from chronic inflammatory changes (increased airway mucus and irritant receptor sensitivity) exacerbated by viral or bacterial infections that impact airway smooth muscle constriction. A wide variety of species and agents has been used to induce a state of enhanced mucus production, epithelial changes and airway inflammation leading to obstruction and cough, diagnostic endpoints commonly associated with chronic bronchitis (see Tab. 1 for a representative list). The animals include rodent-like species such as rats, hamsters and guinea pigs (not a true rodent [10]), as well as larger animals such as cats, dogs, sheep, pigs and monkeys. Clearly, there are as many differences as similarities in response to a particular manipulation among species. Experimental exposures include irritants (gases, particulates and smokes), infectious agents (viral and bacterial) and selected pharmacologic agents (Tab. 1).

Inhaled irritant gases, such as SO_2 and NO_2, and oxidants, such as chlorine and ozone, have been used effectively to produce a chronic bronchitic state in laboratory animals. By far, the greatest number of studies of irritant-induced COPD-like pathology utilize SO_2 because: (1) it is a very soluble gas, easily delivered into the airways [11]; (2) it is a common atmospheric pollutant [12]; and (3) the histologic changes associated with exposure in animals has been reported to produce conditions similar to those found in human bronchitis [3]. Generally, successful studies in rats have involved chronic exposure to SO_2 (250–700 ppm) for 5–7 days a week, 2–6 h/day for 2–6 weeks. Regardless of the precise protocol, there are several consistently demonstrable changes: (1) increased concentration of secretory cells in large and small airway epithelium and extension into peripheral regions where they are normally absent [5, 6]; (2) increased ratio of acid (alcian blue staining) to neutral (PAS staining) mucins [5, 6, 12, 13]; (3) thickening of airway epithelium with a loss or stunting of cilia [5]; (4) reduced tracheal mucus flow rates [13]; (5) increased airway mucin gene expression [14]; and (6) changes in airway responsiveness to exogenous spasmogens [15–17]. Corresponding increases in tracheal responsiveness in the tissue bath are not evident, suggesting mechanisms involving different generations of airways where aspects of permeability of spasmogen, the impact of mucus upon airflow or sensory innervation may play a more important role [15, 18]. Interestingly, dogs more often develop a hypo-responsiveness in corresponding bronchi-

Table 1 - Representative animal models of bronchitis

Species	Agent or insult	Measurement parameters	Ref.
Rat	Sulphur dioxide	Morphology and mucus	[6]
Rat	Sulphur dioxide	Mucus flow rate	[13]
Rat	Metabisulphite	Morphology and mucus	[20]
Rat	Sulphur dioxide	Airway responsiveness	[15]
Rat	Nitrogen dioxide	Morph., mucus and resp. rate	[21]
Rat	Nitrogen dioxide	Morph., mucus and resp. rate	[22]
Rat	Acrolein	Lung vol., diffusion, mechanics	[23]
Rat	Ozone	Morphology and ventilation	[24]
Rat	Ozone	Morphology and mucus	[25]
Rat	Cigarette smoke	Bronchoconstrict., vascular leakage	[26]
G. pig	Tol. diisocyan.	Airway responsiveness	[27]
G. pig	Cigarette smoke	Bronchoconstrict., vascular leakage	[28]
G. pig	Cigarette smoke	Airway hyperresponsiveness	[29]
G. pig	Cigarette smoke	Pulm. hypertens.,vent. hypertrophy	[30]
G. pig	Metabisulphite	Bronchoconstriction	[31]
G. pig	Metabisulphite	Bronchoconstrict., vasc. leakage	[32]
G. pig	Polymyxin B	Cough exacerbation	[33]
Hamster	Nitric acid	Morphology and matrix content	[34]
Hamster	Sulphur dioxide	Morphology and airway mechanics	[35]
Hamster	Nitrogen dioxide	Morph., mucus and resp. rate	[22]
Dog	Cigarette smoke	Morphology and mucus	[36]
Dog	Sulphur dioxide	Morphology and mucus	[37]
Dog	Sulphur dioxide	Airway obstruction/responsiveness	[16]
Dog	Sulphur dioxide	Airway responsiveness/morphology	[39]
Dog	Sulphur dioxide	Response to i.v/inhaled spasmogen	[17]
Dog	Sulphur dioxide	Airway mucus hypersecretion	[38]
Lamb	Tobacco smoke	Morphology	[40]
Monkey	Ozone	Morphology and neutrophilia	[41]

tis models [16, 17, 19]. Other insults or toxicants which induce similar COPD-like manifestations include (Tab. 1): metabisulphite [20], nitrogen dioxide [21, 22], acrolein [23], ozone [24, 25] and cigarette smoke [26] in rats; toluene diisocyanate [27], cigarette smoke [28–30], metabisulphite [31, 32] and polymyxin B [33] in the guinea pig; nitric acid [34], sulphur dioxide [35] and nitrogen dioxide [22] in the hamster; tobacco smoke [36] and sulphur dioxide [16, 17, 19, 37–39] in the dog; tobacco smoke [40] in lambs; and ozone in monkeys [41].

Experimental emphysema

Emphysema is characterized by abnormal permanent enlargement of airspaces distal to the terminal bronchiole resulting from non-uniform destruction of alveolar septal walls without obvious fibrosis. Although certain diagnosis is made only through histologic examination of inflated, fixed whole lung, demonstrable airway obstruction may be detected due to severely diminished elastic recoil of affected tissues, which allows the airways to collapse during expiration [3]. Salient differences from chronic bronchitis include a later age of diagnosis, greater dyspnea with less complications associated with infection, production and clearance of sputum and pulmonary vascular manifestations such as cor pulmonale and pulmonary hypertension. Centriacinar emphysema, with the disease centered in the bronchiole-alveolar duct junction consistent with the pattern of deposition of inhaled toxicants, is the most common form in humans [42]. Panlobular emphysema, associated with an imbalance of enhanced elastase burden relative to a deficiency of its natural inhibitor α-1 antiprotease, permanently enlarges airspaces distal to the terminal bronchioles [42]. Regardless of its classification or distribution, airspace enlargement resulting from destruction of alveolar walls constitutes the morphological hallmarks of this disease.

Experimental animal models of emphysema have been reviewed previously [43, 44], and an overview is presented in Table 1. Generally, emphysema is induced by: (1) a single intratracheal instillation of elastase or papain into the lung of rat [45, 46], mouse [47], hamster [48–51], guinea pig [52], rabbit [53] or dog [54]; (2) cadmium [55, 56], tobacco smoke [57], hyperoxia [58], starvation [59] or nitrogen dioxide [60] in the rat; (3) copper deficiency in the hamster [61]; (4) cigarette smoke in the guinea pig [62]; and (5) endotoxin administration to hamsters [63, 64], dogs [65] or monkeys [66]. These models most resemble the panlobular type of disease [42]. Figure 1 demonstrates the substantial changes in lung capacity that occur in emphysema (adapted from [67]), and a similar depiction of data in the elastase-induced emphysematous state produced by intratracheal instillation of elastase in the hamster (adapted from [50]). Papain or purified preparations of bovine or human pancreatic elastase or neutrophil elastase have been most commonly used to produce emphysemic lesions in animals [42]. Because of their relatively greater sensitivity to elastolytic enzymes, hamsters have been used more often as an emphysema model. A single instillation of elastase into the upper airways results in an acute alveolitis characterized by an early neutrophillic invasion and results in degradation of elastin, rupture of the alveolar epithelium, pulmonary edema, hemorrhage and airspace enlargement. The acute inflammatory response generally resolves within a

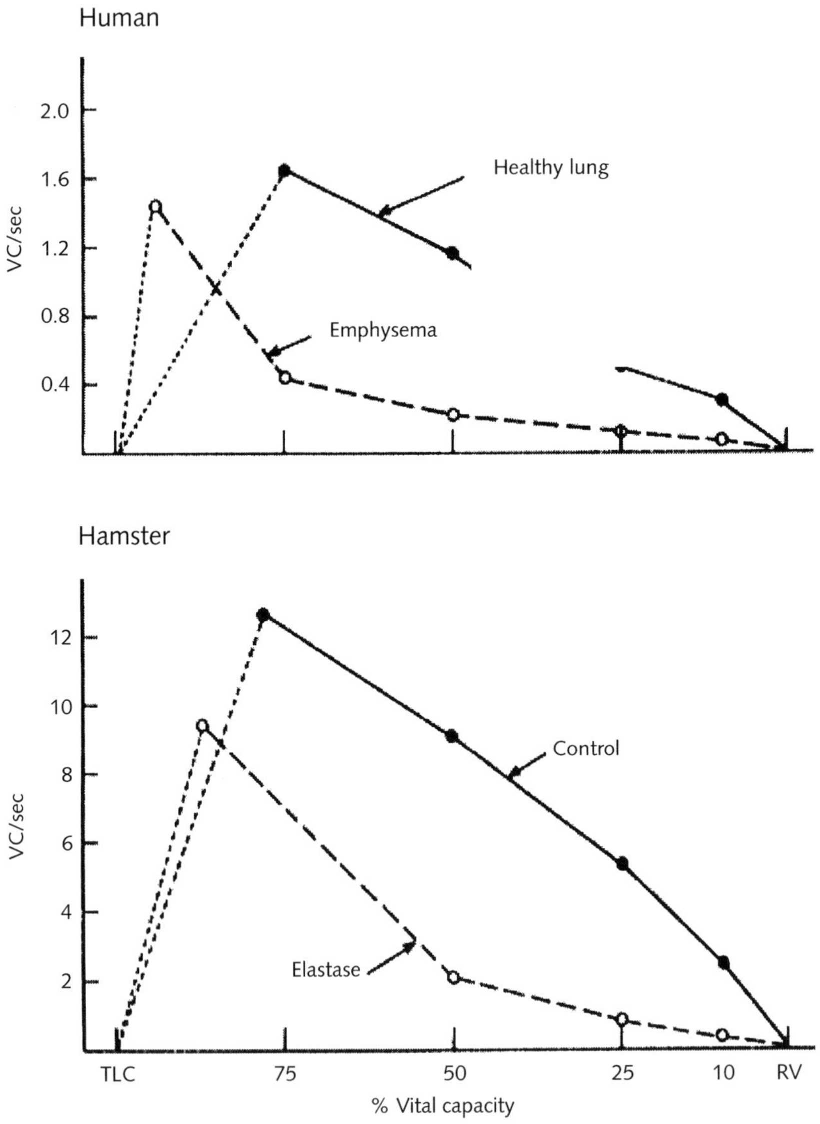

Figure 1
The upper graph demonstrates the differences in lung vital capacity that are evident with emphysema (adapted from [67]), and a similar depiction (lower graph) of data in the emphysematous state produced by intratracheal instillation of human neutrophil elastase in the hamster (adapted from [50]).

week. The rapid synthesis of new elastin and collagen results in a permanent distortion and derangement of alveolar structures [41].

A few genetically emphysema-prone mouse strains of emphysema are available. The emphysema of the "tight-skinned" and "pallid" mice results from a hereditary deficiency in α-1 antiprotease [68, 69]. The results of the imbalance of elastin/elastolytic activity are discernible within the first 2 months of life. There is a spontaneous loss of total lung elastin over time, and leukocyte recovery by bronchoalveolar lavage is generally unremarkable suggesting that the characteristic inflammatory response as demonstrated in models of exogenous elastase administration is not evident in these animals. A genetic malabsorption of copper which is critical for lysyl oxidase, an essential connective tissue cross-linking enzyme, is the basis for panlobular emphysema formation in another strain, the blotchy mouse [70]. While the lung remodeling that occurs has some resemblance to emphysema, it does not involve a dysfunction of α-1 antiprotease, generally believed to be critical in the pathogenesis in humans. Few functional differences have been noted in these animals, possibly due to the lack of underlying inflammation.

Inflammatory cell localization to the lung

It must be recognized that localization of leukocytes at an inflammatory site in the lung involves three separate, but related, didactic phenomena: (1) recruitment (a multistep process in itself including circulatory cytosis from bone marrow or spleen, adhesion and diapedesis to and through endothelial surfaces); (2) proliferation, persistence and activity state at the site (complicated by cell-cell interactions and feed-forward mechanisms); and (3) resolution (by diapedesis into the airway lumen, apoptosis, phagocytosis and rare return to the circulation). Polymorphonuclear (PMN) leukocytes (i.e.neutrophils) are major contributors to lung cell injury in many forms of COPD, ARDS and ischemia-reperfusion injury. Invading PMN's release cytotoxic humoral mediators including oxygen-derived free radicals, elastase, cytokines and chemokines. Although many acute and chronic lung injury models are associated with increases in neutrophil or mononuclear cell numbers, perhaps the most common animal model of non-allergic inflammatory cell chemotaxis or proliferation in the lung is intratracheal or aerosol exposure to lipopolysaccharide (LPS). Techniques commonly used to study ARDS are geared toward symptoms: (1) microvascular lung injury leading to pulmonary hypertension and edema, and (2) impairment of lung mechanics leading to inappropriate gas exchange. It has generally been accepted, but not exclusively proven, that the neutrophil produces much of the damage resulting from an airway insult and, indeed, neutrophilia was common in many of the bronchitis and emphysema models listed in Tables 1 and 2. However, as depicted in Figure 2 where inhaled LPS produces a dose- and time-

Table 2 - Representative animal models of emphysema

Species	Agent or insult	Ref.
Rat	Elastase	[45]
Rat	Elastase	[44]
Rat	Elastase	[46]
Rat	Cadmium	[55]
Rat	Cadmium	[56]
Rat	Tobacco smoke	[57]
Rat	Hyperoxia	[58]
Rat	Starvation	[59]
Rat	Nitrogen dioxide	[60]
Mouse (A/J)	Elastase	[47]
Mouse (tight-skinned)	Genetic elastase imbalance	[68]
Mouse (blotchy)	Genetic copper deficiency	[70]
Mouse (pallid)	Genetic elastase imbalance	[69]
Hamster	Elastase	[48]
Hamster	Elastase	[49]
Hamster	Elastase	[50]
Hamster	Papain	[51]
Hamster	Copper deficient diet	[61]
Hamster	Endotoxin	[63]
Hamster	Endotoxin	[64]
Guinea pig	Elastase	[52]
Guinea pig	Cigarette smoke	[62]
Rabbit	Elastase	[53]
Dog	Endotoxin	[65]
Dog	Elastase	[54]
Rhesus monkey	Endotoxin	[66]

Figure 2
demonstrates the concentration response (A) and time-course (B) of leukocytosis, predominantly neutrophils (recovered by bronchoalveolar lavage), produced by inhaled nebulized aerosol of LPS (10 μg/ml for 15 min) in the guinea pig. Edema formation (C), expressed as the ratio of wet/dry lung weights required a greater concentration and exposure time (1 mg/ml for 30 min) compared to the neutrophilic response (from [71]).

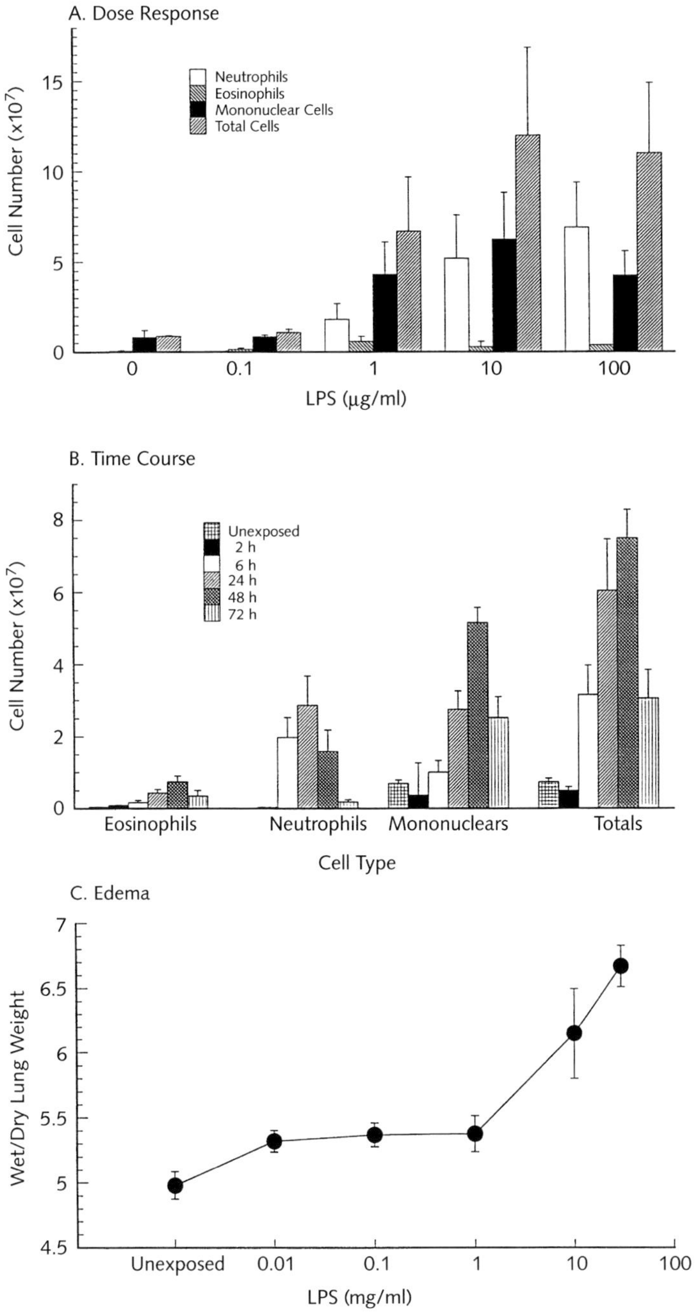
A. Dose Response
Cell Number (x10^7)
15
10
5
0
Neutrophils
Eosinophils
Mononuclear Cells
Total Cells
0
0.1
1
10
100
LPS (μg/ml)
B. Time Course
Cell Number (x10^7)
8
6
4
2
0
Unexposed
2 h
6 h
24 h
48 h
72 h
Eosinophils
Neutrophils
Mononuclears
Totals
Cell Type
C. Edema
Wet/Dry Lung Weight
7
6.5
6
5.5
5
4.5
Unexposed
0.01
0.1
1
10
100
LPS (mg/ml)

dependent airway neutrophilia and edema in the guinea pig, the neutrophilia generally resolves within three days [71]. Therefore, the aspect of repetitive nature of the exposure, characteristic of cigarette smoking, occupational exposure or sleep apnea may loom as the key part of the manifestations of COPD.

Fibrosis

Pulmonary fibrosis is a proliferative reaction to lung injury characterized by altered lung architecture and function resulting from excessive deposition of matrix macromolecules and generally associated with alveolar inflammation (for review, see [72]). Idiopathic pulmonary fibrosis is rare, but occupational exposures to various metals, mineral and organic dusts and drugs have been commonly linked to the disease [72]. Pulmonary fibrosis represents a repair of injury that is reversible upon removal of the injurious stimulus provided basic tissue scaffolding remains intact. Therefore, as in many animal models, the magnitude of the injurious exposure may dictate differential activation of distinct cell populations or mechanisms. Fibrosis may be evident in both chronic bronchitis and emphysema disease states and the animal models that reflect them. A variety of stimuli have been employed in fibrosis models, including drugs (bleomycin and amiodarone), infectious agents, radiation, pesticides (paraquat), particulates (silica and asbestos) and oxidants. The general pattern that the disease and the models follow is an initial injury followed by a repair process, usually involving fibrogenesis, depending on the agent, intensity and route of exposure [73]. The most extensively used pulmonary fibrosis model is that induced by bleomycin. The antineoplastic agent, bleomycin, when administered systemically or by intratracheal instillation into rats or hamsters results in an alveolar fibrosis that becomes evident approximately 2 weeks after a single exposure [72, 74]. Damage is routinely assessed by morphology and increased hydroxyproline content in the whole lung, reflecting enhanced collagen deposition. The fibrosis is generally dose- and time-dependent, but uneven distribution of the bleomycin instillation may dictate a nonhomogeneous occurrence. Pulmonary hypertension and right ventricular hypertrophy may accompany bleomycin-induced fibrosis.

Pulmonary hypertension

Pulmonary hypertension develops from an increase in the pulmonary vascular resistance, primarily localized to the precapillary arteries and arterioles. Although the etiology in a small percentage of patients is idiopathic, pulmonary hypertension is more commonly a severe secondary complication of chronic hypoxia and acidosis resulting from chronic obstructive pulmonary disease, congestive heart failure, respiratory distress syndrome, cystic fibrosis and hypoventilation syndrome [75, 76].

Right ventricular hypertrophy, increased vascular resistance and enhanced vascular remodeling, including hyperplasia (increase in myocardial and smooth muscle cell number), hypertrophy (increase in muscle cell size) and muscle extension (appearance of new smooth muscle in previously less muscularized arterioles), are hallmarks of pulmonary hypertension which combine to produce an anatomic resistance to flow [77]. Two animal model approaches have been extensively characterized. The first is monocrotaline-induced pulmonary hypertension, in which a single parenteral injection of monocrotaline (50–100 mg/kg) results in progressive development of pulmonary vascular endothelial injury, hypertension and right ventricular hypertrophy [78–80]. Monocrotaline produces a significant inflammatory response resulting in alveolar injury which somewhat resolves while pulmonary vascular remodeling persists [78–80]. The model has also been characterized as reflective of adult respiratory distress syndrome [81]. The second type of insult is chronic hypoxia which elicits a variety of cardiopulmonary responses, including pulmonary hypertension, increased vascular resistance (especially in the pulmonary circulation), right ventricular hypertrophy, polycythemia and structural changes in the pulmonary vascular beds, collectively known as vascular remodeling [81, 82]. Physiological, biochemical and histological responses to chronic hypoxia and/or hypobaric exposure have been studied in a variety of species, including rats [83, 84], mice [85, 86] and guinea pigs [87–89]. Such preclinical models of hypoxia-induced pulmonary hypertension are utilized routinely to explore the potential therapeutic efficacy of various classes of compounds. Among the many published reports of models of hypoxia-induced pulmonary hypertension from different laboratories, as expected, the experimental protocols generally differ with respect to several parameters: species and strain utilized, exposure time, specific gas concentration and atmospheric pressure. Our laboratory has endeavored to standardize protocols of chronic hypoxia in rats and guinea pigs, by using a custom-built chamber to optimize the environment for these laboratory animals [90]. There appear to be measurable differences in cardiopulmonary responses to hypoxia among different strains and vendors of rats, and Figure 3 represents a side-by-side comparison of Sprague-Dawley, Wistar and High altitude-sensitive rats from the same vendor [91]. In this study it is apparent that the high altitude-sensitve strain has an exagerated pulmonary artery pressor response to chronic hypoxia, while the Wistar develops slightly greater right ventricular hypertrophy. Interestingly, only the Sprague-Dawley strain develops an increased responsiveness to methacholine-induced bronchoconstriction.

Future directions

Animal research in chronic obstructive pulmonary disease has proceeded in two general directions which are not mutually exclusive: (1) models to help discover and

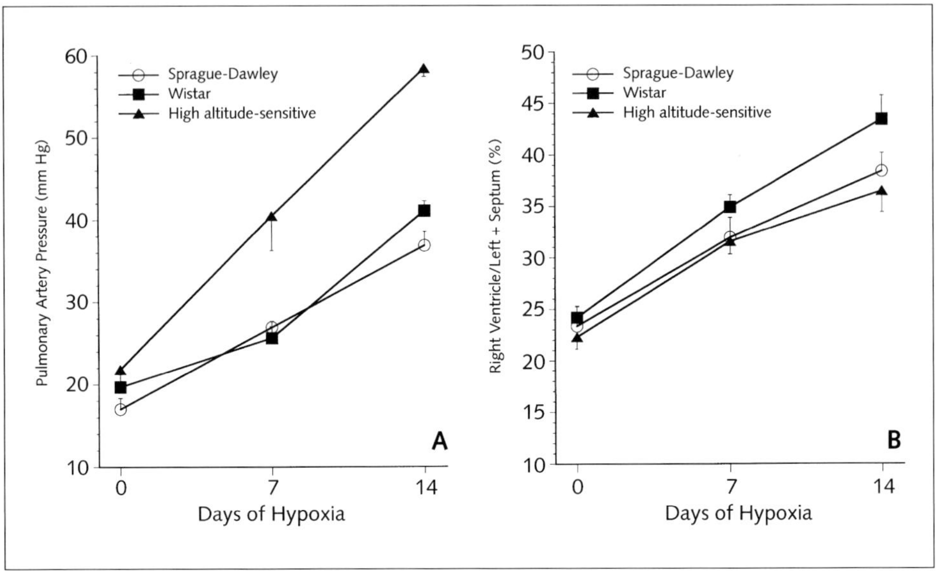

Figure 3
demonstrates the effects of 7- or 14-day hypoxic (9% O_2) exposure on pulmonary artery pressure (A) and right ventricular hypertrophy (B), in Sprague-Dawley, Wistar and High-Altitude Sensitive rats (from [91]).

describe the etiology of the disease, and (2) models in which to test therapeutic agents. Both approaches lend a hand to each other by helping to discern new targets for therapeutic research and by providing pharmacological tools to test hypotheses. As the need for therapeutic approaches for COPD is immediate, many patients require agents to improve their quality of life in a disease which has already manifested itself as a presumably irreversible state. When acute exacerbations of their disease by stress, excercise or viral and bacterial infections occur, these patients are often in an emergency situation with a poor long-term prognosis. Therefore, modeling for this situation may require maintenance of animals in an essentially plateaued state of obstruction followed by measurement of *in vivo* and *ex vivo* parameters before and after challenge with spasmogens, hypoxia or respiratory infection. Astute attention to this approach would help establish the quality of life assessment in animals as it is so often characterized in the human disease.

Concerning pharmacological characterization of drugs which may be effective in the treatment of COPD, just as there is no single animal model, we must also remember that no single drug has provided consistent efficacy in the clinical treatment of the disease. Therefore, a potential therapeutic drug regimen must be assessed in a model which appropriately reflects a particular aspect of the disease (i.e. inflammatory cell infiltration, changes in airway or pulmonary vascular pressure responsiveness, histopathologic evidence of smooth muscle or matrix remodeling). Because some agents (agonists, inhibitors or antagonists) may work in only certain animals, the appropriate stimulus and the particular therapeutic drug standard to which it should be compared may be species- and strain-dependent. Whenever possible, careful *in vitro* or *ex vivo* coordination and comparison of the activity found in tissues from the species and models selected to healthy and diseased human tissues should be made. Although rational therapeutic approaches based on inhibitory activity in a number of these models may increase the level of confidence in finding efficacy in the disease state, one should not over-simplify the etiology of the disease to fit the overall profile of the drug.

Acknowledgements
The author wishes to thank the scientists Ruth Osborn, Steven Bochnowicz, Charles Kotzer and Douglas W.P. Hay for their help in the expansion of our laboratory from primarily asthma into the interesting world of COPD.

References

1 Cherniack NS, Altose MD (1996) Diagnosis of chronic obstructive pulmonary disease. In: AR Leff (ed): *Pulmonary and critical care pharmacology and therapeutics*. McGraw-Hill, New York, 813–820
2 Snider GL (1996) Epidemiology and natural history of COPD. In: A. R. Leff (ed): Pulmonary and Critical Care Pharmacology and Therapeutics. McGraw-Hill, New York 821–828
3 Thurlbeck WM (1990) Pathology of chronic obstructive pulmonary disease. *Clin Chest Med* 11: 389–403
4 Kodavanti UP, Costa DL, Bromberg PA (1998). Rodent models of cardiopulmonary disease: their potential applicability in studies of air pollutant susceptibility. *Environ Health Persp* 106(Suppl. 1): 111–130
5 Reid L, Jones R (1983) Experimental chronic bronchitis. *Int Rev Exp Pathol* 24: 335–384
6 Lamb D, Reid L (1968) Mitotic rates, goblet cell increase and histochemical changes in mucus in rat bronchial epithelium during exposure to sulphur dioxide. *Br J Exp Pathol* 44: 437–445
7 Mullen JBM, Wiggs BR, Wright JL, Hogg JC, Pare PD (1986) Nonspecific airway reactivity in cigarette smokers. Relationship to airway pathology and baseline lung function. *Am Rev Respir Dis* 133: 120–125

8 Parker CW, Bilbo RE, Reed CE (1965) Methacholine aerosol as a test for bronchial asthma. *Arch Intern Med* 115: 452–458
9 Ramsdale EH, Hargreave FE (1990) Differences in airway responsiveness between asthma and chronic airflow obstruction. *Med Clin N America* 74: 741–751
10 D'Erchia AM, Gissi C, Pesole G, Saccone C, Arnason U (1996) The guinea-pig is not a rodent. *Nature* 381: 597–600
11 Frank NR, Speizer FE (1965) SO_2 effects on the respiratory system in dogs; changes in mechanical beahvior at different levels of the respiratory system during acute exposure to the gas. *Arch Environ Health* 11: 624–34
12 Higgins BG, Francis HC, Yates CJ, Warburton CJ, Fletcher AM, Reid JA, Pickering CA, Woodcock AA (1995) Effects of air pollution on symptoms and peak expiratory flow measurements in subjects with obstructive airways disease. *Thorax* 50: 149–55
13 Lightowler NM, Williams JRB (1969) Tracheal mucus flow rates in experimental bronchitis in rats. *Br J Exp Pathol* 50: 139–149
14 Jany B, Gallup M, Tsuda T, Basbaum C (1991) Mucin gene expression in rat airways following infection and irritation. *Biochem Biophys Res Commun* 181: 1–8
15 Shore S, Kobzik, Long NC, Skornik W, Van Staden CJ, Boulet L, Rodger IA, Pon DJ (1995) Increased airway responsiveness to inhaled methacholine in a rat model of chronic bronchitis. *Am J Respir Crit Care Med* 151: 1931–1938
16 Drazen JM, O'Cain CF, Ingram RH Jr. (1982) Experimental induction of chronic bronchitis in dogs. Effects on airway obstruction and responsiveness. *Am Rev Respir Dis* 126: 75–79
17 Shore S, Kariya ST, Anderson K, Skornik W, Feldman HA, Pennington J, Godleski J, Drazen JM (1987) Sulfur dioxide-induced bronchitis in dogs. Effects on airway responsiveness to inhaled and intravenously administered methacholine. *Am Rev Respir Dis* 135: 840–847
18 Moreno RH, Hogg JC and Pare, PD (1986) Mechanics of airway narrowing. *Am Rev Respir Dis* 133: 1171–1180
19 DeSanctis GT, Kelly SM, Saetta MP, Shiner RJ, Stril JL, Hakim TS, Cosio MG, King M (1987) Hyporesponsiveness to aerosolized but not to infused methacholine in cigarette-smoking dogs. *Am Rev Respir Dis* 135: 338–344
20 Pon DJ, van Staden CJ, Rodger IW (1994) Hypertrophic and hyperplastic changes of mucus secreting epithelial cells in rat airways: assessment using a novel, rapid and simple technique. *Am J Respir Cell Mol Biol* 10: 625–634
21 Freeman A, Cane SC, Furioso NH, Stephens RJ, Evans JJ, Moore, WD. (1972) Experimental induction of chronic bronchitis by nitrogen dioxide. *Am Rev Respir Dis* 106: 563–568
22 Foster JR, Cottrell RC, Herod IA, Atkinson HA, Miller K (1985) A comparative study of the pulmonary effects of NO_2 in the rat and hamster. *Br J Exp Pathol* 66: 193–204
23 Costa DL, Kutzman RS, Lehmann JR, Drew RT (1986) Altered lung function and structure after subchronic exposure to acrolein. *Am Rev Respir Dis* 133: 286–291
24 Tepper JS, Costa DL, Lehmann JR, Weber MF, Hatch GE (1989) Unattenuated struc-

tural and biochemical alterations in the rat lung during functional adaptation to ozone. *Am Rev Respir Dis* 140: 493–501
25 Harkema JR, Hotchkiss JA (1993) Ozone- and endotoxin-induced mucous cell metaplasias in rat airway epithelium: Novel animal models to study toxicant-induced epithelial transformation in airways. *Toxicol Letters* 68: 251–263
26 Rogers DF, Jeffery PK (1978) Inhibition by oral N-acetylcysteine of cigarette smoke-induced 'bronchitis' in the rat. *Exp Lung Res* 10: 267–283
27 Thompson JE, Scypinski LA, Gordon T, Sheppard D (1987) Tachykinins mediate the acute increase in airway responsiveness caused by toluene diisocyanate in guinea pigs. *Am Rev Respir Dis* 136: 43–49
28 Lei YH, Barnes PJ, Rogers DF (1995) Mechanisms and modulation of airway plasma exudation after direct inhalation of cigarette smoke. *Am J Respir Crit Care Med* 151: 1752–1762
29 Dusser DJ, Djokic TD, Borson DB, Nadel JA (1989) Cigarette smoke induces bronchoconstrictor hyperresponsiveness to substance P and inactivates airway neutral endopeptidase in the guinea pig. Possible role of free radicals. *J Clin Invest* 84: 900–906
30 Berry B, Wright JL (1991) Long-term pulmonary hypertension produced by cigarette smoking is associated with subendocardial fibrosis and inflammation of the right ventricle: a morphometric analysis in the guinea pig model. *Exp Pathol* 43: 163–172
31 Lotvall JO, Skoogh BE, Lemen RJ, Elwood W, Barnes PJ, Chung-KF (1990) Bronchoconstriction induced by inhaled sodium metabisulfite in the guinea pig. Effect of capsaicin pretreatment and of neutral endopeptidase inhibition. *Am Rev Respir Dis* 142: 1390–1395
32 Sakamoto T, Elwood W, Barnes PJ, Chung KF (1992) Pharmacological modulation of inhaled sodium metabisulphite-induced airway microvascular leakage and bronchoconstriction in the guinea-pig. *Br J Pharmacol* 107: 481–487
33 Ogawa H, Fugimura M, Saito M, Matsuda T, Akao N, Kondo K (1994) The effect of the neurokinin antagonist FK-224 on the cough response to inhaled capsaicin in a new model of guinea-pig eosinophilic bronchitis induced by intranasal polymyxin B. *Clin Auton Res* 4: 19–27
34 Coalson JJ, Collins JF (1985) Nitric acid-induced injury in the hamster lung. *Br J Exp* Pathol 66: 205–215
35 Goldring IP, Greenburg L, Park S-S, Ratner IM (1970) Pulmonary effects of sulfur dioxide exposure in the Syrian hamster. II. Combined with emphysema. *Arch Environ Health* 21, 32–37
36 Frasca JM, Auerbach O, Carter HW, Parks VR (1983) Morphologic alterations induced by short-term cigarette smoking. *Am J Pathol* 111: 11–20
37 Chakrin LW, Saunders LZ (1974) Experimental chronic bronchitis. Pathology in the dog. *Lab Invest* 30: 145–154
38 Bhaskar KR, Drazen JM, O'Sullivan DD, Scanlon PM, Reid LM (1988) Transition from normal to hypersecretory bronchial mucus in a canine model of bronchitis: changes in yield and composition. *Exp Lung Res* 14: 101–120

39 Seltzer J, Scanlon PD, Drazen JM, Ingram RH Jr., Reid L (1984) Morphologic correlation of physiologic changes caused by SO_2-induced bronchitis in dogs. The role of inflammation. *Am Rev Respir Dis* 129: 790–797
40 Mawdesley-Thomas LE, Healey P (1970) The effects of cigarette smoke on the bronchial tree of the lamb (*Ovis aries*). *J Pathol* 101: P18
41 Hyde DM, Hubbard WC, Wong V, Wu R, Pinkerton K, Plopper CG (1992) Ozone-induced acute tracheobronchial epithelial injury: relationship to granulocyte emigration in the lung. *Am J Respir Cell Mol Biol* 6: 481–497
42 Snider GL (1992) Emphysema: The first two centuries and beyond. A historical overview, with suggestions for future research: part 2. *Am Rev Respir Dis* 146: 1615–1622
43 Snider GL, Lucey EC, Stone PJ (1986) Animal models of emphysema. *Am Rev Respir Dis* 133, 149–169
44 Karlinsky JB, Snyder GL (1978) Animal models of emphysema. *Am Rev Respir Dis* 117: 1109–1133
45 Kaplan PD, Kuhn C, Pierce JA (1973) The induction of emphysema with elastase. 1. The evolution of the lesion and the influence of serum. *J Lab Clin Med* 82: 349–356
46 Busch RH, Lauhala KE, Soscutoff SM, McDonald KE (1984) Experimental pulmonary emphysema induced in the rat by intratracheally administered elastase: morphogenesis. *Environ Res* 33: 497–513
47 Valentine R, Rucker RB, Chrisp CE, Fisher GL (1983) Morphological and biochemical features of elastase-induced emphysema in strain A/J mice. *Toxicol Appl Pharmacol* 68: 451–461
48 Martorana PA (1976) The hamster as a model for experimental pulmonary emphysema. *Lab Animal Sci* 26: 352–354
49 Hayes JA, Christensen TG, Snider GL (1977) The hamster as a model of chronic bronchitis and emphysema in man. *Lab Anim Sci* 27: 762–70
50 Lucey EC, Stone PJ, Christensen TG, Breuer R, Snider GL (1988) An 18-month study of the effects on hamster lungs of intratracheally administered human neutrophil elastase. *Exp Lung Res* 14: 671–686
51 Ackerman NR, Corkey R, Perkins D (1978) Pathogenesis of papain-induced emphysema in the hamster. *Inflammation* 3: 49–58
52 Loscutoff SM, Cannon WC, Buschbom RL, Busch RH, Killand BW (1985) Pulmonary function in elastase-treated guinea pigs and rats exposed to ammonium sulfate or ammonium nitrate aerosols. *Environ Res* 36, 170–180
53 Fonzi L, Lungarella G (1980) Correlation between biochemical and morphological repair in rabbit lungs after elastase injury. *Lung* 158: 165–171
54 Milton DK, Godleski JJ, Feldman HA, Greaves IA (1990) Toxicity of intratracheally instilled cotton dust, cellulose and endotoxin. *Am Rev Respir Dis* 142: 184–192
55 Snider GL, Hayes JA, Korthy AL, Lewis GP (1973) Centrilobular emphysema experimentally induced by cadmium chloride aerosol. *Am Rev Respir Dis* 108, 40–48
56 Snider, GL, Lucey, EC, Faris B, Jung-Legg Y, Stone PJ (1988) Cadmium chloride-induced

airspace enlargement with interstitial pulmonary fibrosis is not associated with destruction of lung elastin Implications for the pathogenesis of human emphysema. *Am Rev Respir Dis* 137: 918–923
57 Heckman CA, Dalbey WE (1982) Pathogenesis of lesions induced in rat lung by chronic tobacco smoke inhalation. *J Natl Cancer Inst* 69: 117–129
58 Riley DJ, Kerr JS, Yu SY, Berg RA, Edelman HH (1983) Pulmonary oxygen toxicity. Connective tissue changes during injury and repair. *Chest* 83 (Suppl): 98–99
59 Harkema JR, Mauderly JL, Gregory RE, Pickrell JA (1984) A comparison of starvation and elastase models of emphysema in the rat. *Am Rev Respir Dis* 129: 584–591
60 Glasgow JE, Pietra GG, Abrams WR, Blank J, Openheim DM, Weinbaum G (1987) Neutrophil recruitment and degranulation during induction of emphysema in the rat by nitrogen dioxide. *Am Rev Respir Dis* 135: 1129–1136
61 Soskel NT, Watanabe S, Sandberg LB (1984) Mechanisms of lung injury in the copper-deficient hamster model of emphysema. *Chest* 85 (Suppl): 70–73
62 Wright JL, Churg A (1990) Cigarette smoke causes physiologic and morphologic changes of emphysema in the guinea pig. *Am Rev Respir Dis* 142: 1422–1428
63 Rudolphus A, Kramps JA, Kijkman JH (1991) Effect of human antileucoprotease on experimental emphysema. *Eur Respir J* 4: 103–108
64 Stolk J, Rudolphus A, Davies P, Osinga D, Dijkman JH, Agarwal L, Keenan KP, Fletcher D, Kramps JA (1992) Induction of emphysema and bronchial mucus cell hyperplasia by intratracheal instillation of lipopolysaccharide in the hamster. *J Pathol* 167: 349–356
65 Guenter CA, Coalson JJ, Jaques J (1981) Emphysema associated with inravacular leukocyte sequestration: a potential mechanism of lung injury. *Am Rev Respir Dis* 123: 79–84
66 Wittels EH, Coalson JJ, Welch MH, Guenter CA (1974) Pulmonary intravascular leukocyte sequestration: a potential mechanism of lung injury. *Am Rev Respir Dis* 109: 502–509
67 Bates DV, Macklem PT, Christie RV (1971) *Respiratory function in disease.* WB Saunders, Toronto
68 Szapiel SV, Fulmer JD, Hunninghake G W, Elson NA, Kawanami O, Ferrans VJ, and Crystal RG (1981) Hereditary emphysema in the tight-skin (Tsk/+) mouse. *Am Rev Respir Dis* 123: 680–685
69 de Santi MM, Martorana PA, Cavarra E, Lungarella G (1995) Pallid mice with genetic emphysema. Neutrophil elastase burden and elastin loss occur without alteration in the bronchoalveolar lavage cell population. *Lab Invest* 73: 40–47
70 Ranga V, Grahn D, Journey TM (1993) Morphologic and phenotypic analysis of an outcross line of blotchy mouse. *Exp Lung Res* 4: 269–279
71 Underwood DC,Osborn RR, Bochnowicz S, Hay DWP, Torphy TJ (1998) The therapeutic activity of SB 207499 (Ariflo™), a second generation phosphodiesterase 4 (PDE4) inhibitor, is equivalent to that of prednisolone in models of pulmonary inflammation. *Am J Rep Crit Care Med* 157: A827
72 Crouch E (1990) Pathobiology of pulmonary fibrosis. *Am J Physiol* 259: L159–L184

73 Hepleston A G (1991) Minerals, fibrosis and the lung. *Environ Health Perspect* 94: 149–168
74 Raghow R, Lurie S, Seyer JM, Kang AH (1985) Profile of steady state levels of messenger RNAs coding for type I procollagen, elastin, and fibronectin in hamster lungs undergoing bleomycin-induced interstitial pulmonary fibrosis. *J Clin Invest* 76: 1733–1739
75 Zapol WM, Snider MT (1977) Pulmonary hypertension in severe acute respiratory failure. *New Eng J Med* 296: 476–480
76 MacNee W (1994) Pathophysiology of cor pulmonale in chronic obstructive pulmonary disease, state-of-the-art: Parts one and two. *Am J Respir Crit Care Med* 150: 833–852; 1158–1168
77 Reid LM (1979) The pulmonary circulation: Remodeling in growth and disease. *Am Rev Resp Dis* 119: 531–546
78 Roth RA, Ganey PE (1988) Platelets and puzzles of pulmonary pyrrolizidine poisoning. *Toxicol Appl Pharmacol* 93: 463–471
79 Meyrick B (1991) Structure function correlates in the pulmonary vasculature during acute lung injury and chronic pulmonary hypertension. *Toxicologic Pathol* 19: 447–457
80 Huxtable RJ (1993) Metabolic activation and toxicity of chemical agents to lung tissue and cells. In: TE Gram (ed): *Hepatic nonaltruism and pulmonary toxicity of pyrrolizidine alkaloids.* Pergamon Press, United Kingdom, 213–237
81 Rabinovitch M (1991) Investigational approaches to pulmonary hypertension. *Toxicol Pathol* 19: 458–469
82 Marayuma K, Ye C, Woo M, Venkatacharya H, Lines LD, Silver MM, Rabinovitch M (1991) Chronic hypoxic pulmonary hypertension in rats and increased elastolytic activity. *Am J Physiol* 261 (*Heart Circ Physiol* 30): H1716–H1726
83 Rabinovitch M, Gamble W, Nadas AS, Miettinen OS, Reid L (1991) Rat pulmonary circulation after chronic hypoxia: hemodynamic and structural features. *Am J Physiol* 236: H818–H827
84 Abraham AS, Kay JM, Cole RB, Pincock AC (1971) Haemodynamic and pathological study of the effect of chronic hypoxia and subsequent recovery of the heart and pulmonary vasculature of the rat. Cardiovasc Res 5: 95–102
85 Hunter C, Barer GR, Shaw JW, Clegg EJ (1974) Growth of the heart and lungs in hypoxic rodents: a model of human hypoxic disease. *Clin Sci Molec Med* 46: 375–391
86 James WRL, Thomas AJ (1968) The effect of hypoxia on the heart and pulmonary arterioles of mice. *Cardiovasc Res* 3: 278–283
87 Janssens SP, Thompson BT, Spence CR, Hales CA (1991) Polycythemia and vascular remodeling in chronic hypoxic pulmonary hypertension in guinea pigs. *J Appl Physiol* 71: 2218–2223
88 Thompson BT, Hassoun PM, Kradin RL, and Hales CA (1992) Acute and chronic hypoxic pulmonary hypertension in guinea pigs. *Am J Physiol* 66: 920–928
89 Underwood DC, Bochnowicz S, Osborn RR, Luttmann, Hay DWP (1997) Nonpeptide endothelin receptor antagonists. X. Inhibition of endothelin-1- and hypoxia-induced

pressor responses in the guinea pig by the endothelin receptor antagonist, SB 217242. J *Pharmacol Exp Ther* 283: 1130–1137
90 Bochnowicz S, Osborn RR, Hay DWP, Underwood DC (1997) Hypoxia-induced pulmonary hypertension in an optimized environment for the guinea pig. *Lab Animals* 31: 347–356
91 Underwood DC, Bochnowicz S, Osborn RR, Louden CS, Hart TK, Ohlstein EH, Hay DWP (1998) Chronic hypoxia-induced cardiopulmonary changes in three rat strains: Inhibition by the ET receptor antagonist, SB 217242. *J Cardiovasc Pharmacol* 31 (Suppl. 1): S453–S455

Skin inflammation

Kenneth M. Tramposch

Bristol-Myers Squibb Pharmaceutical Research Institute, 100 Forest Avenue, Buffalo, NY 14213, USA

Introduction

Inflammation is a component of many skin disorders including psoriasis, atopic dermatitis, contact sensitivity and acne [1–3]. These diseases collectively affect over 20% of the population and dermatologists have developed many different treatment plans. Topical corticosteroids have been mainstay therapies since their introduction in 1952 [4, 5]. While these powerful anti-inflammatory drugs produce good resolution of both acute and chronic skin inflammation, long-term treatment in chronic diseases such as psoriasis are not feasible due to side-effects. Topical steroids induce skin atrophy which limits the time of treatment [6]. Also, application of topical steroids to large body surface areas results in hypothalamic-pituitary-adrenal suppression due to systemic absorption thus requiring the withdrawal of treatment. In these cases the disease rebounds sometimes in a form more severe than before the treatment.

There is a definite need to develop novel non-steroidal anti-inflammatory drugs with efficacy equal to or better than topical steroids but without the steroid-associated side-effects. Many different molecular targets have been, and continue to be, the focus of research at many pharmaceutical research laboratories. Potential targets involved in inflammatory cascades are plentiful and biochemical screens to identify inhibitors and antagonists can be established. Commonly, molecular targets are selected based on the observation of increased expression or activity of the target in human disease. Cytokines, lipid mediators, protein kinases, adhesion molecules and growth factors have all been identified as possible candidates. Inhibitors or antagonists are then identified and tested for activity in cell-based and animal models to determine proof-of-principle. That is, if the target plays a key role in inflammation then anti-inflammatory activity will be observed. For example, the lipid mediators of inflammation, collectively known as eicosanoids, are one class of inflammagens that have been studied in this way. Eicosanoids have been known to be elevated in psoriatic skin since 1975 [7, 8]. Many of these compounds have pro-inflammatory actions and their biosynthesis is well-characterized [9]. Since arachidonic acid is the precur-

In Vivo Models of Inflammation, edited by D.W. Morgan and L.A. Marshall

sor for leukotrienes and prostaglandins, inhibition of the phospholipase A2 activity responsible for the release of arachidonate from membrane phospholipids, would potentially result in the blockade of this whole family of mediators [10]. Enzyme screens have found inhibitors which have been shown to block arachidonic acid release in cells [11]. This *in vitro* proof-of-concept suggests that such compounds may be useful in skin diseases where leukotrienes and prostaglandins are elevated, e.g. psoriasis. However, psoriasis does not appear to be naturally occurring in animals nor is there general agreement that induced models suitably mimic the human disease. To overcome this deficiency researchers have developed models which mimic some aspects of the pathophysiology that occurs in the human condition. Activity in such models has been used in decision-making concerning selection of compounds for clinical trials. Of course, the ultimate test for establishing the therapeutic efficacy of novel anti-inflammatory agents for the treatment of skin inflammation is activity in the human disease. However, since clinical evaluations are costly and have many potential variables which require control, they should be reserved for the evaluation of candidates which show activity in the most stringent models available.

This review will discuss the bioassays which have been commonly used in the screening of potential anti-inflammatory agents for skin inflammation. Each model discussed in this report has one or more deficiencies that impact the reliability of the method to provide estimates of therapeutic potential. In addition, several promising new models will be discussed which will likely receive more scrutiny in the years to come. It seems clear that the combined use of several models is necessary for the complete characterization of novel anti-inflammatory agents prior to selection of the best candidate for clinical trial.

Models of acute skin inflammation

Non-immune mediated models

Acute skin inflammation can readily be induced in animals and humans by the topical application of a contact irritant. A number of methods have been developed using naturally occurring substances which were known to produce irritation including croton oil, tetradecanoylphorbol acetate(TPA, the active component of croton oil), carrageenan, cantharidin and arachidonic acid (reviewed in [12]). Inflammation induced by UV light has also been used. For the chemical agents, cutaneous inflammation is usually induced by applying the irritant in a suitable organic solvent to the surfaces of the mouse ear. This results in an edematous response which can be measured in the intact animal using a micrometer to quantitate thickness or by sacrificing the animal and weighing full thickness ear biopsies. Other endpoints such as epidermal hyperplasia and cellular infiltration can also be quantitated by standard methods.

TPA-induced skin inflammation

The croton oil or TPA-induced skin inflammation is the most commonly used test for both steroidal and non-steroidal anti-inflammatory agents. The croton oil test in rat ears was first described for the evaluation of corticosteroids by Tonelli et al. [13]. The use of croton oil has been supplanted by the active component TPA [14]. In the case of TPA, a single topical dose to the ears of mice results in an edematous response that reaches a maximum level at 6 h [15]. Phorbol ester also induces an infiltration of neutrophils in which peak levels are observed by 20–24 h [15–17]. These responses quickly diminish as the acute lesion resolves. The mechanism by which phorbol ester causes inflammation is not completely clear, but seems to be related in part to the release of eicosanoid mediators. TPA induces an early and transient increase in skin prostaglandin levels [18]. LTB_4 levels increase in parallel with changes in vascular permeability and cell infiltration [16]. The effects of oral and topically administered anti-inflammatory drugs on phorbol ester-induced mouse ear inflammation has been reported. Carlson et al. surveyed the activity of a wide variety of pharmacological agents in this model [19]. A more recent confirmation of this work has been published [16]. A summary of the anti-inflammatory activity in this model by various classes of agents is shown in Table 1. Topical steroids such as betamethasone strongly block phorbol ester-induced edema. Classical NSAIDS given topically, such as indomethacin, diclofenac and piroxicam are also very effective. Leukotriene synthesis inhibitors such as phenidone BW 755c, and zileuton are also active. An LTB_4 antagonist has also been shown to be effective. Compounds which block phospholipase A_2 have been shown to block phorbol ester-induced edema and cell infiltration [21, 22]. In general, compounds that modulate eicosanoid biosynthesis, either cyclooxygenase and/or lipoxygenase or phos-

Table 1 - Survey of drug classes evaluated in the TPA-induced mouse ear edema assay

Class	Oral/IP[a]	Topical
NSAIDs	–/+	+++
DMARDs	–	+
Lipoxygenase inhibitors and leukotriene antagonists	++	++
Phospholipase A2 inhibitors	–	++
Anti-histamines	–	+++
Anti-serotonin	+	++
Immunophillin-ligands	–	+++
Glucocorticoids	++	++++

[a]*Relative activity: + indicates active, ++++ indicates most active, – indicates not active.*

pholipase A_2 inhibitors, show topical activity in this model. In view of these data, this model appears to be a reasonable screen for the *in vivo* evaluation of leukotriene/cyclooxygenase inhibitors that have been selected based on enzyme and cell-based assays. However, it is evident that other mediators besides eicosanoids are operative in this type of inflammation given the fact that anti-histamines and serotonin antagonists are topically active [19]. The use of mast cell deficient mice has shown that full expression of the increase in vascular permeability and cell infiltration induced by TPA requires the participation of mast cells [23]. The fact that antihistamines can block the full development of the inflammatory reaction suggests that mast cells play an important interacting role in the development of this acute inflammation. Other compounds also not recognized as anti-inflammatory agents which have shown activity include hydroclorothiazide, chlorpromazine, haloperidol and nifedipine [19]. The activity of these latter compounds put into question the predictive value of this model in the ability to identify novel anti-inflammatory agents for human skin inflammation. The activity of the classical NSAIDS is also troublesome. Indeed, indomethacin has been demonstrated not to be active in the oral or topical treatment of psoriasis [24, 25]. In fact, one report claims that indomethacin exacerbates the disease [26]. Therefore, the use of secondary models to eliminate these false positives should be considered before considering active compounds as clinical trial candidates.

Arachidonic acid induced skin inflammation

Topical application of arachidonic acid to mouse skin results in a rapid onset of edema which resolves quickly [15]. The peak edema occurs at 1 h post-application. Topical and oral corticosteroids are generally active. In contrast to the TPA model, very little cell infiltration is observed. However, quantitation of cellular infiltrate using biochemical markers has been reported [16]. Prostaglandin and leukotriene levels in the skin increase with kinetics that parallel the increase in vascular permeability [16]. However, in contrast to the TPA model, arachidonic acid appears to induce a greater increase in prostaglandins than leukotrienes. Consistent with this feature, NSAIDS administered by the intraperitoneal route have been found to be active in this model [16]. In another study, topical indomethacin is also active, but piroxicam, ibuprofen, and aspirin appear to be inactive [19]. One group measured the effect of systemic NSAIDS on cell infiltration and found that this class did not affect cell infiltration as measured by tissue myeloperoxidase content despite the strong effect on tissue swelling [16]. A summary of compounds tested in this model is shown in Table 2. Lipoxygenase inhibitors appear to be consistently active in the AA model and activity of these compounds such as zileuton suggests a role for 5-lipoxygenase products in the inflammation [16, 19, 27]. This class of compounds blocks both swelling and cell infiltration. LTB_4 antagonists and a FLAP inhibitor

Table 2 - Survey of drug classes evaluated in arachidonic acid-induced mouse ear assay

Class	**Oral/IP[a]**	**Topical**
NSAIDs	++	+
DMARDs	–	+
Lipoxygenase inhibitors and leukotriene antagonists	++	++
Phospholipase A_2 Inhibitors	–	++
Anti-histamines	–	++
Anti-serotonin	–	–
Immunophillin-ligands	–	++
Glucocorticoids	+	++

[a]Relative activity: + indicates active, ++++ indicates most active, – indicates not active.

have also been shown to be active [16]. As with the acute TPA-induced mouse ear edema model, the value of this assay appears to be in the ability to evaluate the *in vivo* activity of compounds which have been selected based on their effectiveness in blocking leukotriene production or function in cell-based or enzyme assays.

Many agents known to not have an anti-inflammatory effect on human skin are active in the AA model. Compounds like anti-histamines, haloperidol and verapamil were judged to be active after topical application [19]. Topical cyclosporin A has been shown to block AA-induced edema in mouse skin but did not affect the levels of prostaglandins or leukotrienes [28]. The authors suggested that the observed activity on swelling might be due to a vasoconstriction activity since a higher degree of inhibition of dye extravasation compared to edema inhibition was noted. In general, a high rate of false positives is likely to be generated when using this model as a selection criteria for higher level tests and more stringent models should be employed before further consideration as clinical candidates.

Persistent chronic-like TPA-induced inflammation

The acute skin inflammation models described above test the ability of the test compound to prevent inflammation. Therefore, the test compound must be applied prior to or at the same time as the inflammagen. The inherent weakness of this approach is obvious. In reality, anti-inflammatory drugs are generally not used prophylactically. In clinical conditions, the patient usually presents with an ongoing inflammatory reaction or has a chronic disease. Therefore, useful clinical drugs will have to be able to resolve an ongoing inflammation. For preclinical evaluation of anti-inflammatory agents, it would be useful to have a model of more prolonged inflam-

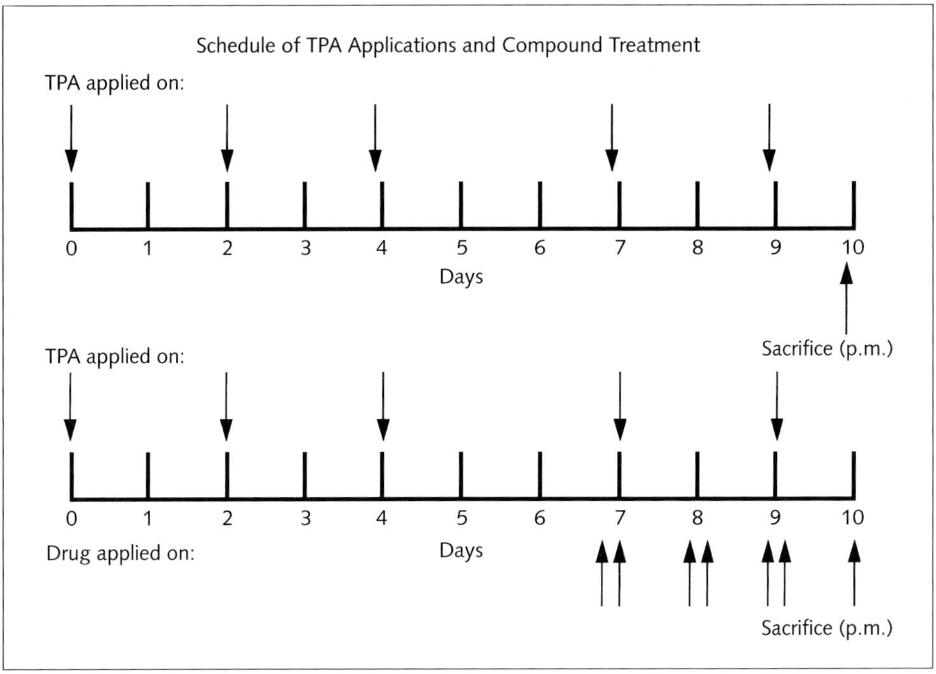

Figure 1
Schedule of TPA and drug applications. Upper line: experiments where TPA only was applied; lower line: experiments where TPA and drug or vehicle were applied. On days where two drug doses were applied the applications were made 6 h apart. Sacrifice was performed at 6 h after last drug application.

mation in which the test compound is applied to an already established inflammatory skin lesion. This type of animal model may be more relevant to the clinical situation and be more suitable for predicting useful drug candidates.

Multiple applications of the phorbol ester TPA to mouse ears has been found to induce a prolonged chronic like inflammatory response characterized by edema, cell infiltration and epidermal hyperplasia [29]. In this model TPA is applied on alternate days during the first week (days 0, 2 and 4). Additional phorbol ester is applied on days 7 and 9. Animals are then sacrificed on day 10 (see Fig. 1). Ear biopsies are taken and weighed to estimate edema, epidermal thickness measurements are made from histological sections and cell infiltration is quantitated by measurement of tissue myeloperoxidase content. The time-course of each inflammatory parameter is shown in Figure 2. A two-fold increase in ear weight was apparent by 6 h which was maintained throughout 10 days. MPO content increased to 160 fold higher than

Table 3 - Survey of drug classes evaluated in the multiple application TPA-induced ear skin inflammation assay

Class	Topical Application[a]
NSAIDs	–
Anti-histamines	–
Lipoxygenase Inhibitors	++
Phospholipase A2 Inhibitors	+++
Glucocorticoids	+++

[a]Relative activity: + indicates active, ++++ indicates most active, – indicates not active.

normal skin at the peak (day 3). In comparison, a single application of phorbol ester induces a 40-fold increase at 24 h. The initial wave of MPO activity was then replaced with a more modest but still greatly elevated MPO level that was 60-fold higher than in non-treated controls. In order to characterize the cell infiltrate during the establishment of this inflammatory reaction, immunohistochemistry was performed to determine the number of macrophages and T cells [30]. Figure 3 shows that by day 4, the number of macrophages increased 2.9-fold over the nontreated control and by day 10 were elevated 6.0-fold. In comparison, the first significant elevation of T cells was not observed until day 7 at 9.5-fold over nontreated controls and peaked at day 8 with a 19-fold elevation. The relative course of the infiltration of each of the cell types is consistent with a transition from an acute inflammation dominated by neutrophils to a chronic inflammation characterized by a dominant macrophage and T cell infiltrate. Interestingly, the unique temporal pattern of the cell infiltrate suggests that phorbol ester is not attracting the leukocytes directly but that factors generated subsequent to phorbol stimulation are responsible.

A survey of the types of compounds reported to be tested in this model is shown in Table 3. Topical steroids such as hydrocortisone valerate and betamethasone dipropionate effectively resolved the inflammation caused by multiple applications of phorbol ester. Of the compounds tested only corticosteroids had an effect on all three inflammatory parameters: ear weight, MPO and epidermal thickness. The pleotropic effect of these agents make it difficult to decipher which inflammatory mediators may be most important in this model. However, corticosteroids have been shown to reduce PLA_2 activity in psoriatic skin [31]. Phorbol ester does increase PLA_2 activity in mouse skin [32]. Therefore, it is possible that at least one of the mechanisms by which steroids modulate the persistent inflammation in this model is by lowering PLA_2 activity and the resultant production of eicosanoids. Indeed, A phospholipase inhibitor, BMS 181162, has also been shown to resolve this chronic

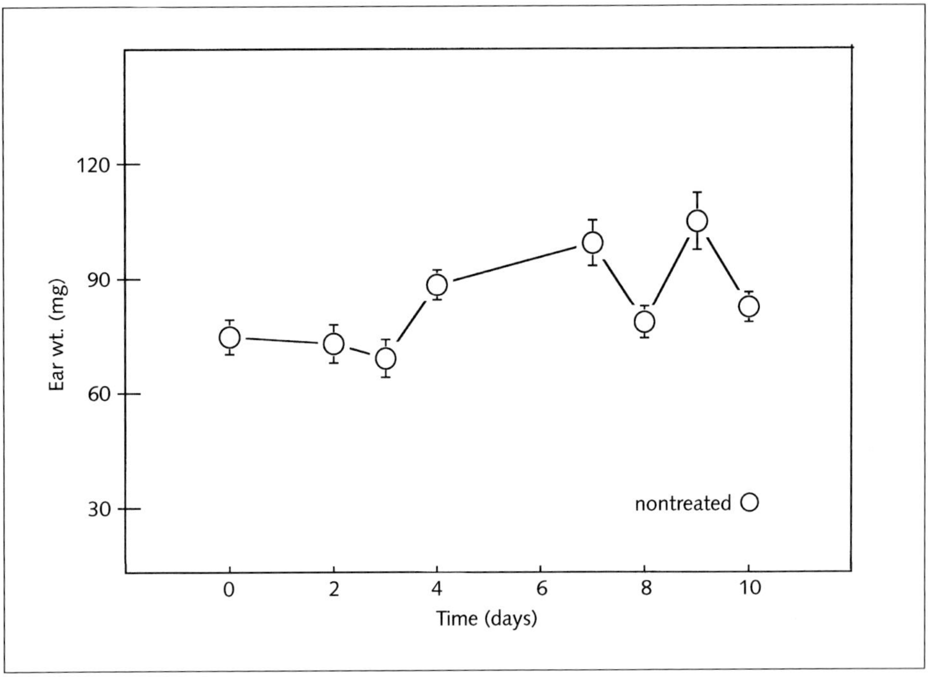
120
90
60
30
Ear wt. (mg)
nontreated
0
2
4
6
8
10
Time (days)

A

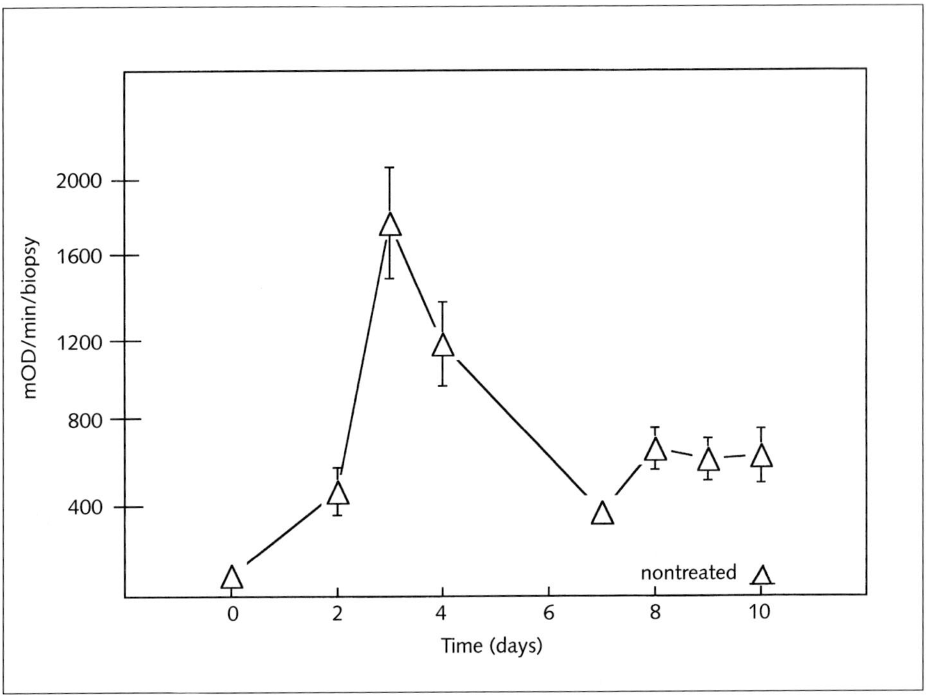
2000
1600
1200
800
400
mOD/min/biopsy
nontreated
0
2
4
6
8
10
Time (days)

B

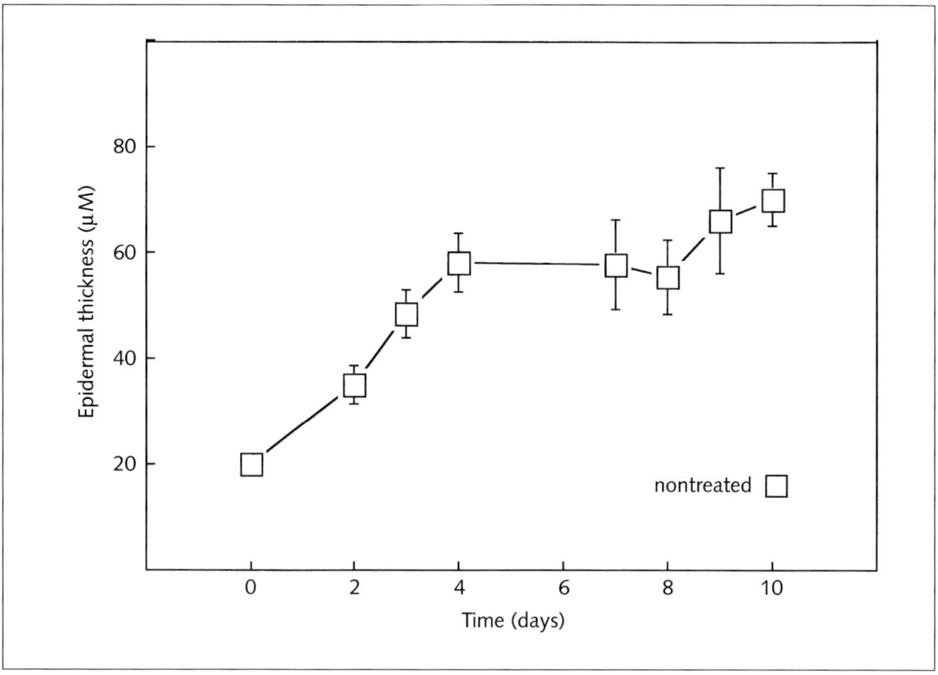

C

Figure 2
Effect of multiple applications of TPA on ear weight (A), MPO (B) and epidermal hyperplasia (C). 20 µl of TPA in acetone/pyridine/water (97:2:1) was applied to each ear of 10 mice according to the schedule shown in Figure 1. (A) Ear weight: each point represents the mean (± SD) of 10 pairs of ears in mg (normal = 35.2 ± 2.6 mg). (B) Myeloperoxidase (MPO) content: each point represents mean (± SD) of 10 biopsies in mOD/min/biopsy (normal = 15.0 ± 4 mOD/min/biopsy). (C) Progression of epidermal thickness: each point represents the mean (± SD) of histological sections from 10 ears in µM (normal = 15.7 ± 3.9).

inflammation. This compound dose dependently resolved the cell infiltrate and completely blocked the increase in LTB_4 and PGE_2 that was observed with TPA alone [33].

Application of 0.1% hydrocortisone valerate significantly reduced the number of T cells and neutrophils present in the tissue compared to phorbol ester alone. In contrast, this steroid had no effect on the number of macrophages in the tissue at day 10.

The lipoxygenase inhibitors lonapalene, NDGA and phenidone were all effective in resolving the chronic inflammation (Table 3). These results suggest a prominent

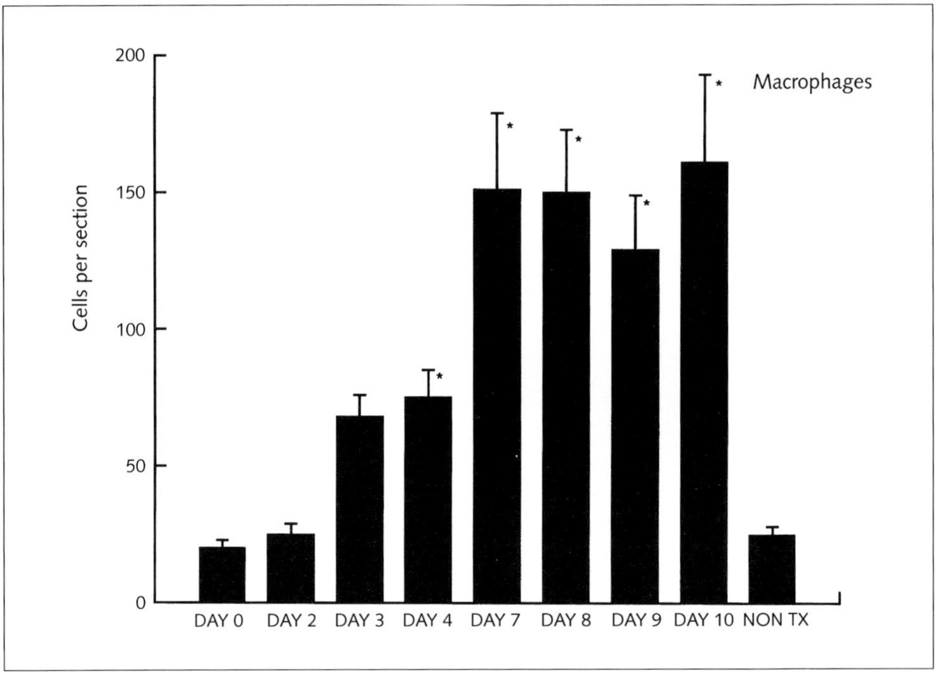
Cells per section
200
150
100
50
0
Macrophages
DAY 0
DAY 2
DAY 3
DAY 4
DAY 7
DAY 8
DAY 9
DAY 10
NON TX

A

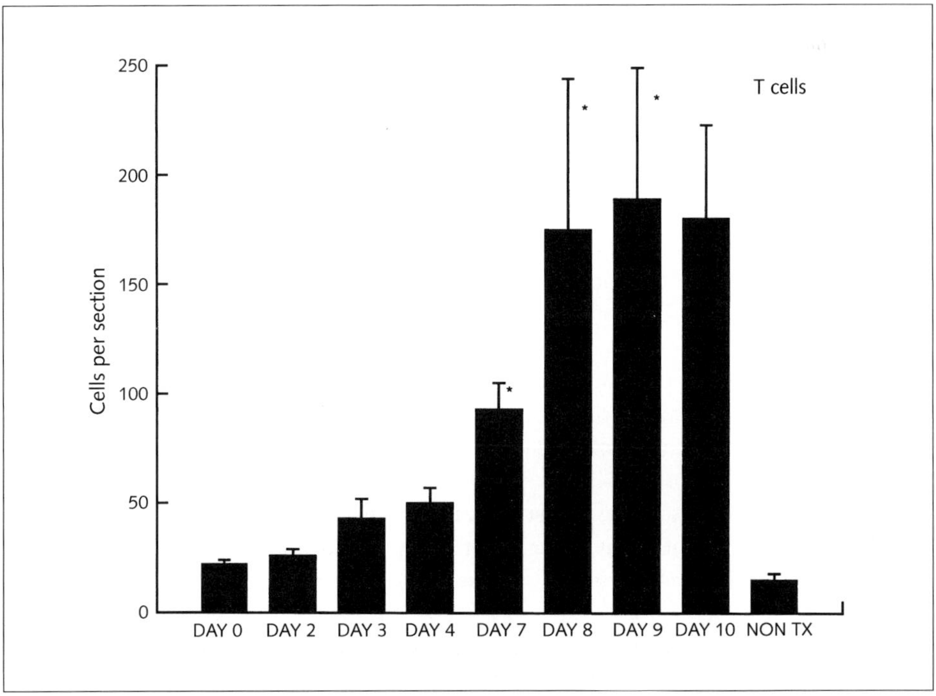
Cells per section
250
200
150
100
50
0
T cells
DAY 0
DAY 2
DAY 3
DAY 4
DAY 7
DAY 8
DAY 9
DAY 10
NON TX

B

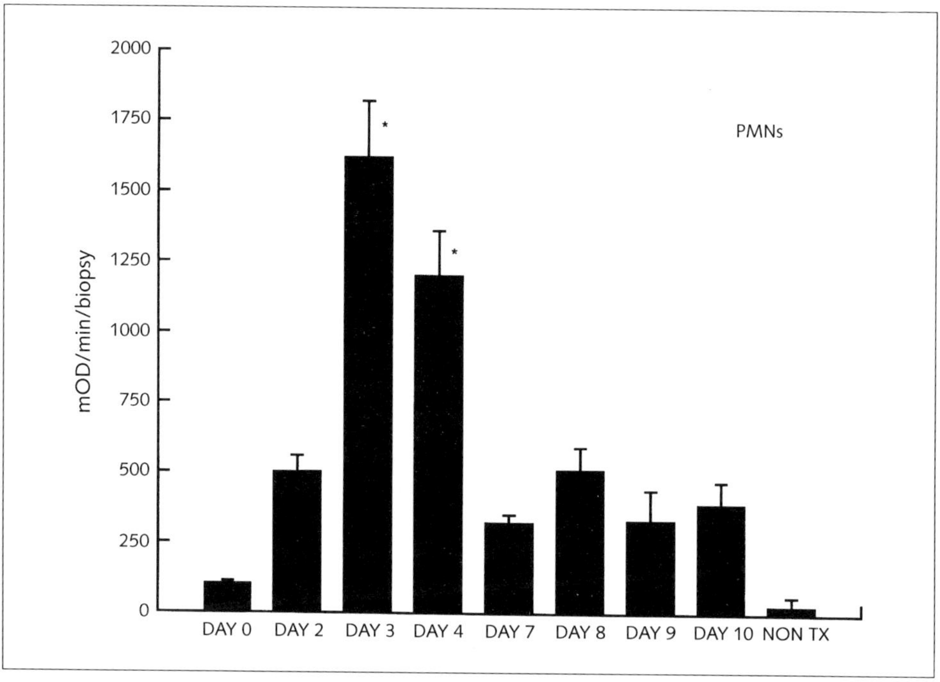

C

Figure 3
Effect of multiple applications of TPA on cell infiltration. TPA in aetone/pyridine/water (97:2:1) was applied to mouse ears according to the schedule shown in Figure 1. The bars represent mean numbers of cells ± SD (n = 0). For each animal, five 5 μm vertical sections were counted from end to end under 200× magnification. The bars with asterisks (*) are significantly different from non-treated controls (p < 0.05), within each graph. (A) The mean number of immunostained macrophages using MOMA-2 antibody. (B) The mean number of immunostained T cells using Thy-1 antibody. (C) The mean number of PMNs as reflected by tissue myeloperoxidase content.

role of leukotrienes in this model. In contrast, cyclo-oxygenase inhibitors were uniformly inactive. Anti-histamines of the H_1 and H_2 antagonist classes have been demonstrated to be active in the acute phorbol ester model. This finding implies that histamine is involved in the acute reaction. Methdilazine, a H_1 antagonist was not active in the chronic model which is consistent with the lack of anti-inflammatory activity for this class in humans.

Multiple applications of phorbol ester to mouse skin results in a prolonged inflammation and produces a pattern of cell infiltration in the skin in which T cells and macrophages are predominant. Unlike in acute models, cyclooxygenase and anti-histamines are not active. Because test compounds are applied after the inflammatory lesion is well established, this model may be more relevant to clinical dermatology and may be more suitable for the screening of new compounds for topical anti-inflammatory activity.

Immune-driven models of skin inflammation

Accumulating evidence exists to support the idea that psoriasis and atopic dermatitis are genetically determined, immunologic diseases, where T cells and cytokines play a major role [34]. Recently developed treatments include immunosuppressive agents such as methotrexate, cyclosporin A, FK-506 and Interleukin-2-diphtheria toxin fusion protein. The success of these drugs in clinical trials have stimulated the search for immunosuppressive agents which inhibit early T cell activation and can act as local immunosuppressants after topical application. Other than transplantation models, most of the frequently used *in vivo* assays for the assessment of drug-induced immunosuppression measure the changes in inflammatory skin responses in delayed-type hypersensitivity(DTH) models. These models are prototypical responses of cellular immunity. Classic DTH occurs in humans after infection with Mycobacterium tuberculosis (reviewed in [35]). Mice, guinea pig and swine can be made sensitive to a variety of antigens. In all of these animals the essential histological feature is the infiltration of lymphocytes which occurs 4–6 h after challenge with antigen. The inflammatory cell infiltrate is accompanied by extravasated plasma which accounts for the swelling and induration seen in most DTH responses. Several popular protocols and species have been used to evaluate inflammatory and immuno-modulating agents.

Oxazalone-induced delayed-type hypersensitivity in mice

The mouse is a popular species for evaluating the effect of novel anti-inflammatory agents on DTH responses [36]. DTH responses can easily be assessed in sensitized animals by micrometer measurements of ear thickness or by measuring the increase in weight of ears or paws after sacrifice. Cell infiltrate can also be measured by histological evaluation of tissue biopsies or by use of a marker of neutrophils such as myeloperoxidase. In general, animals are sensitized to the antigen oxazolone by applying a solution (1–3%, acetone) to the shaved belly. Four or 5 days later the animals are challenged with a low dose of oxazolone (0.5–1%) to elicit a DTH response. Careful selection of the challenge dose is needed to ensure that the dose is not so high that a irritant response is also induced.

The pattern of cell infiltration observed in the mouse DTH response has been well characterized [37]. At 24 h after challenge the edema is severe and the cell infiltration is moderate. The ratio of mononuclear cells to PMNs is about equal (31:27). At the 48 and 72 h timepoints the predominance of mononuclear cells over PMNs is clear (40:12 and 26:7, respectively). The early and prominent PMN cell infiltrate distinguishes the mouse DTH from other animal species and humans. The essential microscopic feature of classic human DTH is the diapedesis of lymphocytes with variable participation of monocytes and macrophages [35]. Lymphocytes are the predominating infiltrating cell. In addition studies have shown that the majority of these lymphocytes are of the helper/inducer phenotype, although some killer/suppressor cells are also present. Interestingly, mutant mice lacking the CD4 gene show marked hyporesponsiveness in DTH protocols [38].

Murine $CD4^+$ T helper clones have been categorized into two subsets on the basis of differences in their profiles of cytokine production. The Th 1 subset secretes INFγ and IL-2. The Th 2 cells produce IL-4, IL-5, IL-6, IL-10, and IL-13. IL-3, GM-CSF and TNFα are produced by both subsets [39]. Cytokine responses have been studied in murine DTH responses. INFγ and IL-2 mRNA levels were dramatically upregulated 12–24 h after challenge in dinitrofluorobenzene sensitized mice [38]. These data support the concept that Th 1 cells are important in DTH responses. However, the importance of Th 2-derived cytokines has also been suggested. Increased IL-4 levels have been observed by several groups studying murine DTH responses [40, 41] Reduced levels of all three cytokines has been observed in the hyporesponsive $CD4^-$ mutant mice [38].

Several groups have used murine DTH responses to rank-order the potency of topical corticosteroids [42, 43]. A good correlation has been observed between the potency of steroids in this model and the potency observed in humans. Steroids in pure solvents and in complex topical formulations have been evaluated. While some topical formulations may be suitable for evaluation, some may not because of very high activities observed for vehicle treatment alone. The effect of drug-free vehicle must be determined in each case to ensure that a true drug effect can be determined.

Anti-histamines and anti-serotonin drugs are ineffective against mouse DTH responses [36]. Classical NSAIDS are variably effective against edema and cell infiltrate with some compounds affecting the vascular phase and not the cell infiltrate [44]. The poor activity of these compounds is consistent with the well-known lack of effect of these compounds in inflammatory dermatoses such as psoriasis and atopic dermatitis.

Immune modulating drugs have been found to inhibit mouse DTH responses. Drugs affecting T cell function such as methotrexate, azathioprine, and thioquanine are all active [36]. All of these agents have an inhibitory effect on proliferating cells and have their main therapeutic use in cancer chemotherapy. Immunosuppressive drugs such as cyclosporin A and FK-506 which are considered non-cytotoxic have been shown to be effective in blocking mouse DTH [45, 46]. This is of great interest since oral cyclosporin A therapy has been found to be useful in the treatment of psoriasis [47]. Similarly, topical FK-506 has been reported to be active in both psoriasis and atopic dermatitis [48, 49]. In addition, topical FK-506 can block induced DTH responses in man [50]. The relatively good correlation between activity in the mouse DTH model and in the treatment of dermal inflammatory conditions in man make this model valuable despite the noted differences in the histopathology.

Guinea pig DTH reactions

DTH reactions can be raised in guinea pigs in a manner similar to mouse and man [51]. In contrast to the mouse where the contact sensitivity is transient, guinea pig DTH responses can be raised throughout the life of the sensitized animal. This feature makes the sensitivity more like that observed in humans. In general, animals are sensitized to the hapten (oxazolone and dinitrofluorobenzene are popular) by applying one or two topical doses spaced 1–4 days apart. The animals are then challenged on day 10–14 on shaved flank skin or on ears, with a lower dose of hapten. The inflammatory endpoints can be measured at 24, 48 and/or 72 h after challenge. The macroscopic endpoints such as erythema and edema are semi-quantitated using a visual grading scale. Alternatively, photometric or laser Doppler measurements can be made to estimate objectively the increase in blood flow in the inflamed tissue [52]. For edema a more objective measure of skin fold thickness has been used. While these latter measurements are more objective than visual observations using the naked eye, it is not clear if these measurements are more precise. Histological endpoints have also been used. The skin infiltrate in guinea pig DTH is predominantly mononuclear cells [51]. Basophils and eosinophil numbers are also increased significantly. Neutrophil infiltration is much less pronounced compared to the latter cell types. This feature clearly distinguishes the guinea pig DTH from the mouse response and makes the guinea pig response more akin to man.

The guinea pig DTH model has been used to evaluate the immunosuppressive effects of mycophenolic acid, cyclosporin A and FK-506 [53–55]. All of these agents can suppress the DTH response. FK-506 has been compared to rapamycin and cyclosporin A after topical or oral administration [53]. FK-506 significantly inhibited T cell infiltration and erythema by both routes. Cyclosporin A was inactive topically and suppressed only the erythema response after oral administration. In another report cyclosporin A was active after topical application [56]. Rapamycin was not active by any route on either endpoint. Unlike FK-506 or cyclosporin A, rapamycin does not block cytokine biosynthesis, but instead blocks cytokine stimulated signal transduction [57]. The lack of consistent activity of topical cyclosporin A is likely reflective of its lack of topical bioavailability. Guinea pig skin is less permeable than mouse skin and this difference may account for the discrepancy between the two animal models. However, since guinea pig skin is much more permeable to drugs than human skin, prediction of topical activity in humans should be done with caution. While FK-506 has been shown to be active topically in the treatment of psoriasis and atopic dermatitis, topical cyclosporin A has not been therapeutically successful [58].

Swine DTH reactions

Because of the marked differences in skin morphology between small laboratory animals such as mice, rats and guinea pigs compared to human skin, several labs have searched for a model which may be more clinically relevant and therefore more useful in the evaluation of novel drugs for treating skin inflammation. The close morphologic and functional similarities between swine skin and human skin have long been recognized [59]. The thickness of swine epidermis varies from 70–120 μm and compares well to human epidermis which varies from 75–125 μm [60]. The tritiated thymidine labeling pattern, index and turnover time of epidermal cells is similar to humans [61]. Some of the differences observed in swine skin include a thicker stratum corneum, absence of eccrine sweat glands and high levels of alkaline phosphatase in the epidermis [62]. A key feature desired in a novel model for evaluating topical compounds is drug permeability similar to human skin. Compared to other animal models, swine skin has permeability properties more similar to human skin than rodent skin [63].

A swine model of DTH has been developed [64]. Domestic pigs were sensitized to dinitroflourobenzene by application of a 10% solution to the ears and groin on day 1. A second application was made on day 4. Animals were then challenged on day 12 on sites on the back, and erythema was measured by reflectometry. The difference in reddening of the drug treated sites compared to the vehicle treated sites was then calculated. Blood flow increases were also assessed using a laser Doppler perfusion monitor. The peak inflammatory response occurred at 24 h. The immunosuppressive macrolides were evaluated in this model and compared to corticosteroids. In one study, test compounds were applied topically at 0.5 and 6 h after challenge [64]. As in guinea pigs, rapamycin was not active in inhibiting DTH responses. Cyclosporin A was inactive when given at a concentration of 10%. The topical steroid clobetasol propionate was very active (87%) in blocking the visual reddening of the DTH site compared to vehicle sites. Dexamethasone, which is less active in humans than the latter compound, was less active in swine at an equivalent dose (38% inhibition). FK-506 at doses as low as 0.04% caused a marked reduction in the inflammatory signs. These data are consistent with findings in humans. Corticosteroids are potent suppressors of DTH responses in humans [65]. Oral Cyclosporin A and topical FK-506 have also been shown to block human DTH responses but topical cyclosporin A has not shown activity presumably due to poor skin penetration [66]. In another study, an ascomycin derivative SDZ ASM 981 has been shown to be as potent as clobetasol propionate in blocking a swine DTH response and was recently reported to be active topically in psoriasis, atopic dermatitis, and human DTH [67].

In view of the fact that validated animal models of psoriasis and atopic dermatitis still do not exist, the DTH reaction in animals may be a key model in identifying novel immunosuppressants for this disease. The inflammatory response in DTH has some common features with psoriatic lesions including the activation of T cells and the localization to the dermal/epidermal compartment. For steroids and immunosuppressive macrolides a good correlation between blocking DTH and activity in psoriasis exists and these observations strengthen the potential predictive value of this model as a preclinical evaluation tool.

Future models

The models described above are well-used and time tested preclinical assays. These models of local inflammatory reactions elicited by the topical application of chemicals have provided valuable insights in the study of inflammation. However, no clear examples exist where novel compounds that were subsequently proved to be therapeutically active in man have been identified solely using these models. The common major flaw these models share is that they are induced and are only of transient nature. In other words they are acute models which resolve spontaneously, giving rise to completely restored tissue. The diseases for which novel anti-inflammatory drugs are needed are of a chronic nature which remit and relapse over many years. To be able to mimic the chronic nature of skin diseases animal models are needed which have a persistent inflammatory component and also have most of the key pathological features of the human condition.

Psoriasis is one of the most common chronic dermatoses seen in the clinic affecting roughly 2% of the population. Because of this high incidence, many new anti-inflammatory agents are first evaluated in this "marker" disease. Clinical success in psoriasis might then foretell potential utility in many inflammatory skin diseases. Psoriasis is a complex disorder involving marked epidermal hyperplasia and altered differentiation leading to the production of thick scale on the skin. This hyperplasia is accompanied by a mixed leukocytic infiltration of the dermis and epidermis composed of T cells, neutrophils and mast cells [34]. The exact underlying pathological cause of psoriasis has not been determined but many potential targets have been identified. The most recent studies have focused on the role of T cells. Although the role of T cells has not been proven, evidence is accumulating that this disease has a major T cell component in its pathogenesis. One possibility that has strong support is that autoreactive T cells initiate an inflammatory reaction which causes other skin cells such as keratinocytes to secrete a variety of cytokines and growth factors. Many pro-inflammatory cytokines are candidates for the mediation of the skin changes seen in psoriasis including TNFα, INFγ, IL-1, IL-6, IL-8, and VEGF. Any new models of psoriasis should have three main components: (1) acanthosis and hyperkeratosis of the epidermis, (2) a dense leukocytic infiltration composed of T cells and neutrophils, and (3) a pattern of cytokine and other biochemical marker expression that resembles the human disease. In addition, as mentioned above, the inflammation should be persistent.

Several mouse models have been recently described which are reported to have a psoriaform phenotype. Some of these are the result of genetic engineering and others are spontaneous mutations. The asebia mouse (ab/ab) represents a spontaneous mutation in BALB/c mice resulting in hyperplasia of the epidermis [68]. The sebaceous glands of these animals are small and the sebaceous cells are abnormal. The dermis shows signs of chronic inflammatory changes including a mixed cell infiltrate, edema and elevated mast cell numbers. Asebia mice have been shown to have constitutively elevated ICAM-1 mRNA expression [69]. Cyclosporin A treatment was able to restore the wild-type phenotype by decreasing the epidermal hyperplasia, dermal cell infiltration and edema. ICAM-1 mRNA levels were also reduced. Other psoriatic treatments including coal tar and UVB-light were not effective in reducing the epidermal thickness, but instead increased the epidermal hyperplasia [70]. More studies aimed at evaluating other anti-inflammatory agents in this model are needed to determine the potential as a valuable screen.

The flaky skin mouse (fsn/fsn) is characterized by a progressive hyperproliferative epidermis [71]. Affected mice are normal at birth except for a hypochromic anemia. Subsequently, they develop hyperkeratotic plaques and acanthosis with an elongation of rete ridges. The infiltrating cells in the skin are primarily mononuclear but intraepidermal neutrophils have been observed through electron microscopic evaluation. At the dermal-epidermal junction numerous macrophages and mast cells are present. Full-thickness skin grafts from fsn/fsn mice to nude mice (nu/nu) recipients maintain the fsn phenotype thus ruling out a systemic immune abnormality [72]. As in human psoriatic lesions, EGF receptor immunostaining is elevated in fsn skin lesions [73]. Treatment with two known anti-psoriatic therapies, UVB light and cyclosporin A, results in a remission of the fsn phenotype as well as a normalization of the EGF-receptor immunostaining patterns. Topical EGF treatment also was effective in reducing the inflammation and epidermal thickness. While data on topical EGF in humans is not available, this treatment has been reported to resolve the psoriatic phenotype in human psoriatic skin xenografts on nude mice [74].

A spontaneous mutation in C57BL/Ka mice (cpdm/cpdm) has been reported which gives rise to a chronic proliferative dermatitis [75]. This dermatitis is characterized by intense erythema, severe hair loss, and skin scaling beginning at about 5 weeks of age and persists throughout the animal's lifetime. Histologically, epidermal hyperplasia and dermal infiltration of mast cells, macrophages and granulocytes (primarily eosinophils) are observed. Pronounced proliferation and dilation of capillaries were also present. The expression of selected adhesion molecules has been investigated in this model. Strong skin immunostaining of ICAM-1 was observed [76]. However, E-selectin and VCAM-1 staining was absent. Very little of the cellular infiltrate was due to T cells. The relative paucity of T cells and the marked infiltration of eosinophils is very interesting. This model may be more closely related to allergic inflammatory skin diseases such as atopic dermatitis where eosinphil infiltration and mast cell hyperplasia are prominent. Interestingly, the corticosteroid triamcinolone caused complete regression of the lesions while cyclosporin A was without effect [75]. The lack of effect by the latter compound suggests a minimal role for activated T cells. Indeed, hematopoietic cell transfer from cpdm donors to lethally irradiated syngeneic C57BL/Ka mice did not result in skin lesions after 2 months.

A recent murine psoriasis-like disorder has been developed by reconstitution of scid/scid mice with minor histocompatibility mismatched naive $CD4^+$ T cells [77]. These cells expressed high levels of CD45RB and were derived from F_2 progeny of BALB/cj *129/Svj donor mice. This reconstitution resulted in skin alterations that resembled psoriasis macroscopically, histologically and in cytokine expression. Skin lesions appeared 3–4 weeks after transfer and affect 100% of the animals after 8 weeks. These lesions were characterized by erythema and scaling that ranged from local lesion on the ears and head to widespread body involvement. A heavy lymphocytic infiltration was present in the dermis and epidermis after 3 weeks accompanied by a hyperplastic epidermis. The cytokines TNFα, INFγ, and IL-1 were observed to be elevated in the psoriaform lesions. Two immunosupressive treatments which improve psoriasis, namely cyclosporin A and UV-B light exposure both markedly improved the disease severity over 20 days of treatment. This model appears to represent the first psoriasis animal model with a proven T cell pathological basis. It will be of interest to evaluate other known treatments of psoriasis to determine the potential predictive value of this model.

Topical drug delivery: Pitfalls in the translation from animal models to man

For the treatment of skin disease, topical delivery of pharmacological agents seems like a strightforward and sensible approach. In this way the highest concentration of drug can be achieved in the target organ and systemic exposure that could result in unwanted side-effects can be minimized. However, the skin is a formidable barrier aginst the entry of foreign chemicals. The skin is often viewed as a protective covering that isolates the animal from the environment and limits systemic exposure to the effects of harmful external agents. An effective topical drug has to overcome this barrier and be bioavailble at the target site.

The number of effective topical drugs is quite low. There are numerous antibacterial and anti-fungal agents on the market. However, these agents are primarilily targeted at surface infections and do not need to penetrate the stratum corneum barrier. The most widely used topical drugs fall in the category of ligands for the steroid/retinoid nuclear receptor superfamily. Corticosteroids, retinoids, and vitamin D analogs represent successful applications of the topical treatment approach. Even though the penetration of these drugs is relatively low they are effective due to very high potency. Levels as low as picogram/g of skin can produce pharmacological effects. Since the mechanism of action of these compounds involves the regulation of gene expression, the effects are amplified many fold. Compounds that act as enzyme inhibitors or receptor antagonists may be at a disadvantage in that much higher target concentrations must be maintained to expect activity.

The proven therapeutic activity of the anti-psoriatic lipoxygenase inhibitor lonapalene is an example of a good translation of anti-inflammatory activity in animals to a clinical application [78]. This compound is effective in the arachidonic acid- and TPA-induced mouse ear inflammation models and has been shown to be active in resolving an established TPA-induced inflammation [29]. However, there are other examples involving compounds which can block eicosanoid generating enzymes where activity in humans could not be demonstrated. Not because the biochemical target was inappropriate, but due to the lack of skin penetration resulting in insufficient concentrations of the drug at the target site, the epidermis and dermis.

As mentioned previously cyclosporin A is an example of a compound which fails as a topical agent due to insufficient skin penetration. Another example of a failed translation of efficacy in animals to activity in humans can be found in the clinical development of the topical phospholipase A_2 inhibitor, BMS 188184 [79]. The structure of this compound is shown in Figure 4. A summary of the biochemical and pharmacological data used to support the elevation of this compound to clincal development status is shown in Table 4. In mouse inflammation models the applied topical dose ED_{50} values ranged from 0.1% to 1.6%.

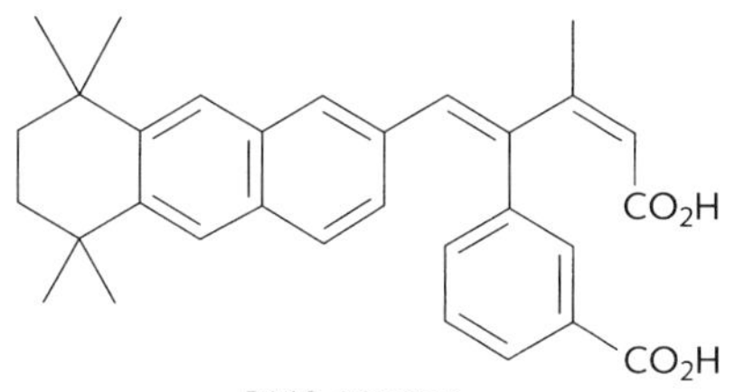

Figure 4
Structure of BMS-188184.

Table 4 - Preclinical profile of the type II phospholipase A_2 inhibitor BMS-188184

Test	Activity (IC_{50} or ED_{50})
Biochemistry and cell biology	
Type II 14 kDA PLA_2	14 μM
Arachidonic acid release in human PMNs	8 μM
LTB_4 biosynthesis inhibition in human PMNs	13 μM
Platelet-activating factor biosynthesis inhibition in human PMNs	10 μM
Antigen-stimulated arachidonic acid release in mouse mast cell	5 μM
Pharmacology	
Single dose TPA-induced mouse ear inflammation (topical)	0.5%
Mouse delayed-type hypersensitivity reaction (topical)	1.6%
Multiple dose TPA-induced mouse ear inflammation (topical)	0.1%

A single 12 h application of two formulations containing 4 or 5% w/w ^{14}C-BMS 188184 to hairless rats and measurement of radioactivity in excreta showed 3–4% of the dose was absorbed systemically. In addition, drug concentrations in the skin were 25–50 μg/g (60–120 μM) 6 h after a single topical dose. These formulations were also tested in normal human volunteers using a single topical dose applied for 12 h. Radioactivity in the excreta was below 0.1% of the dose for both formulations and radioactivity in the blood was below detection limit(< 5 ng/ml). Autoradiograms of vertical sections of drug treated skin biopsies that had been stripped of stratum corneum had radioactivity levels below detection (< 6 μM). Autoradiograms of sections from unstripped biopsies revealed that the radioactivity was only associated with the surface of the skin, the stratum corneum. Because of this lack of penetration and skin localization, a proof-of-principle could not be obtained for this compound.

While it is not the only pitfall in translation of topical activity in animals to activity in man, skin permeation must be put high on the list of criteria in a topical drug discovery program. Skin absorption is a complex phenomenon and the fundemental concepts have been reviewed [80, 81]. In most case the inherent physical properties of the compound class will dictate the feasibility of topical delivery to the skin. Properties such as moleular weight, melting point, aqueous solubility, ionization state and partition coefficient are all contibuting factors that will determine skin permeation. Taking these properties into account during the structure-activity relationship phase of a drug discovery project may improve the chance of developing novel agents for treating skin inflammation. Topical activity in animal models where the permeation barrier is low can be misleading and can often lead to total failure in humans if permeability differences are not considered.

Summary

Routine screening of compounds in animal models is a key component in any discovery effort aimed at identifying novel anti-inflammatory drugs. While true skin disease models which reflect all aspects of the human condition have not yet been found, models do exist which mimic some, but not all, of the pathophysiology. Skin pharmacologists have relied on building correlations between the effects of compounds with known human anti-inflammatory activity and observed efficacy in an animal model to rationalize the selection of *in vivo* screens. This can be very fruitful when working within a class of compounds with a well-defined mechanism of action such as the corticosteroids. However, for compounds with novel mechanisms of action and unproven human efficacy the choice of screening models can often result in high rates of false positive and false negative activities. To reduce these false results the coupling of a gross inflammatory endpoint such as leukocyte infiltration with a biochemical marker relevant to the drug candidates mechanism of action may be a fruitful approach. Ideally, the marker should also be found in the human disease and then could potentially serve as an evaluation endpoint in a clinical trial. At minimum, the effect of a drug on the biomarker reveals that the compound is capable of working *in vivo* by the intended molecular mechanism and that the compound is bioavailable in the target organ.

The combination of multiple animal models in a tiered screening system employing both gross and biochemical endpoints as well as paying special attention to skin permeability seems to be the best current approach for discovering novel anti-inflammatory drugs for skin diseases. Hopefully, future work will bring forth new models in which the human disease is more completely mimicked, resulting in clinically relevant and predictive models for drug discovery.

References

1 US National Health Survey (1971–1974) *Vital and Health Statistics Series II*, No. 212

2 Schultz LF (1993) The epidemiology of atopic dermatitis. *Monogr Allergy* 31: 9–28

3 Menne T, Christopherson J, Maibach HI (1987) Epidemiology of allergic contact sensitization. *Monogr Allergy* 21: 132–161

4 Sulzberger MB, Witten VH (1952) Effect of topically applied compound F in selected dermatoses. *J Invest Dermatol* 19: 101–102

5 Schlagel CA (1965) Comparative efficacy of topical anti-inflammatory corticosteroids. *J Pharm Sci* 54: 335–354

6 Kligman AM (1988) Adverse effects of topical corticosteroids. In: E Christophers, E Schopf, AM Kligman (eds): *Topical corticosteroid therapy. A novel approach to safer drugs*. Raven Press, New York, 181–187

7 Hammerstrom S, Hamberg M, Samuelsson B, Duell E, Voorhees JJ (1975) Increased

concentrations of non-esterified arachidonic acid, 12-HETE, prostaglandin E_2, prostaglandin $F_{2\alpha}$ in epidermis of psoriasis. *Proc Natl Acad Sci USA* 72: 5130–5134
8 Brain S, Camp RDR, Dowd P, Black A, Woollard P, Mallet A, Greaves M (1982) Psoriasis and leukotriene B_4. *Lancet* 2: 762–763
9 Samuelsson B (1987) An elucidation of the arachidonic acid cascade. Discovery of prostaglandins, thromboxane and leukotrienes. *Drugs* 33: 2–9
10 Irvine RF (1982) How is the level of free arachidonate controlled in mammalian cells. *Biochem J* 204: 3–16
11 Marshall LA, Mayer RJ (1995) Phospholipase A_2 isoforms as novel drug targets. In: RR Ruffolo, MA Hollinger (eds): *Inflammation, mediators and pathways*. CRC Press, Boca Raton, 1–22
12 Trancik RJ, Lowe NJ (1985) Evaluation of topical non-steroidal anti-inflammatory drugs. In: HI Maibach, NJ Lowe (eds): *Models in dermatology*, Vol 2, Karger, Basel, 35–42
13 Tonelli G, Thibault L, Ringler I (1965) A bioassay for the concomitant assessment of antiphlogistic and thymolytic activities of topically applied corticosteroids. *Endocrinology* 77: 625–634
14 Kuehl FA, Humes JL, Egan RW, Ham EA, Beveridge GC, Van Arman CG (1977) Role of prostaglandin endoperoxide PGG_2 in inflammatory processes. *Nature* 265: 170–173
15 Young JM, Wagner BM, Spires DA (1983) Tachyphylaxis in 12-O-tetradecanoylphorbol acetate and arachidonic acid induced ear edema. *J Invest Dermatol* 80: 48–52
16 Rao TS, Currie JL, Shaffer AF, Isakson PC (1993) Comparative evaluation of arachidonic acid and TPA-induced dermal inflammation. *Inflammation* 17: 723–741
17 DeYoung LM, Kheifets JB, Ballaron SJ, Young JM (1989) Edema and cell infiltration in the phorbol ester-treated mouse ear are temporally separate and can be differentially modulated by pharmacologic agents. *Agents Actions* 26: 335–341
18 Fuerstenberger G, Marks F (1982) Early prostaglandin E synthesis is an obligatory event in induction of cell proliferation in mouse epidermis *in vivo* by TPA. *Biochem Biophys Res Comm* 92: 749–756
19 Carlson RP, Oneill-Davis L, Chang J, Lewis AJ (1985) Modulation of mouse ear edema by cyclooxygenase and lipoxygenase inhibitors and other pharmacologic agents. *Agents Actions* 17: 197–204
20 Rao TS, Yu SS, Djuric SW, Isakson PC (1994) Phorbol ester-induced dermal inflammation in mice: evaluation of inhibitors of 5-lipoxygenase and antagonists of LTB_4 receptor. *J Lipid Mediat Cell Signal* 10: 213–228
21 Tramposch KM, Steiner SA, Stanley PL, Nettleton DO, Franson RC, Lewin AH, Carroll FI (1992) Novel inhibitor of phospholipase A_2 with topical anti-inflammatory activity. *Biochem Biophys Res Comm* 189: 272–279
22 Burley ES, Smith B, Cutter G, Ahlem JK, Jacobs RS (1982) Antagonism of TPA-induced inflammation by the natural product manoalide. *Pharmacologist* 24: 117
23 Wershil BK, Murakami T, Galli SJ (1988) Mast cell-dependent amplification of an immunologically non-specific inflammatory response. Mast cells are required for the full

expression of cutaneous acute inflammation induced by TPA. *J Immunol* 140: 2356–2360
24 Kern AB (1966) Indomethacin for psoriasis. *Arch Dermatol* 93: 239–240
25 Green CA, Schuster S (1987) Lack of effect of topical indomethacin on psoriasis. *Br J Clin Pharmacol* 24: 381–384
26 Katayama H, Kawada A (1981) Exacerbation of psoriasis induced by indomethacin. *J Dermatol* 8: 323–327
27 Crummey A, Harper GP, Boyle EA, Mangan FR (1987) Inhibition of arachidonic acid-induced ear edema as a model for assessing topical anti-inflammatory compounds. *Agents Actions* 20: 69–76
28 Puignero V, Queralt J (1997) Effect of topically applied cyclosporine A on arachidonic acid- and TPA-induced dermal inflammation in mouse ear. *Inflammation* 21: 357–369
29 Stanley PL, Steiner S, Havens M, Tramposch KM (1991) Mouse skin inflammation induced by multiple topical application of 12-O-tetradecanoyphorbol-13-acetate. *Skin Pharmacol* 4: 262–271
30 Alford JG, Stanley PL, Todderud G, Tramposch KM (1992) Temporal infiltration of leukocyte subsets into mouse skin inflamed with phorbol ester. *Agents Actions* 37: 260–267
31 Norris JFB, Iderton E, Yardley HJ, Summerly R, Forster S (1984) Utilization of epidermal phospholipase A2 inhibition to monitor topical steroid action. *Br J Dermatol* III (Suppl 27): 195–203
32 Bresnick E, Bailey G, Bonney RJ, Wightman P (1981) Phospholipase activity in skin after application of phorbol ester and 3-methyl cholanthrene. *Carcinogenesis* 2: 1119–1122
33 Tramposch KM, Chilton FH, Stanley PL, Franson RC, Havens MB, Nettleton DO, Davern LB, Darling IM, Bonney RJ (1994) Inhibitor of phospholipase A_2 blocks eicosanoid and platelet activating factor biosynthesis and has topical anti-inflammatory activity. *J Pharmacol Expt Therap* 271: 852–859
34 Christophers E (1996) The immunopathology of psoriasis. *Int Arch Allergy Immunol* 110: 199–206
35 Dvorak HF, Galli SJ, Dvorak AM (1986) Cellular and vascular manifestations of cell mediated immunity. *Human Pathology* 17: 122–137
36 Crowle AJ (1975) Delayed hypersensitivity in the mouse. *Adv Immunol* 20: 197–264
37 Chapman JR, Ruben Z, Butchko GM (1986) Histology of and quantitative assays for oxazolone-induced allergic contact dermatitis in mice. *Am J Dermatopathol* 8: 130–138
38 Kondo S, Beissert S, Wang B, Fujisawa H, Kooshesh F, Stratigas A, Granstein RD, Mak TW, Sauder DN (1996) Hyporesponsiveness in contact hypersensitivity and irritant contact dermatitis in CD4 gene targeted mouse. *J Invest Dermatol* 106: 993–1000
39 Mosmann TR, Coffman RL (1980) Th1 and Th2 cells: differential patterns of lymphokine secretion lead to different functional properties. *Ann Rev Immunol* 17: 145–173

40 Mohler KM, Butler LD (1990) Differential production of IL-2 and IL-4 mRNA *in vivo* after primary sensitization. *J Immunol* 145: 1734–1739
41 Thomson JA, Troutt AB, Kelso A (1993) Contact sensitization to oxazolone: involvement of both interferon γ and IL-4 in oxazolone specific Ig and T-cell responses. *Immunology* 78: 185–192
42 Bailey SC, Asghar F, Przekop PA, Kurtz ES (1995) A novel contact hypersensitivity model for rank ordering formulated corticosteroids. *Inflamm Res* 44 (Suppl 2): 5162–5163
43 Back O, Egelrud T (1985) Topical glucocorticoids and suppression of contact sensitivity. A mouse bioassay of anti-inflammatory effects. *Br J Dermatol* 112: 539–545
44 Cavey D, Bouclier M, Burg G, Delamadeleine F, Hensby CN (1990) The pharmacological modulation of delayed-type hypersensitivity reactions to topical oxazolone in mouse skin. *Agents Actions* 29: 65–67
45 Rullan PP, Barr RJ, Cole GW (1984) Cyclosporine and murine allergic contact dermatitis. *Arch Dermatol* 120: 1179–1183
46 Meingassner JG, Stutz A (1992) Anti-inflammatory effects of macrophilin-interacting drugs in animal models of irritant and allergic contact dermatitis. *Int Arch Allergy Immunol* 99: 486–489
47 Powles AV, Baker BS, Fry L (1989) Cyclosporin A and skin disease. In: AW Thomson (ed): *Cyclosporin: mode of action and clinical application.* Academic Publishers, Dordrecht, Kluwer, 191–212
48 Abu-Elmagd K, Van Thiel D, Jegasothy BV, Ackerman CD, Todo S, Fung JJ, Thomson AW, Starzl TE (1991) FK506: a new therapeutic agent for recalcitrant psoriasis. *Transplant Proc* 23: 3322–3324
49 Ruzicka T, Bieber T, Schopf E, Rubins A, Dobozy A, Bos JD, Jablonska S, Ahmed I, Thestrup-Pedersen K, Daniel F et al (1997) A short-term trial of tacrolimus ointment for atopic dermatitis. *N Engl J Med* 337: 816–821
50 Lauerma AI, Maibach HI, Granlund H, Ercko P, Kartamma M, Stubb S (1992) Inhibition of contact allergy reactions by topical FK-506. *Lancet* 340: 556
51 Anderson C (1985) The effect of selected immunomodulating agents on experimental contact reactions. *Acta Derm Venerol Suppl* 116: 1–48
52 Nilsson G, Otto U, Wahlberg JE (1982) Assessment of skin irritancy in man by laser Doppler flowmetry. *Contact Dermatitis* 8: 401–406
53 Duncan JI (1994) Differential inhibition of cutaneous T-cell mediated reactions and epidermal cell proliferation by cyclosporin A, FK506 and rapamycin. *J Invest Dermatol* 102: 84–88
54 Lauerma AI, Stein BD, Homey B, Lee CH, Bloom E, Maibach HI (1994) Topical FK-506: suppression of allergic and irritant contact dermatitis in the guinea pig. *Arch Dermatol Res* 286: 337–340
55 Shoji Y, Fukumura T, Kudo M, Yanagawa A, Shimeda J, Mizushima Y (1994) Effect of topical preparation of mycophenolic acid on experimental allergic contact dermatitis of guinea pigs by dinitrofluoro-benzene. *J Pharm Pharmacol* 46: 643–646

56 Nakagawa S, Oka D, Jinno Y, Bang D, Veki H (1988) Topical application of cyclosporin on guinea pig allergic contact dermatitis. *Arch Dermatol* 124: 907–910
57 Dumont FJ, Staruch MJ, Koprak SL, Melino MR, Sigal NH (1990) Distinct mechanisms of suppression of murine T-cell activation by the related macrolides FK-506 and rapamycin. *J Immunol* 144: 251–255
58 Griffiths CEM, Powles AV, Baker BS, Fry L, Valdimarrson H (1987) Topical cyclosporin and psoriasis. *Lancet* I: 806
59 Montagna W, Yun JS (1964) The skin of the domestic pig. J Invest Dermatol 43: 11–21
60 Meyer W, Schwartz R, Neurand K (1978) The skin of domestic mammals as a model for human skin, with special reference to the domestic pig. *Current Probl Dermatol* 7: 39–52
61 Weinstein GD (1965) Autoradiographic studies of the turnover time and protein synthesis in pig epidermis. *J Invest Dermatol* 44: 413–419
62 Bissett DL, McBride JF (1983) The use of the domestic pig as an animal model of human dry skin and for comparison of drug and normal skin properties. *J Soc Cosmet Chem* 34: 317–326
63 Barteck MJ, LaBudde JA, Maibach HI (1972) Skin permeability *in vivo*: comparison of rat, rabbit, pig and man. *J Invest Dermatol* 58: 114–123
64 Meingassner JG, Stutz A (1992) Immunosuppressive macrolides of the type FK506: a novel class of topical agents for the treatment of skin diseases? *J Invest Dermatol* 98: 851–855
65 Kaidbey KH, Kligman AM (1976) Assay of topical corticosteroids by suppression of experimental rhus dermatitis in humans. *Arch Dermatol* 112: 808–810
66 Hermann RC, Taylor RS, Ellis CN (1988) Topical cyclosporin for psoriasis *in vitro* skin penetration and clinical study. *Skin Pharmacology* 1: 246–249
67 Meingassner JG, Grassberger M, Fahrngruber H, Moore HD, Schuurman H, Stutz A (1997) A novel anti-inflammatory drug, SDZ ASM 981, for topical and oral treatment of skin diseases: *in vivo* pharmacology. *Br J Dermatol* 137: 568–576
68 Brown WR, Hardy MH (1988) A hypothesis on the cause of chronic epidermal hyperproliferation in asebia mice. *Clin Exp Dermatol* 13: 74–77
69 Oran A, Marshall JS, Kondo S, Paglia D, McKenzie RC (1997) Cyclosporin inhibits ICAM-1 expression and reduces mast cell numbers in the asebia mouse model of chronic skin inflammation. *Br J Dermatol* 136: 519–526
70 Brown WR, Rogozinski TT, Ramsey CA (1988) Anthralin and tar with UVB increase epidermal cell proliferation in asebia mice. *Clin Exp Dermatol* 13: 248–251
71 Morita K, Hogan ME, Nanney LB, King LE, Manabe M, Sun TT, Sunberg JP (1995) Cutaneous ultrastructural features of the flaky skin mouse mutation. *J Dermatol* 22: 385–395
72 Sunberg JP, Dunstan RW, Roop DR, Beamer WG (1994) Full thickness skin grafts from flaky skin mice to nude mice: maintenance of psoriaform phenotype. *J Invest Dermatol* 102: 781–788

73 Nanney LB, Sunberg JP, King LE (1996) Increased epidermal growth factor receptor in fsn/fsn mice. *J Invest Dermatol* 106: 1169–1174
74 Nanney LB, Yates RA, King LE (1992) Modulation of epidermal growth factor receptors in psoriatic lesions during treatment with topical EGF. *J Invest Dermatol* 98: 296–301
75 HogenEsch H, Gijbels MJJ, Offerman E, vanHooft J, vanBekkum DW, Zurcher C (1993) A spontaneous mutant characterized by chronic proliferative dermatitis in C57BL mice. *Am J Pathol* 143: 972–982
76 Gallardo Torres HI, Gijbels MJJ, HogenEsch H, Kraal G (1995) Chronic proliferative dermatitis in mice: neutrophil-endothelium interactions and the role of adhesion molecules. *Pathobiology* 63: 341–347
77 Schon M, Detman M, Parker CM (1997) Murine psoriasis-like disorder induced by naive $CD4^+$ T-cells. *Nature Med* 3: 183–188
78 Lassus A, Forsstrom S (1985) A dimethoxy-nepthalene derivative (RS43179) compared with 0.025% fluocinolone acetonide gel in the treatment of psoriasis. *Br J Dermatol* 113: 103–106
79 Springer DM, Lu BV, D'Andrea SV, Branson JJ, Mansuri MM, Burke JR, Gregor KR, Stanley PL, Tramposch KM (1997) Dicarboxylic acid inhibitors of phospholipase A_2. *Bioorg Med Chem Lett* 7: 793–795
80 Barry BW (1983) *Dermatological formulation: percutaneous absorption.* Marcel Dekker, New York
81 Schaefer H, Zesch A, Stuttgen G (1982) *Skin permeability.* Springer Verlag, New York

Animal models of inflammatory bowel disease

Sreekant Murthy and Anne Flanigan

Division of Gastroenterology and Hepatology, Krancer Center for IBD Research, MCP Hahnemann University, Philadelphia, PA 19102-1192, USA

Introduction

Inflammatory bowel diseases (IBD) are genetically complex and multifactorial diseases. They consist of two major illnesses: ulcerative colitis (UC) and Crohn's Disease (CD). These are two of the most debilitating chronic diseases of unknown etiology, characterized by periods of quiescence and relapse. IBD can be differentiated from infectious, ischemic and collagenous colitis since these diseases do not cause a chronic relapsing and remitting disease. They affect the lifestyle and functional capabilities of those afflicted. Epidemiological studies (Tab. 1) show apparent differences in the occurrence of IBD based on age, sex, race and ethnicity. In general, these diseases affect children, adolescents, and young adults, and the majority of those afflicted are between the ages 16 and 45. They are the major causes of morbidity in this very young population.

Crohn's disease, popularized by Crohn, Ginzburg and Oppenheimer [1] was once considered a disease of the ileum. It is now recognized that CD can affect any part of the digestive tract from mouth to anus, although more frequently the ileum and colon are involved. The distinctive clinical and histological features of CD are enumerated in Table 2.

Ulcerative colitis, first described in 1859 by Wilkes [2], on the other hand, is considered primarily a mucosal disease. commonly restricted to the colon. The disease usually originates in the distal colon, invariably involves the rectum and may eventually affect the entire colon to develop into pancolitis. If pancolitis is the initial diagnosis, the patients are at very high risk (30%) for developing colitis-associated colo-rectal cancer and the risk increases with the duration of disease [3]. As shown in Table 2, the clinical features of UC show significant overlap with those of CD, suggesting a commonality between these two diseases. Conversely, the macroscopic and microscopic features of UC differ from CD.

IBD has been recognized for several decades, yet its etiology remains obscure. Advances in genetics, immunology and pharmacology have improved our understanding of the etiology of IBD. IBD is now recognized as a genetically complex

In Vivo Models of Inflammation, edited by D.W. Morgan and L.A. Marshall

Table 1 - Epidemiology of IBD

	Crohn's Disease	Ulcerative colitis
Age	Late adolescence or early adulthood	Late adolescence or early adulthood
Gender	Equal or slightly higher in females	Equal or slightly higher in males
Incidence	1–10/100 000 population	1–16/100 000 population
Prevalence	10–20 times incidence	10–14 times incidence
Genetics	Familial	Familial
Ethnicity	Increased in Jewish population	Increased in Jewish population

multifactorial disease involving a number of overlapping genetic, environmental and immunological factors as shown in Figure 1. It is now increasingly clear that genetic susceptibility may be driven by many environmental factors which ultimately may aggravate the immune response, resulting in chronicity. Thus, regardless of the mechanism of induction, the full spectrum of cellular and humoral elements of the immune system are activated and they coordinately contribute to disease pathology.

There is overwhelming evidence that IBD is both a neutrophil and T cell driven-disease. However, over the last few years, the focus of attention has drifted from neutrophils to T cells that, upon activation, secrete powerful cytokines. These cytokines regulate the recruitment of inflammatory cells and the immune system. Thus, these cytokines have become one of the primary targets of future therapy of IBD.

Medical knowledge, treatment, and research that are involved in understanding the etiopathology of IBD have used laboratory animals for decades. These studies have shown that producing an ideal IBD animal model that entirely mimics human disease is problematic. The disease process is multifactorial and involves several background genes. Therefore, the axiom for producing a true model of IBD should be that the model must express many phenotypic variations that resemble human disease. A promising animal model of IBD is one in which the relative frequency of the overlapping clinical and biochemical features are near the peak of human disease. It is reasonable to believe that current animal models of IBD do not possess all of the characteristics of human disease; however, all of the models express inflammation and utilize the same inflammatory cascades making it difficult to select the best model for preclinical trials. Thus, it is reasonable to assume that these are "models of intestinal inflammation" that have great potential for testing the efficacy of a variety of pharmacological agents.

Selection of an animal model for preclinical testing must meet several fundamental criteria as listed in Table 3. Until today, these models have been empirically selected since the mechanisms of injury and mediator involvement in several of these models have remained unknown. Sometimes, the model is selected based upon

Table 2 - Clinical features of IBD

	Crohn's Disease	**Ulcerative colitis**
Location	Entire GI tract Ileocolitis (40–55%) Ileitis (30–40%) Colitis (less common)	Colon Proctosigmoiditis (40–50%) Left-sided colitis (30–40%) Pancolitis (20%) Backwash ileitis (<10%)
Clinical features	Abdominal pain Diarrhea Occult blood (less common) Low grade fever	Abdominal pain Bloody diarrhea Weight loss Fever
Macroscopic features	Fibrotic and stenotic bowel (strictures present) Deep, serpignous and aphthous ulcers Skip lesions "Cobble stone" appearance Fissures and fistulas Anus is involved	Strictures absent Presence of punctate ulcers Pseudopolyps Shortened colon Rectum always involved Anus free of disease
Microscopic features	Transmural inflammation Focal cryptitis Crypt abscess Granulomas Neuronal hyperplasia	Primarily mucosal inflammation Cryptitis, crypt abscess, dilated crypts and crypt branching and mucin depletion Granulomas absent
Extraintestinal manifestations	Ankylosing spondolytis Psoriasis Iritis and uveitis Erythema nodosum	Ankylosing spondolytis Primary sclerosing cholangitis Pyoderma gangrenosum
Risk of cancer	Rare	High especially in pancolitis and long standing disease (>20% after 30 years)

investigator's experience with the model and the ease of producing and measuring end points. This chapter provides a comprehensive list (Table 4) of animal models that have been applied for preclinical evaluation and models that have contributed to significant breakthroughs in understanding the etiopathology of IBD. The models are classified into different groups based upon how the disease is produced. The disease induction process, the clinicopathological features and success of preclinical therapies in these models are listed in Table 4.

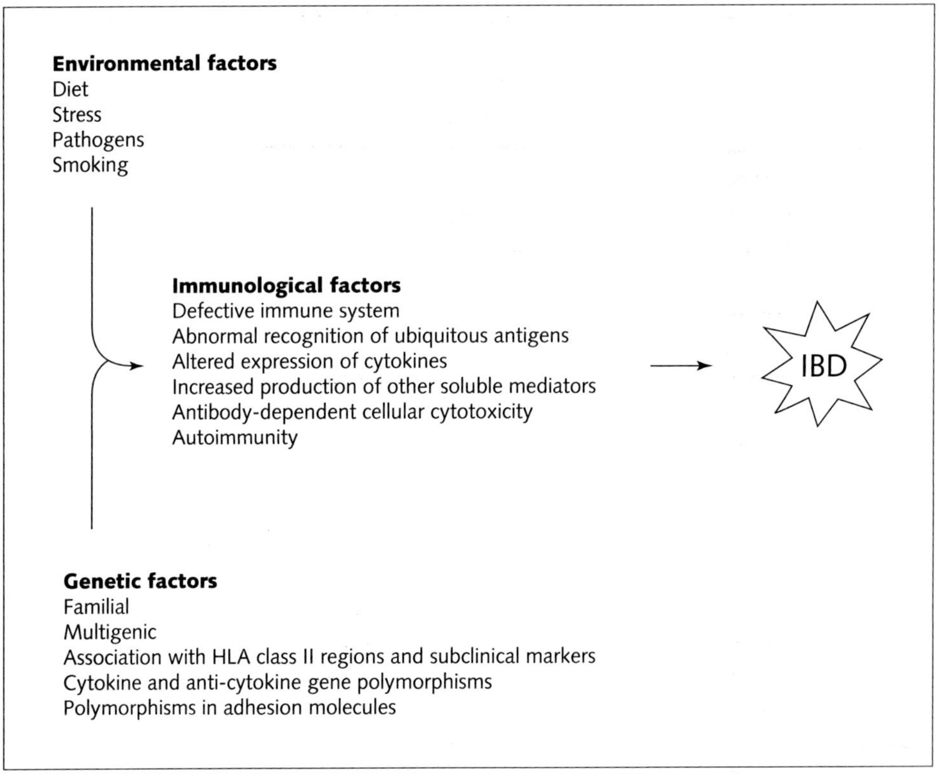

Figure 1
Current concept of the etiology of IBD.

Table 3 - Criteria for selecting an animal model of IBD

1. The model should accurately reproduce some of the phenotypical changes, such as histopathological features and clinical disease spectrums, which are common to human disease, and are modulatable by diet, environmental factors and treatment.
2. The model must be economical, polytocous and convenient to house.
3. The model should be available in multiple species and genetic backgrounds for second or third species verification to rule out interspecies differences and to predict its efficacy in human IBD.
4. The disease should be sufficiently long-lasting and animals should have adequate survival to predict appropriate outcome of therapy.
5. Up-to-date information must be available on the extent of mediator involvement to inhibit its synthesis, release or to antagonize its action.

Spontaneous or natural models

As the etiology of IBD still remains unclear, natural models of intestinal inflammation continue to be of significant interest. Over the years, many natural models of IBD have been described. These include animals that exhibit non-specific inflammation caused by infection, breeding or stress due to domestication, particularly in juvenile animals [4]. Since colitis observed in these animals is primarily due to bacterial infection, in most instances, the disease may be self-limiting. The sporadic nature of the appearance of the disease, self-limiting inflammation, expense in procuring some of these animals and their husbandry contribute to their limited use in preclinical trials.

C3H/HeJ Model

A substrain of C3H/HeJ mice which develops spontaneous cecitis and right-sided colitis in 60% of mice when they are 4–5 weeks old is reported (Tab. 4). These mice are lipopolysaccharide (LPS) resistant due to a defect in the LPS locus on chromosme 4. A further substrain which develops spontaneous colitis with perianal ulcers and occult blood loss in 98% of animals has also been reared and named C3H/HeJBir mice [5]. These mice show remarkable reactivity to enteropathogenic bacterial antigens and dextran sulfate, resulting in fulminant colitis. The model is now extensively used in genetic mapping studies to identify genes that are susceptible or resistant to colitis.

Cotton top tamarin model

This next group consists of captured animals, which develop spontaneous colitis in captivity. This is exemplified by *Sanguinas oedipus* or cotton top tamarins (CTT) belonging to the marmoset family (Tab. 4) [6]. These are new world monkeys native to northcentral Columbia, South America. When these animals are in their natural environment, they show little or no disease. In a confined environment, they develop severe colitis resembling human UC, including heightened occurrence of colon cancer [7].

Colitis in CTT is associated with anorexia, watery diarrhea or bulky soft stools, dehydration and weight loss. Histologically, acute colitis shows mucin depletion, an influx of poly-morphonuclear leukocytes and crypt abscesses. Lymphocytes, plasma cells and eosinophils are seen, but gross ulcers are less common. In the chronic state, neutrophils and granulomas are not seen, but the lamina propria contains monocytes and plasma cells. Dysplasias are commonly seen in areas of chronic inflammation suggesting that severe inflammation may predispose these animals to neo-

Table 4 - Animal models of inflammatory bowel disease

Model	Species	Method of induction	Time-course	Disease location	Pathology
I. Spontaneous					
1. C3H/HeJBir	Mice	Natural	4–6 weeks	Cecum and colon	A
2. Cotton top tamarins *oedepus oedepus*	*Sanguinas*	Natural, in a confined environment	<1 year	Colon	A and C
3. SAMP1/YIT	Mice	Natural	30 weeks	Ileum	C
II. Immunological models					
1. Immune complex-mediated	Rabbits and rats	Injection of albumin-antialbumin Complex + a formalin enema	3 h–8 weeks	Colon	A
2. Dinitorchlorobenzene (DNCB)	Rats, Guinea pigs	Application on skin DNCB enema in ETOH		Colon	A
III. Bacteria and bacterial products models					
1. Bacteria models					
A. *Helicobacter hepaticus*	SCID Mice	Infection		Colon	A and C
B. *Chlamydia trachomatis*	*Macaca fasicularis*	Rectal inoculation (once or repeat)	1–2 weeks	Proctosigmoid colon	A and C
2. Bacterial products					
A. Peptidoglycan poly-saccharide (PGPS)	Rats	Transmural injection of PGPS	2–90 days	Ileum	C Relapsing
B. Freund's incomplete adjuvent	Rats and Guinea pigs	Transmural injection	4 weeks	Ileum	C
IV. Chemicals or polymers					
1. Acetic acid	Rats	1–10% acetic acid enema	1 day–3 weeks	Colon	A
2. Trinitrobenzene sulfonic acid (TNBS)	Rats, rabbits and mice	TNBS enema (20–30 mg in 30–50% ETOH)	3 days–8 weeks	Small intestine or Colon	A and C
3. Dextran sulfate (DSS)	Hamsters, mice and rats	2–10% DSS feeding	5 days–15 weeks/ longer	Colon	A and C

4. Carrageenan	Rats and Guinea pigs	1–5% carrageenan feeding	6 days–7 weeks	Cecum and right colon	A
5. Indomethacin	Rats	Oral or s.c.once or twice	< 1 day–8 days	Small intestine	A
V. Miscellaneous chemical models					
1. Peroxynitrite	Rats	Enema	< 1 day–3 weeks	Colon	A
2. Mitomycin C	Fischer rats	I.P. 2.75 mg/kg		Colon	A
3. Iodoacetamide	Rats	3% enema		Colon	A and C
4. Phorbal-12-myristate-13-acetate [PMA]	Rabbits	Enema	4 days	Colon	A
VI. Transgenic and mutant models					
1. HLA-B27/b2 micro-globulin	Rats	Transgene	> 10 weeks	Intestines	A and C
2. IL-2 Knockouts	Mice	Gene deletion	> 5 weeks	Colon	A and C
3. IL-10 Knockouts	Mice	Gene deletion	> 6 weeks	Colon	A and C
4. TCR Knockouts	Mice	Gene deletion	> 30 weeks	Colon	A and C
5. Gαi2 Knockouts	Mice	Gene deletion	> 15 weeks	Colon	A and C
6. TNFα 3′ deletion	Mice	3′-AU rich region 589 (69bp) deletion	3–8 weeks	Ileum	A and C
VII. Miscellaneous transgenic and knockout mice					
1. Keratin 8	Mice	Gene deletion		Colon	A and C
2. TGFβ1 Knockouts	Mice	Gene deletion		Colon	A and C
3. IL-7 Transgenic	Mice	Transgene	> 8 weeks	Colon	A and C
VIII. T cell and bone marrow transfer models					
1. CD45RBhi	SCID Mice	Adaptive transfer of CD45RBhi $CD4^+$ T cells		Colon	C
2. Tge 26	Mice	Transfer of syngenic bone marrow		Colon	C

A, acute inflammation; C, chronic inflammation

plastic changes. These animals also develop autoantibodies against epithelial autoantigens [8]. It appears that CTTs do not acclimate well and they are metabolically stressed at temperatures less than 32° C, which may initiate and perpetuate chronic colitis [9]. CTTs exhibit low levels of MHC class I polymorphisms or allelic diversity and therefore, are highly vulnerable for viral attack and possible susceptibility to colitis [10]. The mechanisms of injury, the mediators involved and the efficacy of various therapies are listed in Table 5. Most importantly, CTTs respond very well to 5-aminosalicylic acid and 5-lipoxygenase inhibitor therapies [6, 11]. The success of 5-ASA therapy has contributed to long term remission and survival of CTTs in captivity. It also boosts confidence in the model's ability to predict impending success of novel therapies in human trials. The drawbacks for using this model are pancolitis at the onset of disease, limited availability of animals since there are only a few primate centers in the world, housing, expense and the endangered species status of the animals.

SAMP1/YIT Model

Recently, Kosiewicz et al. [12], in collaboration with Japanese investigators, described a mouse model of spontaneous colitis which closely resembles CD. The model was derived by brother-sister mating of AKR strains that develop spontaneous ileitis. Inflammation is transmural, localized to the terminal ileum with heavy infiltration of inflammatory cells. Crypt abscesses and blunted villous architecture were present. Inflammation in this model did not develop in germfree conditions. The model, when fully characterized, has the potential to elucidate the pathogenesis of ileitis.

Immunological models

Immune complex model

Immune complex colitis has been produced in rabbits by a combination of the intravenous injection of human albumin and anti-albumin complex followed by rectal administration of dilute formalin (Tab. 4) [13]. The purpose of using formalin is to damage the colon and render permeability of systemic albumin and antialbumin complex into the damaged region to elaborate a local immune reaction. The soluble complex (0.5 to 0.75 ml) is given intravenously through the marginal ear vein (rabbits) or tail vein (rats) 2 to 3 h after a 0.4–4% formalin enema. There are three distinct methods described for producing colitis in rats and rabbits. The first method involves the injection of immune complex in non-sensitized animals and the second approach involves first sensitizing the animals with enterobacterial antigen of Kunin

[14]. In the third, rats or rabbits are preimmunized with 100 μg of *Escherichia coli* 014:K7:H-antigen emulsified in Freund's complete adjuvant before a second challenge with albumin-anti-albumin complex and formalin enema [15]. The inflammation resolves after 2 weeks, without clear evidence of chronic colitis except in animals that are presensitized where inflammation can last for several months. Histologically, chronic colitis exhibits many features of acute colitis with an increased influx of neutrophils within 24 h, followed by mixed inflammation, crypt abscesses, crypt distortion, ulceration and necrosis. The mechanism of injury, the mediators involved and results of drug efficacy trials are shown in Table 5. Colitis is associated with increased synthesis of arachidonic acid metabolites which occurs a few hours after the activation of IL-1 gene expression [16]. Pretreatment and treatment with IL-1ra within the first 33 h after the induction of colitis diminishes inflammation scores and necrosis. Sulfasalazine treatment is effective [15, 17], but steroids do not work. The speculated mechanism of injury is not relevant to human disease as the model does not show chronicity, relapse or remission of disease.

Dinitrochlorobenzene (DNCB) model of colitis

Rosenberg and Fisher [18] in 1964 applied, on guinea pig skin, DNCB, a hapten, that binds tissue proteins and produces a delayed-hypersensitivity type of reaction. Seven days later, they intrarectally instilled the chemical to produce T cell mediated colitis that is responsive to immunosuppressants (Tab. 4). This model has been less popular because it produces acute self-limiting colitis and peritonitis, thus making it difficult for long-term preclinical trials.

Bacteria and bacterial products models

Bacteria models

The postulation that an infectious agent is a causative factor of IBD has persisted for many decades. In IBD patients, the presence of many different bacterial agents has been documented; however, none of these pathogens have ever been directly implicated as primary etiopathic agents. Recently, Ward et al. [19] described that 5% of immunodeficient athymic Ncr-nu/nu, BALB/c AnNCr-nu/nu, C57BL/rNCr-nu/nu and C.B17/Icr-scid/ncr mice naturally infected with *Helicobacter hepaticus* develop typhlitis, colitis, proctitis and rectal prolapse (Tab. 4). Histologically, the large intestine shows hyperproliferative atypical epithelial cells. Immunocompetent mice infected with this pathogen do not show large bowel inflammation. The usefulness of this infectious disease model for IBD research is currently under investigation.

Table 5 - Mechanisms, mediators and anti-inflammatory therapies in animal models of experimental colitis

Model	Species	Mechanism	Mediators	Non-effective therapy	Effective therapy
Spontaneous	Cotton top tamarins	Stress, MHC class I polymorphism	AA metabolites	Unknown	Anti-α4β7and anti-TNF antibody, 5-ASA, 5-LO [5, 10, 76, 77]
Immune complex	Rabbits and rats	Bacteria, immunologic	AA metabolites, IL-8, IL-1, IL-8	Steroids [?]	5-ASA, IL-1ra, anti-IL-8 [14, 17, 78, 79]
TNBS	Rats, rabbits and mice	Th1 response, neutrophils, ROM NO, free radicals	AA metabolites, Cytokines	Steroids, 5-ASA	TNFα , fish oil Zileuton, MK 886 EGF, KGF, L-NAME [26, 80–88]
Dextran sulfate	Hamsters, mice and rats	Unclear	AA metabolites, Cytokines	Steroids	IL-1ra, TNFα, TGFβ2, cyclosporin, antioxidants, FLAP-inhibitor [35, 36, 90–92]
Carrageenan	Rats and Guinea pigs	Bacteria	Unclear	5-ASA	Antibiotics [93]
Acetic acid	Rats	Bacteria [?], Neutrophils[?]	AA metabolites,	Antineutrophil antibody	5-ASA, L-NAME
			Cytokines	Unknown	IL-1ra, SOD inhibitors, PGE [94–98]
Indomethacin	Rats	Bacteria, prostaglandin inhibition	Unspecified Bacteria	Unknown	5-ASA, 5-LO and TXB2 inhibitors [99–101]
Peroxynitrite	Rats	Oxidative damage	Nitric oxide	Unknown	L-NAME [53]

Iodoacetamide	Rats	Sulfhydryl blockade	Unknown	Unknown	L-NAME [54]
PG/PS	Rats	Bacterial biproducts, T cells	IL-1, IL-8, IL-1/Ilra		
	Kallikreinin and kinin	IκB, IGF-1mRNA	Unknown		Proteasome inhibitor [102]
IL-2 Knockouts	Mice	Bacteria, T cells	IL-1, TNFα, IFNγ	Unknown	IL-12 mab [103]
IL-10 Knockouts	Mice	Bacteria, T cells	Th1 cytokines	Unknown	Anti-IFNγ mab [63]
CD45Rbhi	SCID Mice	Th1 response, bacteria	TGFβ, IFNγ	Unknown	Anti-TNFα and IFNγ, rIL-10 [73]
Tgε 26	Mice	CD3ε26 cells	TNFα, IFNγ	Unknown	Anti-TNFα and IFNγ [75]

PG/PS, Peptidoglycanpolysaccharide; TNBS, Trinitrobenzene sulfonic acid; 5-ASA, 5-aminosalicylic acid; TNF, tumor necrosis factor; AA, arachidonic acid; EGF, epidermal growth factor; KGF, keratinocyte growth factor; ROM, reactive oxygen metabolites; TGF, transforming growth factor; IGF, insulin like growth facator; IFN, interferon; PGE, prostaglandin E; SOD, superoxide dismutase; TXB, thromboxane B; NO, nitric oxide; L-NAME, L-arginine methyl ester; mab, monoclonal antibody; 5-LO, 5-lipoxygenase; FLAP, 5-lipoxygenase activating protein.
For brevity, only a few representative references are provided.

In addition, certain strains of *Chlamydia trachomatis* when rectally inoculated produce proctitis in Cynomolgus monkeys, *Macaca fasicularis* [20]. Proctitis is seen in 1 to 2 weeks and resolves rapidily within 2 weeks. However, a repeat inoculation results in more severe disease which will last up to 12 weeks. The disease in this model appears to be immunologic. This is an excellent animal model of the primate immune system which has closer homology to humans. The disadvantage is that C. *trachomatis* is not involved in human IBD.

Bacterial products models

Peptidoglycan polysaccharide model

The intestinal lumen contains very high concentrations of exogenous and endogenous antigens derived from food products and bacterial secretory or cell wall products. The bacterial secretory product formyl-methyl-lucyl-phenylanine (FMLP) and their cell products lipopolysaccharides and peptidoglycan polysacchrides (PG/PS) are highly antigenic and they activate every possible cascade of inflammation and cause damage to the colonic mucosa and vascular tissue architecture. Sartor et al. [21] in 1985 showed that a single subserosal injection of group A PG/PS results in ileal, cecal and distal colonic, chronic, spontaneously relapsing, transmural, granulomatous enterocolitis, arthritis, anemia and hepatic granulomas in genetically susceptible Lewis rats (Tab. 4). The acute phase is rapid and resolves after 2 weeks. Chronic granulomatous colitis emerges after 2 weeks; it reaches a peak at 3 weeks after injection. Buffalo rats and Fischer F344 rats which are MHC matched with Lewis rats develop self-limiting inflammation without any signs of extraintestinal or systemic inflammatory reactions. The mechanism of injury, the mediators involved and results of therapeutic studies are given in Table 5. The advantage of using this model is that it resembles CD by showing chronic, relapsing, transmural and granulomatous inflammation including extrainstestinal manifestion and arthritis. The disadvantages are that it is species/strain specific, requires laparotomy and direct serosal injection of PG/PS and produces enterocolitis and inconsistent mucosal ulcerations. PG/PS is expensive and only a few commercial sources are available.

Freund's incomplete adjuvant (FICA)-induced model

A close analogue of the PG/PS model of enterocolitis is the Freund's incomplete adjuvant (FICA)-induced colitis in rats and guinea pigs [22]. The method of producing injury is the same as in the PG/PS model where FICA mixed with heat-killed *Mycobacterium tuberculosis* prepared in mineral oil is injected intramurally into the subserosal regions to produce chronic colitis lasting for 4 weeks. The induction of disease requires surgical intervention. Preclinical trials have shown that 5-ASA and an inhibitor of inducible nitric oxide synthesis provide significant protection in this model.

Chemical injury

Acetic acid model

Acetic acid-induced colonic injury, first introduced by MacPherson and Pfeiffer [23], requires abdominal incision and introduction of 5% acetic acid into the colonic lumen. The method has now been changed to introduce various concentrations of acid as an enema into rats, guinea pigs, rabbits and mice, with or without prepping or buffering the colon. The most commonly used concentration is 3–5% acetic acid given as an enema in volumes less than 1 ml (Tab. 4). At this concentration, animals develop injury at the site of delivery. Colitis is easy to produce and highly reproducible. Injury occurs rather rapidly with the initiation of injury to the mucosal surface altering the mucosal barrier. Peak damage, with severe acute inflammation, ischemia and erosions occurs within 24 h and resolves within 2 weeks with little or no chronic inflammation. The disease is characterized by the loss of surface epithelium, ulcerations, the loss of goblet cells, crypt abscesses, edema and an influx of neutrophils in the lamina propria, occurring within 2 to 3 days. The mechanism of injury is unclear, yet it is known that it involves a protonated form of this acid at a pH of 2–3 since buffered acid and HCl at the same pH do not cause epithelial injury [24]. Oxygen-derived free radicals are also suggested to be involved in producing injury. Colonic injury is associated with increased production of arachidonic acid metabolites produced by infiltrating neutrophils. Many therapies used for the treatment of human IBD also work in this model which help predict beneficial effects of tested drugs for possible human application (Tab. 5). However, the popularity of this model has diminished since the model shows absence of significant chronic inflammation and initial epithelial cell injury, producing a later immune activation which is opposite to that of human IBD.

Trinitrobenzene sulfonic acid

Since it was first described in 1989 by Morris [25], this model has been very popular. Over 100 articles have been published to describe the use of this model for both experimental and therapeutic purposes. Its popularity is well founded because a single application of TNBS in rats, mice, guinea pigs, dogs and rabbits produce rapid, reliable and reproducible disease (Tab. 4) [25–29]. Since the model can be produced in multiple species, it renders the possibility of second or third species verification to rule out false negatives due to species-specific results. The induction of colitis by TNBS is simple, requiring an introduction of 20–80 mg of TNBS dissolved in 30–50% ethanol as an enema in the rat colon, but the degree of disease and time required to produce the injury may vary between laboratories. TNBS induces peak acute inflammation within a week which gradually progresses into chronic inflammation lasting for about 8 weeks. Certain mice are resistant to TNBS [30]. In TNBS-susceptible mice, acute injury can be produced by low concentrations of TNBS and ethanol and the injury can be visualized within 2 to 3 days. The disease may resolve rapidly compared to rats. Repeat enemas of TNBS may accentuate injury, but oral feeding of TNBS results in significant oral tolerance [30].

In both rats and mice, the disease shows some resemblance to CD with "skip lesions", "cobblestoning", linear ulcers and transmural inflammation. Rats show the presence of giant cells, but mice do not form granulomas. The mechanism of injury involves a trinitrophenyl group, which acts as a hapten by covalently bonding with cell surface proteins and presenting the MHC class II peptide aggregates to $CD4^+$ T cells through antigen presenting cells to induce a $CD4^+$ T cell immune response and cytokine production [26, 31]. Inflamed tissues express increased levels of IL-12 and IFNγ. The elevated level of IFNγ has also been suggested to activate macrophages to produce chemotactic factors and proinflammatory cytokines [26]. IL-12 and IFNγ antibody-treated animals not only prevented the induction of TNBS-induced colitis [26], but the IL-12 antibody was therapeutic in established disease [26]. Also, TNFα mab and rIL-4 down-regulate colitis. These data suggest that TNBS colitis is mediated by a Th1 response and agents that modify this Th1 response help down-regulate colitis.

There is also evidence indicating that in TNBS-induced rat colitis, there is exaggerated production of oxygen-derived free radicals suggesting an alternative pathway for injury [32]. The injury is associated with a substantial increase in myeloperoxidase levels, suggesting that neutrophils play a significant role in the production of acute injury.

In the last decade, over 50 papers addressing the application of this model for the preclinical evaluation of drugs have been published. Table 5 summarizes the mechanisms, mediators and results of different classes of compounds that were tested. The success of these therapies provides valuable comparisons for future drugs to be tested (Tab. 6). The advantages of this model is in the simplicity and relative inexpense of producing colitis. The disadvantages are that the reproducibility of the model is dependent upon the dose of TNBS, lot of TNBS, concentration of ethanol, species and the strain of animals used in individual laboratories. The fundamental therapeutic disadvantage of this model is the lack of therapeutic success observed by some investigators with corticosteroids and 5-ASA which are commonly used in IBD therapy, suggesting an indifference of this model to human IBD.

Dextran sulfate model

In 1985, Ohkusa [33], and in 1990, Okayasu et al. [34], reported two new animal models of colitis in Syrian hamsters and BALB/c mice, respectively, by feeding them 3–10% dextran sulfate sodium solution (DSS) dissolved in drinking water (Tab. 4).

In our laboratory, we further characterized the model by producing colitis in outbred female Swiss Webster mice [35, 36]. The disease was induced by feeding 5% DSS (MW 30–40 kDa, ICN Biochemicals) for 5 days to produce "acute" colitis and 7 days of DSS feeding, followed by 21 days of plain water to produce "chronic" colitis. In addition, long term or cyclic administration of DSS for four cycles produces severe colitis which after 3, 6 and 9 months results in the development of dysplasia or adenocarcinoma in mice, rats, hamsters. Published studies show guinea pigs are more susceptible to DSS injury than any other species studied so far [37]. In our studies [35] we have shown that outbred mice develop lesions in the mid-colon first, followed by the distal colon. Fisher 344 rats [38], BALB/c [34] and CBA/j [34] mice develop left-sided colitis. Conversely, hamsters [39], guinea pigs [37] and Wistar rats [40] develop right-sided colitis. These studies suggest species, strain and site specific susceptibility of the colonic mucosa, which probably uncover certain genetic differences in susceptibility to DSS between and within species.

Evaluation of colitis in this model has utilized both functional scores and histology. Functional score utilizes a method of determining disease activity index based on daily measurement of body weight, stool consistency, and testing for occult blood in the stool or bleeding per rectum [35, 36]. The disease in outbred mice begins to develop 3 days after feeding DSS, reaching peak disease on day 7. The most severe disease seen on day 7 is not ideal for preclinical testing since some of the animals become wasted with severe disease. The disease activity index shows an excellent correlation with crypt architectural changes [35]. The disease spares the small intestine. Histopathological changes, which predate clinical symptoms, include a systematic progressive non-inflammatory crypt dropout, hyalination in the lamina propria and in the pericrypt regions. The earliest changes are focal, involving less than 10% of the mucosal surface and involving two to three crypts. After 4 days of DSS, histological changes become more confluent and diffuse with evidence of significant loss of crypts involving 15–75% of the surface area. There are minimal, mixed inflammatory infiltrates of which granulocytes are more abundant. Monocytes and macrophages are rarely seen at this time. After 5 days of DSS, pathological changes become more confluent with the loss of surface epithelium, the formation of erosions, and the emergence of an early hyperplastic epithelium. Inflammation is florid with granulocytes, lymphocytes and plasma cells, with minimal presence of monocytes and macrophages. Some cryptitis is seen, but crypt abscess are not present. Once DSS feeding is stopped, the animals progress to developing severe chronic inflammation and subsequent dysplasia is seen in some animals. Chronic inflammatory infiltrates seen at this point consist of only a few granulocytes, an abundance of lymphocytes, plasma cells, monocytes and macrophages. Many of the histological changes seen at this stage are reminiscent of human chronic UC. Even 60 days after discontinuing DSS, chronic inflammation and erosions continue to be present providing a large enough window for therapeutic intervention without the inherent risk of self healing. Continuous, cyclic or even one-time

Table 6 - Suggested models for preclinical trials

Model	Advantages/Disadvantages	Successful preclinical applications
Cotton top tamarins (CTT)	Reliable Spontaneous with remission and relapses UC-like features Colitis-associated colon cancer Endoscopy and serial biopsies can be taken to determine endpoints of efficacy Difficult to obtain Expensive Endangered species	Inhibitors of arachidonic acid metabolite synthesis and receptors Anticytokine therapy Antiadhesion molecules Anticancer therapy Conventional IBD therapy
Trinitrobenzenesulfonic acid (TNBS)	Easily produced Can be produced in multiple species Gross morphological changes serve as reliable endpoints, but histology is required for final analysis RATS: Reliable, but reproducibility varies from one laboratory to other CD-like linear ulcers Presence of granulomas and giant cells MICE: Reliable, but individual strain responses vary CD-like transmural inflammation	Inhibitors of arachidonic acid metabolite synthesis and receptors Anti- and pro-cytokine therapies Transcription factor (NFKB) inhibitors Immune mechanisms Nitric oxide inhibitors
Dextran sulfate sodium (DSS)	Easily produced DSS is expensive and batch to batch variations could occur Disease can be produced in multiple species Reliable and reproducible Species, strain susceptibility and disease location vary Disease intensity can be manipulated to produce "acute" or "chronic" disease	Inhibitors of arachidonic acid metabolite synthesis and receptors Anticytokine therapy Antiadhesion molecules Anticancer therapy Genetic studies examining the susceptibility genes leading to gene therapy

Dextran sulfate sodium DSS, continued)	Disease activity indices serve as reliable endpoints, but histology is required for final analysis UC-like chronic mucosal disease leading to colitis-associated dysplasia and colon cancer	
Peptidoglycan polysaccharide (PGPS)	Reliable CD-like lesions can be produced Relapsing chronic granulomatous enterocolitis and arthritis in Lewis rats Only transient intestinal inflammation can be produced in MHC-matched Buffalo and Fisher rats PGPS is expensive and difficult to acquire Disease production requires laparotomy and transmural injection	Antiadhesioon molecules Anticytokines Non-steroidal immunosuppressants Nitric oxide inhibitors
HLA-B27	Reliable, CD-like disease demonstrating extraintestinal involvement Expensive to obtain Disease cannot be produced under germ free conditions Gastrointestinal inflammation and arthritis is seen due to overgrowth of anaerobic bacteria	Antibiotics Cytokine therapies Nitric oxide inhibitors
Indomethacin	Easily produced in rats and not expensive Reliable and reproducible CD-like features; however, the mechanism(s) of "acute" disease is different from "chronic" disease. Only chronic disease responds to antibiotics and conventional therapy Disease cannot be produced under germ free conditions	Inhibitors of arachidonic acid metabolite synthesis and receptors Antibiotics Conventional IBD therapy Anti-adhesion molecules Anti-cytokines Nitric oxide inhibitors
$CD45RB^{hi}$	Reliable Difficult to produce since it requires adoptive transfer of pure $CD45RB^{hi}$ T cells into SCID mice Presence of giant cells, chronic inflammation and ulceration leading to wasting syndrome	Antibiotics Anticytokines

administration of DSS results in the development of dysplasia after 3 months. The incidence of dysplasia may vary by the way DSS is administered, but is usually around 20%.

The mechanism by which DSS induces colonic injury has not been thoroughly investigated. Batch to batch variability in disease severity should be expected and controlled with good quality control studies. The degree of sulfation does not seem to be a contributing factor for disease induction. However, since colonic bacteria can dissociate sulfate from DSS, the free sulfate present in the intestine could act as a substrate to produce H_2S which could significantly interfere with cellular metabolism to induce a toxic effect on the epithelium [41]. Other possible mechanisms include alterations in luminal bacterial ecology and the activation of monocytes, macrophages and mast cells. Germ-free animals [42] produce colitis similar to that observed in non-germ free animals. Thus, bacteria are not primarily involved in producing disease. DSS induces inflammation in athymic nude mice, severely combined immune deficient mice (SCID) [43] and antibody-dependent helper T cell depleted mice [44], indicating that lymphocytes are not involved in the disease induction process. Nevertheless, the early appearance of inflammation and the increased mortality of SCID mice exposed to DSS suggest that T lymphocytes may provide protection at the time of induction of disease. TGFβ2 provides significant therapeutic benefit [45]. The lack of trefoil factor and TGFβ1 exaggerates disease production [46, 47]. The mediators involved and the efficacy of various compounds tested in this model are listed in Table 5. While many classes of compounds showed significant efficacy at various disease levels, it is disappointing that corticosteroids are not effective in this model, suggesting a variance from human disease. However, 5-ASA and congeners of 5-ASA show beneficial effects in established disease, but have lower efficacy levels in "chronic" disease.

The advantage of this model also resides in the ease of producing both "acute" and "chronic" disease by simple modification in the DSS feeding protocol. The disease is highly reproducible. Chronic inflammation in this model lasts for a longer period, permitting the evaluation of the efficacy of compounds without any inherent risk of self-healing. The functional end points are easy to measure, but labor intensive. Since disease activity index shows an excellent correlation with architectural changes, it can be used to quickly screen compounds to facilitate decision-making for thorough screening by histology and mediator measurements.

Most importantly, this model is highly applicable for understanding the multistep neoplastic process involved in colitis associated colon cancer and to test investigational drugs that are chemopreventive or therapeutic to combat colon cancer in high risk IBD patients. The disadvantage of the model is that DSS is very expensive. Batch to batch variations in disease severity could occur due to small molecular weight DSS impurities in the DSS preparation. The disease is characterized by progressive crypt dropout suggesting a direct effect of DSS on the epithelial cells as opposed to lamina propria cells as suggested in human IBD. The disease is patchy, crypt abscesses are infrequent, and corticosteroids are ineffective, making this model not fully relevant to human UC.

Carrageenan model

A low molecular weight (20–40 kDa) carrageenan, a sulfated polysaccharide derived from red seaweed given as a 5% solution, produces colitis in guinea pigs (Tab. 4) [48]. The disease spares small intestine, first appears in the cecum within 1 to 2 weeks after the administration and gradually progresses into distal colon and rectum covering the entire colon. The disease usually disappears after discontinuation of carrageenan. The lesions are localized to the mucosa and lamina propria with significant influx of polymorphonucelar cells, crypt dropouts, non-granulomatous crypt abscess and crypt distortion. The mechanism of injury appears to be related to the presence of bacteria, particularly *Bacteroides vulgatus* [49]. The popularity of this model has somewhat diminished due to several reasons. Carrageenan shows significant variability between preparations, but commercially available polygeenan provide some future hope for minimizing batch to batch variability. The lesions take a long time to develop; it is predominantly limited to the cecum and right colon and is self-limiting. Additionally, the disease is dependent on enteric bacteria, and the disease affects mucosal architecture first before inflammation and immune activation.

Indomethacin model

It is commonly known that certain nonsteroidal antiinflammatory agents such as indomethacin produce gastric and intestinal inflammation and ulceration. There is sufficient evidence to support that the mechanism of injury in this model is mediated by inhibition of prostaglandins that are cytoprotective to the intestinal mucosa. Evidence for this alleged mechanism came from studies in which administration of prostaglandins and their analogs prevented formation of lesions [50]. Further studies showed that luminal bacteria and their byproducts contributed to the emergence of lesions since germ free rats failed to develop severe lesions and antibiotics attenuated the disease in this model [50, 51]. The synergic effect of bile and indomethacin has also been documented to be one of the possible mechanisms of injury in this model. The indomethacin-induced injury seems quite different than the effect of aspirin since aspirin does not induce intestinal injury, but inhibits prostaglandin synthesis. Clearly, the mechanisms of acute injury seem different from chronic injury. Acute injury seems non-immune-related since immunosuppressives are not therapeutic and chronic injury seems related to enteric bacteria. The morphological and histological changes of chronic inflammation and ulceration of the small intestine observed in this model show some resemblance to CD. It appears that the model is easy to produce in rats (mice do not develop lesions) and is reproducible. The model also responds to 5-ASA, a conventional therapy used in IBD.

Miscellaneous chemically-induced models

This group includes peroxinitrite [52], mitomycin [53], iodoacetamide [54] and phorbol-myristate acetate (PMA) models produced in rats (Tab. 4) [55]. Peroxinitrite injury is nitric-oxide mediated. The model may help our understanding of the role played by nitric oxide in IBD. The iodoacetamide and N-methylmaleimide (sulfhydryl alkylator that blocks sulfhydryl groups) model [56] is produced by instilling these compounds in the colon which produces multifocal mucosal erosions and ulcerations which resolves within 3 weeks. The mitomycin model of colitis is another relatively easier model to produce. Injury in this model is produced by intraperitoneal injection of mitomycin c at 2.75 mg/kg in Fisher rats. It produces diffuse colitis affecting mainly the superficial epithelium and causing mucosal barrier disruption. The PMA-induced model of colonic inflammation in rabbits was first described by Fretland et al. in 1990 [55]. Like any other experimental colitis model, it is easy to produce by giving an enema comprising of 1.5 to 3 mg/kg dissolved in 10 ml of 20% ethanol. Colitis in this model lasted for at least 4 days. The mechanisms of injury are probably related to up-regulation of protein kinase C. Neutrophils do not seem to be directly involved in precipitating injury in this model. The model is untested for application to preclinical trials.

Transgenic and knockout models

In the past decade, investigators have used several inbred animals to contain the influence of background genes in the development of colitis. Table 4 lists several of these models. These inbred models are now used to facilitate single gene mutations in major histocompatibility complex (MHC) class I polymorphic molecules such as HLA-B-27, b2-macroglobulin and several cytokine genes to understand their role in regulating T cell and immune functions in various diseases. These mutated, inbred models have contributed to significant breakthroughs and will eventually open avenues to exploit pharmacological testing of many genetic, immunological and cytokine-specific targets of inflammation (Tab. 4).

HLA-B27/b2 microglobulin model

In this transgenic model, rats are genetically engineered to over-express HLA B 27/b2 microglobulin using recombinant DNA technology [57]. These animals develop spontaneous spondyloarthropathies and inflammation of multiple organs including gastrointestinal inflammatory disease. The transgenic lines have been maintained in MHC compatible F33 and Lew backgrounds. Only those homozygous animals which bear the highest gene copies of 21-4H and 33-3 lines develop diarrhea, and other clinical symptoms such as arthritis, myocarditis, uveitis and skin inflammation occur at variable frequency [58]. The disease affects the stomach, small and large intestines, showing some characteristics of Crohn's disease and extraintestinal manifestations. In the small intestine, it produces patchy, non-granulomatous inflammation. In the colon, the disease is diffuse with crypt abscesses and heavy influx of mononuclear cells without any evidence of remission. The onset varies between 6 and 20 weeks with male preponderance of disease. The mechanism of injury has many features of MHC class I antigenic responses. There is a suggestion that bacteria is involved in the disease process in this model.

IL-2 knockout mice

In 1993, Sadlack et al. [59] reported that mice homologous for a disrupted IL-2 cytokine gene (IL-2 –/–) grow normally until they are about 4 to 5 weeks old. They, however, develop colitis when they are about 35 days old. Before they are 9 weeks old, nearly 50% of animals die of anemia. The surviving animals will have normal small intestine, but develop severe colitis, gross bleeding per rectum and rectal prolapse and die within 10 to 25 weeks. The disease has much resemblance to human UC since rectum is more inflamed and then the disease progresses proximally, forming pancolitis. Histologically, the colon shows gross ulcerations, epithelial thickening due to hyperplasia, crypt abscesses and crypt distortion. The lamina propria is filled with both acute and chronic inflammatory infiltrates.

The mechanism of injury does not seem to involve many special pathogens, but involves commensal microorganisms since these animals, when derived in a pathogen-free environment, do not develop colitis. It seems to suggest that disruption of IL-2 gene causes a defect in the immune system. The abnormal immune response is associated with an increase in IL-1, IFNγ and TNFα and a decrease in IL-4 and IL-10 transcripts, suggesting that the immune response is shifted to Th2 response which may ultimately lead to breakdown of self tolerance. Backcross of IL-2 –/– with RAG-2 deficient mice suggests the involvement of T lymphocytes [60]. This is further confirmed by backcrossing IL-2 –/– with β2-microglobulin –/– mice which implicates $CD4^+$ as opposed to $CD8^+$ T cells [61]. In IL-2 –/– mice, IL-12 causes abnormal maturation of thymocytes toward the generation of Th1-type thymocytes capable of mediating colitis and anti-IL-12 mab shows improvement in disease (Tab. 5). The model is now extensively used to study the role played by this cytokine in pathological immune cascades.

IL-10 knockout mice

In 1993, Kuhn et al. [62] first produced IL-10 knockout mice that develop enterocolitis characteristic of CD. These mice also develop normally until they are 3 weeks old. However, at about 7 to 11 weeks of age they lose significant body weight, develop anemia and segemental enterocolitis involving duodenum, jejunum and proximal colon with occassional perforated ulcers. The intestine shows varying degree of inflammation with epithelial hyperplasia cells and crypt branching. The lamina propria is filled with a heavy influx of macrophages, neutrophils and occasional multinucleate giant cells. The mechanism of injury involves common enteric microorganisms since animals housed in a germ-free environment develop mild colitis as opposed to enterocolitis seen in animals housed in pathogen containing environments. Abnormal immune activation is common in this model with increased expression of MHC class II molecules and the model also shows suppression of T-cell tolerance to enteric antigens. Anti-IFNγ antibody inhibits inflammation [63] suggesting a Th1 response contributing to the disease process (Tab. 5).

T cell receptor knockout mice

The functional T cell receptor consists of a heterodimeric surface receptor which encompasses an α and β chain or a γ and δ chain. Mice with a disrupted gene at the α or β locus are selectively deficient of the αβ receptor. Likewise, mice with a mutated gene at the δ locus lack the gd receptor. T cell receptor knockout (αβ and γδ –/–) mice were produced by creating a double mutant from crossing β mutant mice with δ mutant mice [64]. Control mice deficient in mature T and B cells were also produced by mutating the recombination activating gene (RAG-1). The TCR α and β mutant mice develop normally for 12 to 16 weeks and they later develop chronic colitis reminiscent of human UC in 30% of animals in a year. They have diarrhea, weight loss and rectal prolapse, leading to high mortality due to wasting syndrome. Gross examination of the colon shows dilatation and thickening, mostly in the rectum and in the most severe cases, the entire colon is affected. The small intestine is not affected. Histologically, particularly in the rectum, crypts are elongated, branched and distorted. Both acute and chronic inflammatory infiltrates are seen in the lamina propria with occasional crypt abscesses. The muscularis mucosa or submucosa is usually spared of inflammation unless the disease is very severe. The mechanism by which these animals develop colitis is unknown. It appears that there is an overproduction of IL-1α and IL-1β, but not TNFα or TGFβ1 in the colonic mucosa, since monoclonal antibodies specific to these cytokines suppressed epithelial cell proliferation and increased influx of T cells in TCRα –/– knockout mice. Lymphocytes with γδ receptors are not involved in inducing the disease. The control nude, RAG-1 deficient mice, housed in the same pathogen-free environment, do not develop disease. A specific pathogen has not yet been isolated from TCR knockout animals.

Gαi2 knockout mice

The α, β and γ chains of G proteins play an important role in signal transduction processes. Mice with disrupted Gαi2 develop pancolitis when they are 8 to 12 weeks old [65]. They show stunted growth and gradually 75% of the animals go into wasting syndrome by the time they are 35 weeks old. The disease is very severe in the distal colon and they commonly develop rectal prolapse. Histologically, these animals develop acute and chronic inflammation with mild crypt elongation, crypt abscess, crypt distortion and gross ulcerations. During a more advanced stage of disease, fibrosis and thickening of the bowel wall become very prominent. During the course of the disease these animals also develop adenocarcinoma involving both mucosa and submucosa. The small bowel is usually normal. The mechanism by which the disease is produced remains unknown. There is evidence to suggest that the genetic background of mice dictates the degree of disease with Gαi2 –/– mutant on 129 sv background producing the most severe disease, whereas, the same mutation on 129sv × C57Bl6 crossbreed produces no disease at all.

TNFα 3′ deletion

Comminelli et al. [66] recently showed that mice carrying an endogenous deletion of the 3'-AU-rich region of the TNFα gene develop a new CD-like phenotype which express disease when they are 3 weeks old, with severe inflammation developing by 8 weeks. Mice developed transmural inflammation particularly in the ileum showing the presence of villous blunting without superficial ulceration and crypt abscess. They also express TNFα in the tissue. This is the first experimental model to show that mutational disregulation of TNFα plays a role in the development of intestinal lesions similar to those seen in CD.

Miscellaneous transgenic and knockout mice

In addition to these well characterized gene knockout models, other mutant cytokine knockout models such as TGFβ1 knockout [67]. IL-7 transgenic [68] and keratin 8 mice have also been produced (Tab. 4) [69]. The TGFβ1 deficient mice produce multiorgan inflammation including stomach and the colon probably due to mucosal barrier disruption caused by lack of TGFβ1. The IL-7 transgenic mice [68] which overexpress IL-7 mRNA in the colonic mucosa and IL-7 receptors on lymphocytes, developed colitis when mice are 8–12 weeks old. These animals show intestinal bleeding and appearance of rectal prolapse. Histologically, the mucosa shows eorosions, crypt abscesses, goblet cell depletion and thickening of the bowel wall. The mechanism involved in producing disease in this model remains to be elucidated, but may involve IL-7 produced from colonic epithelial cells. In the keratin-8 gene disrupted FVB/N mice [69], inflammation, hyperplasia and crypt architectural changes are seen in the cecum, colon and rectum. No homozygous mouse line was established at this time since this gene mutation causes embryonic death in C57Bl/6 × 129sv mice.

Lastly, there is an expensive, complicated and time-consuming autoimmune model of cyclosporine A-induced colitis in mice [70]. Colitis in this model is T cell mediated and appears to occur cooperatively with intestinal bacterial flora.

T cell and bone marrow transfer models

In 1993, Morissey et al. [71] and Powrie et al. [72] demonstrated that adoptive transfer of CD45RBhiCD4$^+$ T cell population from BALB/c or C.B-17 mice into C.B-17 severely combined immune dificicency (SCID) mice results in colitis and wasting syndrome beginning 3 to 5 weeks after transfer of cells. Cotransfer of CD45RBloCD4$^+$ T or unfractionated CD4$^+$ T cells with CD45RBhiCD4$^+$ T cells protects these animals. This is probably mediated by TGFβ, but not by IL-4 [73]. These animals show an enlarged and thickened colon associated with epithelial hyperplasia, goblet cell depletion, loss of crypts, ulceration and multifocal chronic inflammation with granulomas and multinucleated giant cells. This pathologic change associates with increased MHC class II expression and increases IFNγ levels. The model is new and exciting, but difficult to produce in laboratories which are not equipped with a sophisticated FACStar Plus flow cytometer. It requires very high purity (>98%; 10 to 50 000) CD45RBhiCD4$^+$ T cells to be injected intravenously. Anticytokine and cytokine therapies are effective (Tab. 5).

In addition to the CD45RBhiCD4$^+$ T model, mice that transgenically overexpress the human CD3e26 gene (Tgε26) are athymic and are deficient in T and NK cells. When syngenic bone marrow is infused from wild-type animals, Tgε26 mice develop colitis and wasting syndrome [74]. Infusion of NK cells and transplant of neonatal thymus before bone marrow transfer affords protection [75]. These mice also produce high levels of IFNγ and TNFα. Selective inhibition of these cytokines results in complete abrogation of inflammation in this model.

These studies suggest that rats and mice with immunoregulatory defects, either by over-expression of certain cytokines and major histocompatibility molecules or the lack of certain cytokine and signal transduction molecules, develop severe spontaneous colitis.

Conclusion

In the last decade, there has been an explosion of drugs produced to combat inflammation. At the same time, animal models of inflammation that relatively mimic many human diseases have significantly increased, thus enhancing our capability to test the efficacy of drugs in several models. Since all animal models express inflammation after an external insult, they generally seem fit to test any antiinflammatory agent. Unfortunately, most animal models respond to many antiinflammatory therapies, but only a few models help determine whether such observation can be extended to human disease. Animal models that have been successfully used to test different pharmacological agents are included in Table 6. The description of these models clearly suggest that investigators must take into consideration the species, strains, substrains and uniformity in the mode of induction, to reproduce the disease in their laboratory. Only when these criteria are clearly met should they proceed with preclinical investigations. Otherwise, these variables could affect overall therapeutic outcome. Many inducible and genetic models of intestinal inflammation are now available for target-specific therapies. Since they are tailor-made to observe pathological changes when certain specific genes are disrupted, their applicability to test all antiinflammatory agents at present is debatable. Therefore, models of successful therapies in relation to specific mediators are described in Table 6. Application of these therapies has not only helped us understand these models in pharmacological terms, but also helped us discretely select models to determine therapeutic success, second species verifications and to extended successful candidate drugs for human clinical trials.

References

1 Crohn BB, Ginzburg L, Oppenheimer GD (1932) Regional ileitis: a pathologic and clinical entity. *JAMA* 99: 1323–1329

2 Wilks S, Moxon W (1975) *Lectures on pathological anatomy*, 2nd ed. J and A Churchill, London

3 Levin B (1992) Inflammatory bowel disease and colon cancer. *Cancer* 70: 1313–1316

4 Pfeiffer CJ (ed) (1985) *Animal models of intestinal disease.* CRC Press, Boca Raton, FL

5 Sundberg JP, Elson CO, Bedigian CD, Berkenmeir EH (1994) Spontaneous heritable colitis in a new substrain of C3H/HeJmice. *Gastroenterology* 107: 1726–1735

6 Madara JL, Podolsky DK, King NW, Sehgal PK, Moore R, Winter HS (1985) Characterization of spontaneous colitis in cotton top tamarin (*Saguinas oedipus*) and its response to sulfasalzine. *Gastroenterology* 88: 13–19

7 Chalifoux, LV, Bonson RT (1981) Colonic adenocarcinoma associated with chronic

ulcerative colitis in cotton topped marmosets (*Saguinas oedipus*). *Gastroenterology* 80: 942–946

8 Das KM, Squillante L, Henke M, Clapp N (1990) The presence of circulating antibodies in cotton top tamarins (CTT) with spontaneous colitis against an epitope on MR 40,000 protein shared by human and CTT colon epithelial cells. *Gastroenterology* 98: A468

9 Stonerook MJ, Weiss HS, Rodriguez JV, Hernandez JI, Peck PC, Wood JD (1994) Temperature-metabolism relations in the cotton-top tamarin (*Sanguinas oedipus*) model for ulcerative colitis. *J Med Primatol* 23: 16–22

10 Watkins DI, Hodi FS, Levin NL (1988) A primate species with limited major histocompatibility complex class I polymorphism. *Proc Natl Acad Sci USA* 85: 7714

11 Clapp N, Henke M, Hansard R, Carson R, Walsh R, Widomski D, Anglin C, Fretland D (1993) Inflammatory mediator changes in cotton top tamarins (CTT) after SC-41930 anti-colitic therapy. *Agents Actions* 39: C8–10

12 Kosiewicz MM, Krishnan A, Shah M, Bentz M, Matusmoto S, Cominelli F (1998) Characterization of a new spontaneous murine model of inflammatory bowel disease. *Gastroenterology* 114: G4143

13 Hodgson HJF, Potter BJ, Skinner J, Jewell DP (1978) Immune complex mediated colitis in rabbits. *Gut* 19: 225–232

14 Mee AS, McLaughlin JE, Hodgson HGF, Jewell JP (1979) Chronic immune colitis in rabbits. *Gut* 20: 1–5

15 Axelsson LG, Ahlstedt S (1990) Characteristics of immune-complex induced chronic experimental colitis in rats with a therapeutic effect of sulfasalazine. *Scan J Gastroenterol* 25: 203–209

16 Cominelli F, Nast CC, Clark BD, Schindler R, Lierena R, Eysselein VE, Thompson RC (1990) Interleukin 1 (IL-1) gene expression, synthesis, and effect of specific IL-1 receptor blockade in rabbit immune complex colitis. *J Clin Invest* 86: 972–980

17 Cominelli F, Nast CC, Llerena R, Dinarello CA, Zipser RD (1990) Interleukin I suppresses inflammation in rabbit colitis. Mediation by endogenous prostaglandins. *J Clin Invest* 85: 582–586

18 Rosenberg EW, Fisher RW (1964) DNCB allergy in the guinea pig colon. *Arch Dermatol* 89: 99–103

19 Ward JM, Anver MR, Haines DC, Melhorn JM, Gorelick P, Yan L, Fox JG (1996) Inflammatory large bowel disease in immunodeficient mice naturally infected with *Helicobacter hepaticus*. *Laboratory Animal Science* 46: 15–20

20 Quinn TC, Goodell SE, Mkrtichian EE, Schuffler MD, Wangf SP, Stamm WE, Holmes KK (1981) *Chlammydia trachomatis practitis*. *N Engl J Med* 305: 195–200

21 Sartor RB, Cromartie WJ, Powell DW, Schwab JH (1985) Granulomatous enterocolitis induced in rats by purified bacterial cell wall fragments. *Gastroenterology* 89: 587–595

22 Yamada T, Zimmerman T, Specian RD, Grisham MB (1993) Chronic granulomatous colitis induced by intramural injection of Freund's complete adjuvant. *Gastroenterology* 104: A804

23 MacPherson BR, Pfeiffer CJ (1978) Experimental production if diffuse colitis in rats. *Digestion* 17: 135–150

24 Strober W (1985) Animal models of inflammatory bowel disease-an overview. *Dig Dis Sci* 30: 3S–10S

25 Morris GP, Beck PL, Herridge MS, Depew WT, Szewczuk MR, Wallace JL (1989) Hapten-induced model of chronic inflammation and ulceration in the rat colon. *Gastroenterology* 96: 795–803

26 Neurath MF, Fuss I, Kelsall BL, Stuber E, Strober W (1995) Antibodies to interleukin-12 abrogate established experimental colitis in mice. *J Exp Med* 182: 1281–1290

27 Miller MJS, Sadowka-Kowicka H, Chotinnaueml S, Kakkis JL, Clark DA (1993) Amelioration of chronic ileitis by nitric oxide synthesis inhibition. *J Pharmacol Exp Ther* 264: 11–16.

28 Shibata Y, Taruishi M, Ashida T (1993) Experimental ileitis in dogs and colitis in rats with trinitrobenzenesulfonic acid-colonoscopic and histopathologic changes. *Gastroenterologia Japonica* 28: 518–527

29 Goldhill JM, Burakoff R, Donovan V, Rose K, Percy WH (1993) Defective modulation of colonic secretomotor neurons in a rabbit model of colitis. *Amer J Physiol* 264: G671–G677

30 Elson CO, Beagley KW, Sharmanov AT, Fujihashi K, Kiyono H, Tennyson GS, Cong Y, Black, CA, Ridwan BW, McGhee JR (1996) Hapten-induced model of murine inflammatory bowel disease: mucosa immune responses and protection by tolerance. 157: 2174–2185

31 Cavini A, Hackett CJ, Wilson KJ, Rothbard JB, Katz SI (1995) Characterization of epitopes recognized by hapten-specific $CD4^+$ T cells. *J Immunol* 154: 1232–1238

32 Grisham MB, Volkmer C, Tso P, Yamada T (1991) Metabolism of trinitrobenzenesulfonic acid by the rat colon produces reactive oxygen species. *Gastroenterology* 101: 540–547

33 Ohkusa T (1985) Production of experimental ulcerative colitis in hamsters by dextran sulfate sodium and a change of intestinal microflora. *Jpn J Gastroenterol* 82: 1327–1336

34 Okayasu I, Hatekeyama S, Yamada M, Ohkusa T, Inagaki Y, Nakaya R (1990) A novel method in the induction of reliable experimental acute and ulcerative colitis in mice. *Gastroenterology* 98: 694–702

35 Cooper HS, Murthy SNS, Shah RS, Seergran DJ (1993) Clinicopathologic study of dextran sulfate sodium experimental murine colitis. *Lab Invest* 69: 238–249

36 Murthy SN, Cooper HS, Shin H, Shah RS, Ibrahim SA, Sedergran DJ (1993) Treatment of dextran sulfate sodium-induced murine colitis by intracolonic cyclosporin. *Dig Dis Sci* 38: 1722–1734

37 Iwanaga T, Hoshi O, Han H, Fujita T (1994) Morphological analysis of acute ulcerative colitis experimentally induced by dextran sulfate sodium in the guinea pig: possible mechanisms of cecal ulceration. *J Gastroenterol* 29: 430–438

38 Domek MJ, Iwata F, Blackman EI, Kao J, Baker M, Vidrich A, Leung FW (1995) Anti-

neutrophil serum attenuates dextran sulfate sodium-induced colonic damage in the rat. *Scand J Gastroenterol* 30: 1089–1094

39 Yamada T, Ohkudas T, Okayasu I (1992) Occurrence of dysplasia and adenocarcinoma after experimental chronic ulcerative colitis in hamsters induced by dextran sulfate sodium. *Gut* 33: 1521–1527

40 Tamaru K, Kobayashi H, Koshimoto S, Kajiyama G, Shimamoto F, Brown WR (1993) Histochemical study of colonic cancer in experimental colitis in rats. *Dig Dis Sci* 38: 529–537

41 Roediger WE, Moore J, Babdige W (1997) Colonic sulfide in pathogenesis and treatment of ulcerative colitis. *Dig Dis Sci* 42: 1571–1579

42 Bylund-Fellenius AC, Landstrom E, Axelsson LG, Midtvedt T (1994) Experimental colitis induced by dextran sulfate in normal and germfree mice. *Microbial Ecology in Health and Disease* 7: 207–215

43 Dielman LA, Ridwan BU, Tennyson GS, Beakley KW, Elson CO (1994) Dextran sulfate sodium (DSS)-induced colitis occurs in severe combined immunodeficient (SCID) mice. *Gastroenterology* 107: 1722–1734

44 Axelsson L-G, Landstrom E, Goldschmidt TJ, Gronberg A, Bylund-Fellinius A-C (1996) Dextran sulfate sodium (DSS) induced experimental colitis in immunodeficient mice. Effects in CD4+-cell depleted, athymic and NK cell-cell depleted mice. *Inflamm Res* 45: 181–191

45 Murthy SN, Cooper HS, Coppola D, Barrish S, McKibbin R. Cerletti N DiMuzzio J (1992) Transforming growth factor β2, but not epidermal growth factor yields protection against dextran sulfate-mediated colitis in mice. *Gastroenterology* 102: A669

46 Mashimo H, Wu DC, Podolsky DK, Fishman MC (1996) Impaired defense of intestinal mucosa in mice lacking intestinal trefoil factor. *Science* 274: 204

47 Egaer B, Procaccino F, Laksmanan J, Reinshagen M, Hoffman P, Patel A, Reuben W, Gnanakkan S, Liu L, Barajas L, Eyesselein VE (1997) Mice lacking transforming growth factor alpha have an increased susceptibility to dextran sulfate-induced colitis. *Gastroenterology* 113: 825–832

48 Onderdonk AB (1985) The carrageenan model of experimental ulcerative colitis. *Prog Clin Biol Res* 186: 237–245

49 Onderdonk AB, Bronson R, Cisneros R (1987) Comparison of *Bacteroides vulgatus* strains in the enhancement of experimental ulcerative colitis. *Infect Immun* 55: 835–836

50 Yamada T, Deitch E, Specian RD, Perry MA, Sartor RB, Grisham MB (1993) Mechanisms of acute and chronic intestinal inflammation induced by indomethacin. *Inflammation* 17: 641–662.

51 Banerjee AK, Peters TJ (1990) Experimental non-steroidal anti-inflammatory drug-induced enteropathy in the rat: similarities to inflammatory bowel disease and effect of thromboxane synthesis inhibitors. *Gut* 31: 1358–1364

52 Rachmilewicz D, Stanler JS, Karmeli F, Mullins ME, Singel DJ, Loxcalzo J, Xavier RJ, Podolsky DK (1993) Peroxynitrite-induced rat colitis. A new model of colonic inflammation. *Gastroenterology* 105: 1681–1688

53 Keshavarzian A (1992) Mytomicin C-induced colitis in rats: A new model of acute colonic inflammation implicating reactive oxygen species. *J Lab Clin Med* 120: 778–791

54 Rachmilewicz D, Karmeli F, Okon E (1995) Sulfhydryl blocker-induced rat colonic inflammation is ameliorated by inhibition of nitric oxide synthase. *Gastroenterology* 109: 98–106

55 Fretland DJ, Widomski DL, Levin S, Gaginella TS (1990) Colonic inflammation in the rabbit induced by phorbol-12-myristate-13-acetate. *Inflammation* 14: 143–150

56 Satoh H, Sato F, Takami K, Szabo S (1997) New ulcerative colitis model induced by sulfhydryl blockers in rats and the effects of antiinflammatory drugs on the colitis. *Jap J Pharmcol* 73: 299–309

57 Hammer RE, Maika SD, Richardson JA, Tang JP, Taurog JD (1990) Spontaneous inflammatory disease inflammatory disease in transgenic rats expressing HLA-B27 and human beta 2 m: an animal model of HLA-B27 associated disorders. *Cell* 63: 1099–1112

58 Taurog JD, Maika SD, Simmons WA, Breban M, Hammer RE (1993) Susceptibility to inflammatory disease in HLA-B27 transgenic rat lines correlates with the level of B27 expression. *J Immunol* 150: 4168–4178

59 Sadlack B, Merz H, Schorle H, Schimpl A, Feller AC, Horak I (1993) Ulcerative colitis-like disease in mice with a disrupted interleukin-2 gene. *Cell* 75: 253–261

60 Ma A, Datta M, Margosian E, Chen J, Horak I (1995) T cells, but not B cells, are required for bowel inflammation in interleukin deficient mice. *J Exp Med* 182: 1567–1572

61 Simpson SJ, Mizoguchi E, Allen D, Bahn AK, Terhrst C (1995) Evidence that $CD4^+$ but not $CD8^+$ T cells are responsible for bowel inflammation in interleukin-2 deficient colitis. *Eur J Immunol* 25: 2618–2625

62 Kuhn R, Lohler J, Rennick D, Rajewsky K, Muller W (1993) Interleukin-10 deficient mice develop chronic enterocolitis. *Cell* 75: 263–274

63 Berg DJ, Davidson N, Kuhn R, Muller W, Menon S, Holland G, Thompson-Snipes L, Leach MW, Rennick D (1996) Enterocolitis and colon cancer in interleukin-10-deficient mice are associated with aberrant cytokine production and $CD4^+$ Th1-like responses. *J Clin Invest* 98: 1010–1020

64 Mombaeerts P, Mizoguchi E, Grusby MJ, Glimcher LH, Bhan AK, Tonegawa S (1993) Spontaneous development of inflammatory bowel disease in T cell receptor mutant mice. *Cell* 75: 275–282

65 Rudolph U, Finegold MJ, Rich SS, Harriman GR, Srinivasan Y, Brabet P, Boulay G, Bradley A, Bimbauer L (1995) Ulcerative colitis and adenocarcinoma of the colon in Gαi2-deficient mice. *Nature Genet* 10: 141–148

66 Cominelli, F, Kontoyiannis D, Pizzaro TT, Kollias G (1998) Mice carrying an endogenous deletion of the 3'-AU-rich region of the TNFα gene develop a Crohn's disease-like phenotype: A key role of the TNFα in the pathogenesis of chronic intestinal inflammation. *Gastroenterology* 114: G3911

67 Diebold RJ, Eis MJ, Yin MY, Ormsby I, Boivin GP, Darrow BJ, Saffitz JE (1995) TNFα

Doetschman T. Early onset of multifocal inflammation in the transforming growth factor β1-null mouse is lymphocyte mediated. *Proc Natl Acad Sci* 92: 12215–12219
68 Ueno Y, Watanabe M, Yamazaki M, Yajima T, Ishiii H, Uehira M, Nishimoto H, Hata J, Hibi T (1996) Interleukin-7 transgenic mice develop chronic colitis with overexpression of IL-7 mRNA in the colonic mucosa. *Gastroenterolgy* 110: A1033
69 Baribault H, Penner J, Iozzo RV, Wilson-Heiner M (1994) Colorectal hyperplasia and inflammation in keratin 8-deficient mice. *Genes and Development* 8: 2964–2973
70 Bucy RP, Xu XY, Li J, Huang GQ (1993) Cyclosporin A-induced autoimmune disease in mice. *J Immunol* 151: 1039–1050
71 Morissey PJ, Charrier K, Braddy S, Liggit D, Watson JD (1993) $CD4^+$ T cells that express high levels of CD45RB induce wasting disease when transferred into congenic severe combined immune mice. *J Exp Med* 178: 237–244
72 Powrie F, Leach MW, Mauze S, Caddle LB, Coffman RL (1993) Phenotypically distinct subsets of $CD4^+$ T cells induce or protect from chronic intestinal inflammation in C.B-17 SCID mice. *Int Immunol* 5: 1461–1471
73 Powrie F, Carlino J, Leach MW, Mauze S, Coffman RL (1996) A critical role of transforming growth factor-β but not interleukin-4 in the suppression of T helper type-1-mediated colitis by CD45RBlow $CD4^+$ T cells. *J Exp Med* 183: 2669–2674
74 Simspon SJ, Hollander G, Mizoguchi E, Bahn A, Terhorst C (1995) T lymphocytes in murine inflammatory bowel disease. *Clin Immunol Immunopathol* 76: S45–S46
75 Wang B, Shah SA, Simpson SJ, Allen D, Biron CA, Hollander GA, Terhorst C (1996) Protective role of natural killer cells in a mouse model of inflammatory bowel disease. *Gastroenterology* 110: A1042
76 Sterberg PE, Winsor-Hines D, Briskin MJ, Soleer-Ferran D, Merrill C, Mckay CR, Newman W, Ringer DJ (1996) Rapid resolution of chronic colitis in the cotton top tamarin with an antibody to a gut homing integrin alpha 4 beta 7. *Gastroenterology* 111: 1373–1380
77 Watkins PE, Warren BF, Stephens S, Ward P, Foulkes R (1997) Treatment of ulcerative colitis in the cotton top tamarin using antibody to tumor necrosis factor alpha. *Gut* 40: 628–633
78 Axelsson L-G, Ahlstedt S (1990) Characteristics of immune-complex-induced chronic experimental colitis in rats with a therapeutic effect of sulphasalzine. *Scand J Gastroenterol* 25: 203–209
79 Cassini-Raggi V, Herbert C, Monsacchi L, Cominelli F (1994) A specific monoclonal antibody (MoAb) against interleukin-8 (IL-8) suppress inflammation in rabbit immune colitis. *Gastroenterology* 106: A661
80 Neurath MF, Pattesson S, Meyer zum Buschenfelde KH, Strober W (1996) Local administration of antisense phosphorothioate oligonucleotide to the p65 subunit of NF-kappa B abrogates established experimental colitis in mice. *Nature Medicine* 2: 998–1004
81 Neurath MF, Fuss I, Pasparakis M, Alexopoulou L, Haralambous S, Meyer zum Buschenfelde KH, Strober W, Kollias G (1997) Predominant pathogenic role of tumor necrosis factor in experimental colitis in mice. *Eur J Immunol* 27: 1743–1750

82 Bertran X, Mane J, Fernandez-Banares F, Castella E, Bartoli R, Ojanguren I, Esteve M, Gassull MA (1996) Intracolonic administration of zileuton, a selective 5-lipoxygenase inhibitor, accelerates healing in a rat model of chronic colitis. *Gut* 38: 899–904
83 Wallace JM, Keenan CM (1990) An orally active inhibitor of leukotriene synthesis accelerates healing in a rat model of colitis. *Am J Physiol* 258: G527–534
84 Vilaseca J, Salas A, Guarner F, Rodriguez R, Martinez M, Malagelada J (1990) Dietary fish oil reduces progression of chronic inflammatory lesions in a rat model of granulomatous colitis. *Gut* 31: 539–544
85 Wallace JL (1988) Release of platelet activating factor (PAF) and accelerated healing induced by a PAF antagonist in an animal model of chronic colitis. *Can J Physiol Pharmacol* 66: 422
86 Luck MS, Bass P (1993) Effect of epidermal growth factor on experimental colitis in the rat. *J Pharmacol Therap* 264: 984–990
87 Zeeh JM, Procaccino F, Hoffman P, Aukerman SL, McRoberts JA, Soltani S, Pierce GF, Lakshamnan J, Lacey D, Eysseilein VE (1996) Keratinocyte growth factor ameliorates mucosal injury in an experimental model of colitis in rats. *Gastroenterology* 110: 1077–1083
88 Rachmilewicz D, Karmeli F, Okon E, Bursztyn M (1995) Experimental colitis is ameliorated by inhibition of nitric oxide synthase activity. *Gut* 37: 247–255
89 Murthy S, Cooper HS, Coppola D, Shirer R (1992) Interleukin receptor-1 receptor antagonist is effective against dextran sulfate (DSS)-mediated colitis in mice. *Gastroenterology* 102: A669
90 Kojouharoff G, Hans W, Obermeir F, Mannel DN, Andus T, Scholmerich J, Gross V, Falk W (1997) Neutralization of tumor necrotic factor (TNF) but not IL-1 reduces inflammation in chronic dextran sulfate sodium-induced colitis in mice. *Clin Exp Immunol* 107: 353–358
91 Murthy SNS, Fondacaro JD, Murthy NS, Cooper HS, Bolkenius F (1994) Beneficial effect of MDL 73404 in dextran sulfate mediate colitis. *Agents Actions* 41 (special conference): C233–234
92 Murthy S, Murthy NS, Coppola D, Wood DL (1997) The efficacy of BAY y 1015 in dextran sulfate model of mouse colitis. *Inflamm Res* 46: 224–233
93 Onderdonk AB, Cisnaros RL, Bronson RT (1983) Enhancement of experimental ulcerative colitis by immunization with *Bacteriodes vulgatus*. *Infect Immun* 42: 783–788
94 Keshaverzian A. Morgan G, Sedghi S, Gordon JH, Doria A (1990) Role of reactive oxygen metabolites in experimental colitis. *Gut* 31: 786–790
95 Thomas TK, Will PC, Srivatsava A, Wilson CL, Harbison M, Little J (1991) Evaluation of an interleukin-1 receptor antagonist in the rat acetic acid-induced colitis model. *Agents Actions* 34: 187–190
96 Rachmilewitz D, Karmeli F, Okon E, Bursztyn (1995) Experimental colitis is ameliorated by inhibition of nitric oxide synthase activity. *Gut* 37: 245–247
97 Fedorak RN, Empey LR, Macarthur C, Jewell LD (1990) Misoprostol provides a

colonic mucosal protective effect during acetic acid-induced colitis in rats. *Gastroenterology* 98: 615–622

98 Keshavarzian A, Maydek J, Zabihi R, Doria M, D'Astice M, Sorensen JRJ (1992) Agents capable of eliminating reactive oxygen species: Catalase, WR-2721 or Cu(II)2 (3,5-DIPS) decrease experimental colitis. *Dig Dis Sci* 37: 1866–1873

99 Fang W, Broughton A, Jacobson ED (1977) Indomethacin-induced intestinal inflammation. *Dig Dis* 22: 749–760

100 Banarjee AK, Peeters TJ (1990) Experimental non-steroidal antiinflammatory drug-induced enteropathy in the rat: Similarities to inflammatory bowel disease and effect of thromboxane synthetase inhibitors. *Gut* 31: 1358–1364

101 Stenson WF (1986) Role of lipoxygenase products in inflammatory bowel disease. In: D Rachmilewitz (ed): *Inflammatory bowel diseases.* Martinus Nijhoff, The Hague, 95–104

102 Conner EM, Brand S, Davis JM, Laroux FS, Palombella VJ, Fuseler JW, Kang DY, Wolf RE, Grisham MB (1997) Proteasome Inhibition attenuates nitric oxide synthase expression, VCAM 1 transcription and the development of chronic colitis. *J Exp Pharmacol Therap* 282: 1615–1622

103 Ehrhardt RO, Ludviksson BR, Gray B, Neurath M, Strober W (1997) Induction and prevention of colonic inflammation in IL-2 deficient mice. *J Immunol* 158: 566–573

T cell-mediated diseases of immunity

Elora J. Weringer and Ronald P. Gladue

Department of Immunology, Central Research Division, Pfizer Inc., Box 1189, Eastern Point Road, Groton, CT 06340, USA

Introduction

The understanding of the immunological processes involved in T cell-mediated disease in humans has greatly benefited from the study of similar processes in animals [1]. Whether induced by a specific antigen or spontaneously occurring, all have lent insight to and sometimes confused the puzzle of the etiology and pathogenesis of human disease. This chapter emphasizes three major examples of organ-specific and systemic animal models of T cell-mediated disease in mice. Organ-specific diseases are either antigen-induced, as in experimental allergic encephalomyelitis (EAE), experimental allergic neuritis (EAN), transplant allograft rejection, or graft versus host disease (GVHD), or occur spontaneously as for models of insulin-dependent diabetes mellitus (IDDM). The primary animal model of systemic autoimmune disease is a spontaneously occurring syndrome in inbred mice which resembles human systemic lupus erythematosus (SLE). The goal is to provide a descriptive analysis of a few select models with regard to what is known of the etiology and pathogenesis and to afford opportunities for evaluation of new therapeutic targets and mechanisms for the treatment of human immunological disorders which are the result of T cell regulatory processes. Finally, since animal models are constructs either experimentally or from nature as prototypes of human disease, it should be noted that there are as many subtle differences as there are overt similarities. This should always be kept within the investigators' hypotheses and interpretations.

Autoimmune disease states

Utility of animal models

The relevance of laboratory animal models is in the understanding of mechanisms and conditions of development of disease and the ability to perform procedures which for ethical or practical reasons cannot be performed in man. For the investi-

In Vivo Models of Inflammation, edited by D.W. Morgan and L.A. Marshall

gator involved in research to explore unique therapeutic targets and develop novel agents for autoimmune diseases attributed to T cell-mediated immune mechanisms, a wide variety of animal models are available. Not any one model completely mimics human disease [1], as species and strain differences as well as modes of induction (spontaneous vs. induced) introduce differences in the pathogenesis and phenotypic expression of the disease. The choice of one or more animal models to explore and evaluate a disease mechanism can provide pertinent preclinical information on which to base possible strategies and recognize potential pitfalls of new mechanism-based therapies.

An advantage is that the model disease state usually has a predictable course and progression that lend some insight into the mechanism involved. One usually selects a model pertinent to the disease entity to be studied or an area of proposed therapeutic intervention. Models can exhibit antigen-specific, organ-specific, or systemic autoimmune phenomena and hence must be selected with a disease aspect clearly identified. Since many of the autoimmune diseases occurring in man are multiple (polyautoimmune) [2] with probably more than one unidentified antigen, models must be used cautiously and selectively to study the initiation, progression and outcome, and associated pathologies of the disease state. Also, with certain models, species and strain selection can provide genetic insights to study the degree of risk for disease development and identify specific susceptibility traits for developing therapies. Early treatment can intervene before disease symptomology and progression are evident or irreversible.

Thus, the role of animal models of T cell-mediated diseases is to mimic or replicate the process of human diseases and to evaluate possible stages of the disease. They can be instructive as to induction, etiology, pathogenesis and response to therapy as well as lend specific insights into the biological responses of the immune system. The biological responses will be discussed more fully in consideration of genetically altered/modified models whereby a part of the immune system has been modified or deleted and the resultant effects studied.

Advantages and disadvantages are inherent in all animal models and should be considered in the primary objective. The T cell-mediated models are either experimentally induced, antigen-driven or spontaneous, naturally occurring, genetically determined models (Tab. 1). Thus, induced models, such as EAE, EAN, myocarditis, thyroiditis, and nephritis have the advantage of a known, often well-defined antigen (epitope), which are adjusted to appropriate controls and have an established time of onset and a predictable course [3, 4]. The disadvantages are that most of these models require the use of adjuvants with controlled proinflammatory effects and responses that may have arguably less biologic relevance to human disease [1, 4].

With the advent of spontaneous models, a significant advantage is the close approximation to human disease often accompanied by multiple organ system responses. The obvious disadvantages lie in the uncertainty of onset, initiating antigen, disease course, and often a variable and unpredictable pathogenesis. Neverthe-

Table 1 - Animal models of T cell-mediated diseases

Experimentally-induced Antigen-driven organ-specific	Spontaneous, genetically-induced Organ-specific/systemic
Experimental allergic encephalomyelitis	Nonobese diabetic (NOD) mouse
Experimental allergic neuritis	Systemic lupus erythematosus (SLE)
Myocarditis	NZB, (NZB×NZW) MRL/lpr mice
Thryoiditis	BB Rat
Nephritis	
Advantages	**Advantages**
well-defined exogenous antigens	closer to human disease
established time of onset	multiple organ system involvement
predictable, monitored course	genetic relationships
Disadvantages	**Disadvantages**
induction requires stimulus (adjuvant)	undetermined initiating antigen
relevance to human disease uncertain	uncertainty of onset, variable/unpredictable pathogenesis

less, spontaneous autoimmune models both organ-specific (diabetes in the bio-breeding (BB) rat and the nonobese diabetic (NOD) mouse) and systemic (murine lupus in New Zealand black (NZB) mice) disease have been instrumental in the advance of knowledge and therapeutic regimens for similar human diseases and have contributed insight into the multiplicity of events/factors which precipitate autoimmune disease.

The definition of genetic susceptibility, which is silent until triggered by factors and/or agents to induce overt disease phenotype, has emphasized the genetics of autoimmune diseases. The genetics imparts a susceptibility, or predisposition, for disease initiation following an unknown sequence of cumulative effects, which may be related to the immune system, the target organ, or a combination of the two. There are solid data to support the genetic regulation of the immune response through T cell receptor (TCR) and immunoglobulin (IgG) rearrangements, thymic development, major histocompatibility complex (MHC) class I and II molecules, and peripheral tolerance [5, 6]. As yet, genes acting at the target organ are not as well studied or defined. Much of the genetic regulation resides in a delicate balance and is affected by the endogenous and exogenous environment; these modulations can come from infectious or pharmacological agents, components of dietary nutrients, or toxins.

Thus, studies of diseases with a T cell-mediated mechanism, whether experimentally induced or spontaneously occurring, have provided useful insight and information on diseases of the human immune system. The goal of this chapter is to review a few of the most pertinent and utilized models with respect to initiation, etiology (genetics or environment, alone or in combination), pathogenesis, preclinical phenotype, utility for the evaluation of drugs, similarities and differences and importantly, their relevance to human disease. It is for each investigator to select the appropriate models with the knowledge and acceptance of their limitations as well as their utility in a therapeutic or drug discovery effort.

Laboratory animal models

The investigator is challenged to develop procedures which remove pain and distress but still maintain or replicate the biology of the disease condition. The success of models and the relevance to disease in man critically depends on the (1) choice of species and genetic strains, (2) model employed and (3) methods used for analyzing drug action and effects [7]. It is also critical to integrate the mechanisms and conditions observed in the animal models to those which are known or envisaged to occur in man.

Choice of animals

There are response variations which accompany strain differences in both mice and rats following experimental induction and in genetic models of autoimmune disease often between and sometimes also within laboratories. The reasons for high and low responders are not clearly established, but quality (contamination of lines) and degree of breeding, environmental conditions and reagents used for inducing disease are all indicated [8]. Genetic status should be certified to avoid heterogeneity and all models should be evaluated within the conditions utilized for the expression of disease.

What makes a disease autoimmune?

Initially, autoimmune etiology was based on Koch's postulates and required that: (1) the autoimmune response include either autoantibodies or T cells, (2) the autoantigen be identified, (3) a similar autoimmune response be produced in an animal model by the autoantibody or T cells and (4) the animal exhibits a pathogenic course which produces a disease similar to that in human [4]. Years of experimentation on the immune system and autoimmune responses has clarified the nature of

Table 2 - T cell mediated diseases

Pathogenesis: autoantibody and/or autorecreative T cells, resulting in inflammation, functional alterations and anatomic lesions
• disease slowed or prevented by therapeutic immunosuppresion • pathogenesis results in target organ loss • disease produced experimentally by sensitizing vs. target autoantigen • disease transfer by autoantibodies or T cells • association with other autoimmune diseases • association with susceptible MHC haplotype

T cells involved [9] (Tab. 2). As a component of the normal T cell repertoire, self-reactive lymphocytes, particularly those specific for antigens expressed in peripheral tissues, can be isolated from the peripheral blood of normal individuals [3, 10] and normal animals [11, 12]. However, direct proof by the above postulates can only be demonstrated in genetically identical inbred or immunologically modified experimental animals. This has been demonstrated for EAE with myelin basic protein (MBP)-specific CD4 T cells [13] and in diabetes (IDDM) with transfer of T cells or murine bone marrow from NOD mice into F_1 recipients [14].

It is well-established that autoreactive T and B lymphocytes within the blood and lymphoid compartments, if activated beyond a critical threshold can potentially induce an autoimmune response [2]. Plausible perturbations in the immunoregulatory T cell network include environmental and infectious agents [15] as pathogens can induce alterations in the normal immune balance by either eliciting expansion of autoreactive lymphocytes or inducing the deletion of regulatory cells. In an infectious process the immune system must discriminate precisely between self and non-self determinants to eliminate disease-inducing agents and protect self. If successful there is eradication of the pathogen with no consequence to the immune system. Occasionally this process is incomplete (failure to eradicate pathogen precipitating chronic inflammatory state) and escapes immune elimination causing damage and inducing an autoimmune response. There are numerous associations and indirect evidence for pathogens as etiologic agents for autoimmune disease expression, i.e. rheumatoid arthritis (RA), Graves disease, IDDM, SLE, yet the pathogenic antigen remains obscure [1, 2, 16, 17]. The relationship to human disease is an indirect one and requires reproduction of the disease in an animal model by antigen-specific means.

An area of divergence between human disease and many experimentally induced models is that in humans often the initiating autoantigen is unknown and the clinical disease is manifest by an immune response to several autoantigens, as secondary determinant spreading occurs due to release and/or exposure of new epitopes subsequent to the initial injury [18]. Most often the release of new epitopes is responsible for polyclonal activation of T cell clones with heterogeneous antigenic specificity within the lesions which amplify the autoimmune reaction and development of the disease state. Often antigen non-specific T cells outnumber the antigen-specific T cells [19]. Finally, caution dictates that experimental models, induced by discrete, known antigenic epitopes, may be imperfect replicas of human disease. The animals are inbred and genetically-defined; a case not true for human disease. As an example, although MBP-induced EAE, is similar to multiple sclerosis (MS) as a T cell-mediated, demyelinating disease of the central nervous system (CNS) [20], the relationship of MBP as an etiologic autoantigen in MS has yet to be established.

Major histocompatibility complex (MHC)

The degree of concordance in identical twins (IDDM, 35–40%) [21], as well as the occurrence of multiplicity of autoimmune disease among members of the same family, suggests the presence of inherited susceptibility or genetic risk factors. The genes identified for the ability to confer highest risk and association with autoimmunity encode the MHC molecules. This has been reviewed extensively elsewhere [5], but a few points are pertinent to autoimmune models and disease in man. MHC gene products are polymorphic and critical in selecting the T cell repertoire [6] either favoring or preventing autoimmunity. They can process and present autoantigenic peptides with varying efficiencies (accord risk or protection) or can exhibit abnormal expression (increased expression or abnormal presentation) [18]. Thus in humans, MHC associations are important and often striking in certain autoimmune conditions, yet the genetics is often more complicated than a singular MHC association to expression of autoimmunity. There may be associations with other less well-defined genes which contribute to the overall association with autoimmune diseases. MHC typing studies and association statistics serve as useful markers to identify at-risk individuals and offer an opportunity for preventative or therapeutic implementation.

Cytokines

By virtue of their proinflammatory and inflammatory properties, cytokines can direct and modulate the immune response in an autoimmune event and calibrate the resultant pathogenesis for intensity of tissue damage. Whether produced locally or systemically to involve distant responses, cytokines enable activation and proliferation of T cells, induce tissue damage by direct cytotoxicity, and modulate autoantigen presentation by increased expression of MHC class II and adhesion molecules [22]. Inflammatory cytokines are associated with a number of autoimmune diseases including RA, psoriasis, and IDDM [23]. Although a vital component of an intact immune response, overproduction or inappropriate expression is pathogenic. The cytokine network and regulatory mechanism are both synergistic and antagonistic, and cytokines may initiate or exacerbate an inflammatory response (TNFα, IFNγ, IL-1, IL-2, IL-12, IL-18) or downregulate an inflammatory response (IL-4, IL-10, TGFβ) [24] resulting in both tight control and redundancy in homeostasis of the system.

Pathology

It is important to consider the essential features of the disease process in man and the similarities of pathology in the animal model (Tab. 3). Considerations include the principal site of injury in man versus the animal model, similar clinical symptoms, relevance of the pathological changes in the lesion, and extent of injury. Knowledge of disease progression (pathogenesis) is critical to drug testing and evaluation, i.e. if the agent is designed to block the initiation of the immune response and progression of the lesion, the destruction of target tissue, or consequent sequelae. Further, evaluation of the disease state, in many instances, comes from supportive data obtained at the end of the experiment [25] which does not allow for repeated observations within a single animal. Also, mice and rats have distinguishable peculiarities (cellular and proinflammatory responses) in response to experimental induction or in the genetic, spontaneous models (i.e. IFNα/β stimulate human T cells but not murine T cells).

Table 3 - Pathogenesis of T cell-mediated diseases

Markers of disease
• Target cell lesion with mononuclear cell infiltrate • MHC class I and class II hyperexpression • Autoantibodies to target cells and tissues • CD4 and CD8 autoreactive T cells • Inflammatory mediators: IL-1, IL-6, TNFα, INFγ
Environmental factors
• Geography, age, season, viral, toxins, food proteins, stress

Disease induction: Adoptive transfer

Passive transfer experiments, unique to T cell-mediated diseases, are only possible using highly inbred animal strains under carefully controlled conditions. Usually T cell-mediated diseases can be adoptively transferred with populations of purified T cells (CD4 and/or CD8) or lymphoid cells (splenocytes) which have been exposed to the appropriate specific antigens *in vitro* before transfer into the host. Preactivation and expansion of the adoptive T cells facilitates efficient and rapid expansion *in vivo* and subsequent disease induction [26]. Adoptive transfer is used to characterize the phenotype of effector cells and confirm their role of disease induction *in vivo*. At issue in this procedure is reproducibility; thus the numbers must be sufficient to account for the inherent variability in recipients. The advantage is the ability to manipulate and isolate cells *ex vivo* and bioassay their respective pathogenicity and specificity *in vivo* [27]. This strategy is labor intensive and often variability in disease production can impede or obscure reliable conclusions as to disease mechanisms. Evaluation of disease onset and progression in association with drug studies requires repeated observations of a sizable group throughout a dose range followed by appraisal of drug effect (biochemical and efficacy markers) on disease symptoms or metabolic measurements.

Spontaneous, organ-specific autoimmune disease models

Nonobese diabetic (NOD) mouse: Model of insulin-dependent diabetes mellitus (IDDM)

General features

IDDM is an autoimmune disease resulting from the destruction of pancreatic islet beta (β) cells by both cell-mediated and humoral mechanisms. A major advance in the understanding of IDDM was the discovery of the autoimmune origin and development of two rodent models: the spontaneous, genetically determined organ-specific NOD mouse [28] and the BB rat [29] which acquire a disease very similar to human IDDM. The following discussion will focus on IDDM as observed in the NOD mouse with particular similarities and differences which are pertinent in the BB rat model. The NOD mouse is a genetically inbred strain with a unique susceptibility to develop spontaneous, overt autoimmune diabetes by 10–20 weeks of age, shares many features with the human disease (reviewed in [30]), and has been extensively studied as to immunogenetics, pathogenesis, and prevention [31]. In NOD mice, disease manifestations are characterized by a mononuclear cell insulitis and the presence of antibodies against islets, insulin, and glutamic acid decarboxylase (GAD), usually before the abrupt onset of hyperglycemia and overt diabetes. Detection of autoantibodies to several distinct determinants coincides with islet β cell destruction [32, 33]. The disease can be prevented by immunotherapy if initiated at or before onset of clinical symptoms.

The frequency of diabetes achieved and sustained in any NOD colony is dependent on the specific pathogen-free (SPF) quality, as exposure to pathogens can diminish the incidence of IDDM [30]. In most colonies the incidence of IDDM is higher in females as they exhibit higher T cell responses, but IDDM in NOD mice of either sex entails a prolonged preclinical phase during which islet infiltration and β cell destruction occur. Diabetic NOD mice do not become ketoacidotic [34], and do not initially require insulin, but long-term management and chronic studies are difficult and unfavorable [30]. Nonetheless, there is colony-to-colony variation as to the incidence of IDDM and the distinctive insulitis lesions emphasizing the fact that this autoimmune occurrence is a combination of genetic, pathophysiologic, immunologic, and environmental events [30]. In addition to diabetes, NOD mice also present with thyroiditis, sialitis, and, later in life, an autoimmune hemolytic anemia.

Importantly, the temporal characteristics of lesion severity and extent, from peri-insulitis to destruction of β cells within the islets, is a semi-quantitative measurement [30] used to assess the efficacy of various therapeutic modalities. In addition, methods for evaluation of the histopathological stages of insulitis in NOD mice are available [30].

Genetics

The major genetic component conferring IDDM susceptibility in NOD mice is the MHC haplotype $H2g^7$, which is the first *Idd* locus (*Idd1*) for which a specific susceptibility gene is known [30]. The presence of the $I\text{-}Ag^7$ in conjunction with a failure to express the I-E molecule most likely contributes to the immunoregulatory defects observed in NOD mice; both are necessary for autoimmune insulitis and diabetes (reviewed in [35]). There is also strong evidence for T cell activity under the control of non-MHC linked diabetes susceptibility loci. NOD MHC class I alleles, although common, are essential contributors to the haplotype and other alleles are less diabetogenic when congenic on NOD [36].

The BB rat also develops diabetes preceded by insulitis, but early in life experiences a severe lymphopenia, involving a RT6+ (mono ADP-ribosyltransferase) subset of lymphocytes [33]. Not all BB rats develop diabetes, and a subline are diabetes resistant (DR). Immunological studies indicate that the insulitis observed in BB rats is similar to that in NOD mice, and that the autoimmunity is closely linked to the MHC class II region. Several important findings obtained from both of these animal models have provided new insight into the immunological and genetic control of IDDM.

In addition, other models of IDDM which utilize chemical induction with streptozotocin, immunomodulation via neonatal thymectomy, irradiation, and adoptive transfer of immune regulatory T cell subsets have been described [21]. Transgene technology has provided antigen-defined models using the rat insulin promoter (RIP) to create mice which developed IDDM after endogenous or exogenous antigen exposure [37]. Additionally, a number of viruses induce diabetes in several mouse strains by direct or indirect cytotoxic mechanisms [33]. Both NOD mice and the DR strain of BB rats are susceptible to viral induction/enhancement of diabetes [33], however, these phenomenon do not translate to human disease, as there is yet no convincing evidence for a direct pathogenic role of a viral etiology in human diabetes.

Both the NOD mouse and BB rat have allowed the necessary prerequisite experimentation to validate the autoimmune nature of IDDM, the elaboration of the MHC- and non-MHC-linked genetics, the isolation of T cell effector clones, and the evaluation of immunotherapeutic modalities. However, as with all models, representation of a multifactorial disease is difficult and often not a complete reproduction of the human disease. Due to dissimilarities observed between animal models and human disease, more than one model or species is generally useful to study the components of autoimmunity and relationship to human disease. For instance, animal models of IDDM are surrogates of a very complex metabolic imbalance and disease syndrome and lack a complete biological similarity between rodents and humans. Between BB rats, NOD mice and human IDDM, some pertinent dissimilarities exist [38]. With regard to the immune system (BB rats are lymphopenic), gender frequency (disease frequency higher in NOD females) and metabolic status (ketoacidosis is more severe in BB rats and humans) differences are apparent. Also, heterogeneity exists in both human IDDM and in rodent models [38], in that they are subject to genetic drift and environmental effects. These differences, along with an incomplete understanding of the etiology of IDDM in any of these species, limit the assumptions and assertions of identity amongst them.

Immunopathology

The immunopathology of IDDM in NOD mice is involved. Initial investigations firmly established the destruction of β cells as a T cell-mediated process which is characterized by aggregates of mononuclear cells surrounding the islets (peri-insulitis) (Tab. 3). The infiltrate is predominately CD4 T cells but CD8 T cells, B lymphocytes and macrophages are also present. All cell types are believed to contribute to pathogenesis, with CD4 and CD8 T cells the essential mediators [21]. In fact, macrophage infiltration is required for onset of diabetes in NOD mice and BB rats. Recruitment is thought to be a result of the continued stimulation of primary autoreactive β cell specific T cells [39] probably via B7-1 costimulation of CD8 T cells [40]. Macrophages contribute to the progression of the lesion through antigen presentation, cytokine production (IL-6, IL-1, TNFα, TGFβ), adhesion molecule upregulation and direct cytotoxicity by release of free radicals.

Information garnered from studies on T cell clones isolated from the islets of prediabetic or diabetic mice suggest that pathogenicity may be limited to a small population of T cells, some of which may have either a regulatory or suppressor function [41]. Thus, effector cytotoxic function may be due to a shift in the balance of control to the effector cells and a Th1 type cytokine profile. Other immune abnormalities associated with IDDM in NOD mice include deficiency in IL-4 producing T cells, abnormal peripheral lymphoid accumulation (lymphoid and non-lymphoid tissues), defective macrophage IL-1 secretion [30] and deficient functional NK cells [42]. Recently, genetically modified NOD stocks (congenic and transgenic) have been constructed [30] which can be used to study in detail the pathogenesis of this autoimmune disease.

The role of cytokines in the pathogenesis of IDDM in NOD mice

Host production of cytokines has been implicated in IDDM, at all stages, from the initiation of insulitis to progression to diabetes. Mechanisms have yet to be clarified, but cytokines have a definite role in pathogenesis of the lesion as well as secondary ability to modulate the immune system. It is suggested that, initially, peri-insular inflammatory infiltrates are less destructive and, in addition to Th1 cytokines, secrete some Th2 type cytokines (TGFβ, IL-4, IL-10) [43]. Later, the pattern changes to Th1 (IFNγ, IL-2) concomitant with intra-islet infiltration, insulitis, β cell loss and diabetes. Of these, the Th1 cytokine, IFNγ, is suspected to be a key component required for the induction and progression of the autoimmune response [21]. Presumably IFNγ is produced at the site by T cells and is necessary for the upregulation of MHC class I on β cells, MHC class II on antigen presenting cells (APC), and the induction of secondary cytokines. Cytokine regulation and proinflammatory modulation occurs locally within the target organ with cells producing cytokines localized to the lesions. A protective role for IL-4 [44] but a contributory role for IL-1, IL-6 and TNFα has been defined for β cell damage within the localized compartment of the islets [45]. TNFα has a pleiotropic effect on NOD diabetes and may be directly cytotoxic to β cells or may have an indirect effect via the recruitment of autoreactive T cells to the islets. Early in the pathogenesis, as shown in mice bearing TNFα transgenes [46], TNFα prevents induction of β cell loss, but when β cell destruction is underway, TNFα participates in the disease process [47].

Autoantibodies/Autoantigens

Although B lymphocytes are essential as APC [48] there is no evidence that autoantibodies are required in IDDM pathogenesis [21]. The role of autoantibodies to lesion development in NOD diabetes has yet to be proven. Both diabetic BB rats and NOD mice present with a multiple autoantibody response to membrane and cytoplasmic constituents of β cells. As yet, there is no evidence that antibodies have an initial role in the insulitis or are secondary to the autoimmune destruction of the pancreatic β cells, even those that are specific for β cell determinants. Finally, the disease cannot be transferred by serum from diabetic mice [33].

The abnormality leading to autoimmune diabetes is hypothesized to be within the immune system and not due to an abnormal target autoantigen. However, recent studies have identified a putative diabetes-associated T-cell autoantigen using an islet-specific T cell clone which distinguishes antigenic presence on NOD but not on BALB/c [49]. The majority of studies in NOD mice and BB rats suggest that the immune defect precipitating diabetes is most likely at the APC/T cell level and not within the pancreas (reviewed in [21]). Even with a hypothesis of multiple antigens or of β cell derived neoantigens, the etiologic antigen(s) has yet to be verified. The pathogenesis of the lesion is complex and the induction of the autoimmune, T cell antigen-specific response(s) leading to the progression of the disease is unclear.

Experimentally-induced, antigen-driven, organ-specific autoimmune disease models

Experimental allergic encephalomyelitis (EAE)

Experimental allergic encephalomyelitis is a paralytic, autoimmune inflammatory disease of the CNS myelin with pathological similarities to MS [50–52]. The immunological effector cells involved in EAE that are responsible for initiating the disease have been shown, with depletion experiments, to be $CD4^+$ lymphocytes [53, 54]. Once these encephalogenic T cells migrate into the CNS, dependent on the expression of integrins such as VLA-4 [55], they re-encounter antigen presented by perivascular microglia. T cells as well as microglia and astrocytes then release pro-inflammatory cytokines [56] and chemokines [57, 58] which may directly or indirectly recruit inflammatory cells, compromise the integrity of the blood brain barrier, and stimulate the release of mediators which breakdown myelin resulting in impaired nerve conduction and paralysis [50, 52]. Although classical inflammatory cell infiltration in EAE has been reported to consist predominantly of lymphocytes and monocytes, some reports have also indicated the presence of eosinophils [59].

Animal models

EAE can be induced in genetically susceptible animals strains, the most common being SJL/J (H-2^s), PL/J (H-2^u) [20], and B10.PL (H-2^u) [60] mice or the Lewis rat [20], although monkeys have also been used [61]. There are two methods of inducing EAE. A primary disease can be initiated by immunizing animals with myelin proteins in Freund's adjuvant, although this produces the most severe disease in rat [20]. Alternatively, an adoptive transfer system is frequently used where myelin sensitized T cells obtained from immunized animals are cultured in the presence of the antigen for 4 days then injected intraperitoneally into syngeneic naïve animals. T cells in the adoptive transfer model are typically obtained from the draining lymph nodes of immunized animals; however, spleen cells can also be used and sometime result in a higher level of disease intensity [59]. Transgenic mouse models have also been described which in some, but not all cases, develop a spontaneous autoimmune disease similar to EAE [62].

The most common immunogens utilized to induce EAE are MBP, proteolipid protein (PLP) [20, 63], myelin oligodendrocyte glycoprotein (MOG) [64] and the S100 beta protein of astrocytes [65]. In depth T cell epitope studies have identified specific portions of these proteins necessary in particular animals strains [20, 66, 67]. Although most of these proteins induce a paralytic disease, the severity, area of the spinal cord and brain affected, and the extent of the antibody response elicited depends on the specific protein, the animal strain, and the sex of the animal, females being more susceptible [63, 68–70]. Once disease is initiated, paralysis is typically graded on a 1–5 or sometimes 1–6 scale ranging from a mild tail to full hind limb and sometimes forelimb paralysis. Death can also occur in some cases.

The role of cytokines in the pathogenesis of EAE

Once encephalogenic T cells migrate into the CNS, the release of these cytokines directly from the encephalogenic T cells or possibly from microglial cells and astrocytes may contribute to the pathology observed in EAE. Studies using T cell clones have suggested that the Th1/Th2 phenotype determines encephalogenicity [71]. Unfortunately, this relationship has not always been clear. Th2 clones have been shown to induce disease in animals that are immunologically deficient [71]. In addition, administration of IL-10, a Th2 cytokine, did not prevent but rather in some cases worsened disease [72]. Similar discrepancies have been observed with IL-4, another Th2 cytokine. Retroviral gene insertion of IL-4 into Th1 encephalogenic T cells reduced their ability to induce disease [73], however IL-4 deficient mice have been shown to develop EAE with similar severity to their wild type controls [74].

Cytokines that may play a more prominent role in the pathogenesis of EAE include IL-12 and TNFα [75]. Administration of IL-12 to animals has been shown to exacerbate disease [76] and make resistant mice susceptible [77]. This effect of IL-12 may be related to its ability to augment TNFα [78] and enhance the generation of iNOS [68, 79] rather than its effects on IFNγ [60]. Inhibition of TNFα has been shown to decrease paralysis [80]. In addition, CNS-specific expression of

TNFα under control of the MBP promoter resulted in a severe disease that was not self-limiting [81], suggesting that TNFα has the ability to perpetuate the disease.

Chemotactic factors have also been shown to play a role and perpetuate the inflammatory response in EAE. One group of chemotactic factors are the chemokines, a group of 8-10 kDa proteins shown to be expressed in the spinal cord and brain of mice and rats during EAE [57, 82, 83]. Interestingly, the CC chemokine MIP-1α, shown to recruit monocytes and T cells [84] as well as promote a Th1 cytokine profile [58], has been illustrated to play an important role in the acute phase of EAE [58]. In contrast, during the relapsing phase of the disease, MIP-1α appears to be less involved and another CC chemokine, MCP-1, also a chemoattractant for monocytes and T cells [84], but inducing a Th2 cytokine profile, appears to play a more important role [58]. These studies indicate that the factors involved in the acute phase may be different than those in the relapsing disease.

Demyelination mechanisms

Although the recruitment of cells into the CNS is critical to disease progression, the actual mediators responsible for demyelination are unclear. TNFα, as indicated above, appears to be critically important since inhibition can prevent demyelination without affecting the cellular inflammatory response [80]. Apoptosis via FAS/FASL interaction of myelin producing cells [85, 86] has also been shown to be important in the pathogenesis of EAE. In addition, mediators such as matrix metalloproteinases which may facilitate tissue damage and cytokine release [87], iNOS [68], and leukotrienes [59] may also contribute to the end result of paralysis, as can myelin-specific antibodies [88]. Regardless of the mediators involved, the clinical symptom of paralytic disease ensues and although the severity can vary, most animals recover.

Mechanism involved in recovery and relapse

The factors controlling recovery are unclear although some investigators have suggested that glucocorticoid release is important [89, 90]. In addition, experiments have suggested that gamma/delta T cells play a protective role as shown with depleting antibodies [91]. Following the recovery phase, in some but not all cases, a relapsing disease occurs. Sometimes, animals may go through several cycles of recovery and relapse. Even less understood are the factors or conditions responsible for relapse. In MS, relapsing disease has been suggested to relate to viral or bacterial infections [76]. Although these are sometimes associated with relapses in MS patients, no specific MS pathogen has been found. Interestingly, MBP-specific T cell receptor (TCR) transgenic mice have been described which undergo a spontaneous paralytic disease similar to EAE. However, in immunocompetent animals, sponta-

Table 4 - Lessons from immunosuppression trials for T cell-mediated diseases

- Immune-mediated mechanism of target cell loss: mechanism of action of immunosuppression
- Tempo of target cell loss: active and remitting stages of disease
- Therapeutic window: active periods of autoimmunity, T cell mediated target cell destruction
- Rationale for potential experimental therapeutic initiatives
  - preservation of target cell function for individuals at risk (preclinical stage)
  - maintenance of residual cell function in patients with clinical disease

neous paralytic disease only occurs if animals are held in a non-sterile facility unless animals are immunized with MBP [62], suggesting, like in MS, that microbial stimulation is necessary for relapse.

Correlation with MS

Animal models are important tools to study human disease. In the case of EAE, the initial hypothesis that T cells may play a role in the pathogenesis of MS initially came from studies in EAE [88]. This was later supported by the isolation of MBP and PLP specific T cells from MS patients. However, unlike EAE, a direct correlation between autoreactive T cells and MS has not been made [92]. MS is also more complicated than EAE since in addition to $CD4^+$ T cells, myelin specific antibodies [93] and possibly viruses [94] may also contribute to the pathogenesis of MS. In EAE, Th1 clones (IFNγ producing) induce disease whereas Th2 clones (IL-4 producing) generally do not [71]. Yet human autoreactive T cells stimulated with myelin antigens produce IFNγ, IL-4, and IL-10 [54]. Despite these differences, it should be pointed out that therapies, such as IFNβ, which has shown effect in MS patients, is also effective in EAE [95, 96]. Clearly, our knowledge of the pathogenesis of MS is limited and it is likely that EAE will continue to be a valuable tool to study MS and serve as a testing ground for newer therapies.

Opportunities for treatment

The utility of these models is primarily for evaluation of compounds with immunosuppressant mechanisms (Tab. 4). However, other components of lesion pathogenesis may be evaluated including the role of proinflammatory mediators, T cell migration, cytokine and cytokine receptors, mechanisms of T cell costimulation/activation, antigen presentation, B lymphocyte activation/antibody production, and Th1-Th2 cytokine balance (Tab. 5). Recently, a promising approach for MS, RA and IDDM has been the mucosal administration (feeding) of antigens spe-

Table 5 - Potential targets for T cell-mediated diseases

Targets	**Issues**
Antigen presentation	identity of autoantigen unknown peptide recognition
MHC II-TCR interaction	MHC II is extensively polymorphic
	genetic identification
	non-polymorphic, non-specific
T cell activation/effector function	specific targets known
	target of current immunosuppressants
Anergy induction	antigen identification required?
Oral tolerance	antigen required; efficacy questionable

cific to disease. The ability of ingested target autoantigen to mitigate or suppress Th1 dependent autoimmune diseases is most likely due to either the deletion/anergy of specific T cells or to the induction of Th2 type antigen reactive T cells to exert bystander suppression [97]. An important caveat in these models is that perturbations are expressed differentially on the immune response, and that extrapolation to human disease, which is more difficult to immunologically modify, is often not valid.

Spontaneous, systemic autoimmune disease models

Systemic lupus erythematosus (SLE)

The genetically inherited autoimmune disease observed in several strains of mice exhibit a lupus-like syndrome with variations in onset, genetics and immunopathology [98, 99]. New Zealand black (NZB), New Zealand white (NZW) and F_1 hybrids (NZB×NZW) are the best characterized and studied models of human SLE. They allow study of the disease at a preclinical stage whereby polyclonal B cell and T cell activation creates an immunostimulatory environment to enhance autoimmune responses [100, 101]. The mice manifest disease by 6 to 7 months with a severe immune complex glomerulonephritis and high serum levels of IgG anti-nuclear antibodies (ANA). The disease is polygeneic in mice with both MHC and non-MHC linked loci involved [102]. These models have been employed to test new therapies of interest for human lupus and RA. Cyclophosphamide and steroids, often with azathioprine, and more recently mycophenolate [103] have proven effective in delaying the immune complex nephritis. These are long-term studies with daily dosing of drug for 6–8 months. NZB×NZW is considered one of the primary models for SLE (with MLR/lpr mice), yet few investigators have utilized these models for evaluation of pharmacological agents due to difficulty in defining biochemical efficacy markers and the length of time required to detect a beneficial response.

Genetically engineered disease models

Transgenic and knockout mice

In building another tier to understanding the autoimmune phenomenon in man, genetically engineered models of T cell-mediated autoimmune diseases have been created [37] based on the experimental observations of naturally occurring or experimentally-induced autoimmune disease models. Since autoimmunity encompasses many aspects including effector mechanisms, susceptibility factors, cytokine molecules and the *in vivo* activity of lymphocytes, these models were designed to select individual factors or reproduce the autoimmune state. This is done directly by introduction of effector mechanisms, tissue expression of a specific cytokine, or indirectly, by challenging self-tolerance mechanisms [37]. In addition, the utilization of genetically altered mice on disease-prone or resistant and immune deficient [104] backgrounds has contributed to investigations of autoimmune disease mechanisms [105].

The results from these approaches have confirmed the role of inflammatory cytokines in the pathogenesis of autoimmunity, singularly or in an interactive capacity, to recruit cells and effector molecules, and to evoke new questions regarding the nature and understanding of peripheral tolerance [37]. The advantages reside within the animals' response to a transgene insertion or a knockout (deletion) as a single alteration in the entire immune system network or as additions on non-target tissues. However, these new technologies to target a specific mechanism or site does not lead easily to the evaluation of new drug therapies [37] because the constitutive expression of the transgene limits the ability of pharmacological intervention to effect changes in the lesion pathology or the aberrant immune response. These shortcomings are a subtle reminder that experimentation driven primarily by *a priori* assumptions [37] formed into genetically modified models are, though valuable, artificial and must be interpreted with caution and consideration of the immunological universe.

Conclusion

The development of animal models has been the strategy employed to counter the inherent difficulty to study disease in humans, since most often the target organ is

Table 6 - Possible strategies for immune intervention in T cell-mediated diseases

Prevention	identify putative autoantigens which initiate target cell destruction • agents which block antigen presentation • agents which induce tolerance to target autoantigens • agents which mimic autoantigens
Early intervention	identify individuals with active subclinical disease by genetic, immune, or metabolic markers • agents which block T or B cell specific effector function • agents which block proinflammatory signaling and inflammation
Late intervention	prevent complete target cell loss at onset of overt disease with combined immunotherapy and novel therapeutic agents

not accessible and patients cannot be the recipients of unproven therapies. Within animal models, investigators can analyze the etiology, pathogenesis, genetic factors and the effects of selective and specific pharmacological agents. However, as one well recognizes, the lineage of species is well-developed and characterized by its biological diversity thus revealing inherent dissimilarities in the biological pattern of diseases observed in animals and humans. As a principal caution, using these models to design and evaluate therapeutic interventions/modalities, the investigator must use "necessary caution and wisdom" [106] when interpreting the range of immunologic responsiveness and the unique characteristics of the animals immune system. In addition, the ability to replicate these diseases in animal models does not imply that the processes are completely reasoned out. Even as the histopathology reveals a sequence of events, the initiating episodes are elusive as are the molecular events which reside intracellularly and direct the response.

Nevertheless, development of agents for both selective and generalized pharmacological intervention have proven therapeutically useful and have provided an additional view of the complexity of the immune response and the interaction of the humoral and cellular components. Design and use of such agents has been guided by a variety of strategies for immune intervention in T cell-mediated diseases (Tab. 6). Knowledge of their molecular targets and mechanisms of action (Tab. 7) have been elucidated in part through the use of many of the models discussed in this chapter (Tab. 8). This information is certainly useful for determining what models would be appropriate for evaluating potential new therapeutics with similar mechanisms of action. Thus, if the analogies between animal models and human disease continue to provide relevant informaiton, then investigators have before them a creative field of experimentation in which to move forward, cautiously and objectively, into the realm of human disease and clinical application.

Table 7 - Mechanism of action of clincial and experimental immunosuppressive agents[*]

Activity of agent	Molecular target	Mechanism of action	Effect on immune system
Stage: Early T cell activation			
Cyclosporin A	Calcineurin	Inhibit Ca-dependent signaling, IL-2 gene transcription	Inhibition of IL-2 production;
FK-506 (Tacrolimus)	Calcineurin	Inhibit Ca-dependent signaling, IL-2 gene transcription	inhibition of IL-2 dependent T cell proliferation
Stage: Late T cell activation			
Rapamycin (Sirolimus)	Ribosomal p70 S6 Kinase	Cell cycle inhibition; Block growth factor dependent signaling	Inhibits T, B cell proliferation Inhibits macrophage IL-1 secretion and activation
Leflunomide	Dihydroorotate DH (DHODH)	Inhibits *de novo* pyrimidine biosynthesis	Blocks proliferation/ activation of T, B cells, macrophages
Antimetabolites			
Mycophenolate mofetil (CellCept)	Inhibits ionosine monophosphate DH (IMPDH)	Blocks *de novo* pathway of purine biosynthesis	Inhibits T, B cell proliferation inhibits cellular and humoral immunity (cytostatic)
Brequinar sodium	DHODH	Inhibits *de novo* pyrimidine biosynthesis	Blocks proliferation/activation of T, B cells, macrophages
Methotrexate	Dihydrofolate reductase (DHFR)	Inhibits *de novo* purine biosynthesis; inhibits DNA, RNA, protein synthesis	Inhibits proliferation of T and B cells: inhibits all rapidly dividing cells (cytotoxic in S phase)
Cyclophosphamide (Nitrogen mustard)	Purine, pyrimidine bases of DNA	Alkylation of DNA, crosslinking of bases, and inhibition of DNA repair	Inhibits proliferation of T, B and hematopoietic cells: inhibits all rapidly dividing cells (cytotoxic)
Others			
15-Deoxyspergualin	Heat shock proteins	Inhibits T cell cytotoxicity, macrophage activation, IL-1 secretion	Inhibits antigen presentation, T cell activation, antibody production
SKF 105685	?	Suppression of cytokines, inflammation, T, B cells	Inhibits macrophage function, IL-1 and IL-2 synthesis and release, generation of cytotoxic T cells

[*]*Compiled from [107–109]*

Table 8 - Activities of immunosuppressant agents in models of autoimmune disease*

Disease	SLE	SLE	IDDM	IDDM	MS	MS
Model	NZB/NZW	MLR/lpr	NOD	BB/W	SJL/J, PL/J	Lewis
Species	mouse	mouse	mouse	rat	mouse	rat
Cyclosporin A	yes	no	yes	yes	yes	yes
FK-506	no	no	no	no	yes	yes
Rapamycin	yes	yes	yes	yes	yes	yes
Leflunomide	yes	yes	no	no	yes	yes
Mycophenolate	no	no	yes	yes	no	no
SKF 105685	no	yes	yes	yes	no	yes
Outcome	Inhibition protection	Inhibition protection	Inhibition protection	Inhibition protection	Inhibition protection	Inhibition protection
Response	ds-DNA antibodies		Islet cell antibodies,		Hindlimb paralysis, tail drop	
Suppressed	nephritis, proteinuria		insulitis, hyperglycemia			

Compiled from [107–109]; [1]SLE, Systemic lupus erythematosus; [2]IDDM, Insulin dependent diabetes mellitus; [3]MS, Multiple sclerosis

References

1 Rose NR (1989) Pathogenic mechanisms in autoimmune diseases. *Clin Immunol Immunopathol* 53 (Suppl): 57–516

2 Cohen IR (1992) The cognitive paradigm and the immunological homunculus. *Immunol Today* 13: 490–494

3 Bona C (1991) Postulates defining pathogenic autoantibodies and T cells. *Autoimmunity* 10: 169–172

4 Rose NR, Bona C (1993) Defining criteria for autoimmune diseases (Witebsky's postulates revisited). *Immunol Today* 14: 426–430

5 Nepom G, Erlich H (1991) MHC class-II molecules and autoimmunity. *Annu Rev Immunol* 9: 493–525

6 Wicker LS (1997) Major Histocompatibility Complex-linked control of autoimmunity. *J Exp Med* 186: 973–975

7 Greenwald RA, Diamond HS (eds) (1988) *CRC Handbook of animal models for the rheumatic diseases*, Volume II. CRC Press, Boca Raton, 181–183

8 van Gelder M, Mulder AH, van Bekkum DV (1996) Treatment of relapsing experimental autoimmune encephalomyelitis with largely MHC-matched allogeneic bone marrow transplantation. *Transplantation* 62: 810–818

9 Cruse JM, Lewis RE (1988) Cellular interactions in autoimmunity. *Concepts Immunopathol* 6: 1–21

10 Rees AD, Lombardi G, Scoging A, Barber l, Mitchell D, Lamb J, Lechler R (1989) Functional evidence for the recognition of endogenous peptides by autoreactive T cell clones. *Int Immunol* 1: 624–630

11 Fowell D, Mason D (1993) Evidence that the T cell repertoire of normal rats contains cells with the potential to cause diabetes. Characterization of the $CD4^+$ T cell subset that inhibits this autoimmune potential. *J Exp Med* 177: 627–636

12 Saoudi A, Seddon B, Heath V, Fowell D, Mason D (1996) The physiological role of regulatory T cells in the prevention of autoimmunity: the function of the thymus in the generation of the regulatory T cell subset. *Immunol Rev* 149: 195–216

13 Zamvil S, Nelson P, Trotter J, Mitchell D, Knobler R, Fritz R, Steinman L (1985) T-cell clones specific for myelin basic protein induce chronic relapsing paralysis and demyelination. *Nature* 317: 355–358

14 Serreze DV, Leiter EH, Worthen SM, Shultz LD (1988) NOD marrow stem cells adoptively transfer diabetes to resistant (NOD × NOD)F_1 mice. *Diabetes* 37: 252–255

15 Kotb M (1995) Infection and autoimmunity: A story of the host, the pathogen, and the copathogen. *Clin Immunol Immunopathol* 74: 10–22

16 Yoon JW (1991) Role of virus in the pathogenesis of IDDM. *Ann Med* 23: 437–445

17 Oldstone MB (1997) Viruses in autoimmune diseases. *Scand J Immunol* 46: 320–325

18 Sercarz EE, Lehmann PV, Ametani A, Benichou G, Miller A, Moudgil K (1993) Dominance and crypticity of T cell antigenic determinants. *Annu Rev Immunol* 11: 229–266

19 Owens T, Sriram S (1995) The immunology of multiple sclerosis and its animal model, experimental allergic encephalomyelitis. *Neurol Clin* 13: 51–73
20 Swarnborg RH (1995) Experimental autoimmune encephalomyelitis in rodents as a model for human demyelinating disease. *Clin Immunol Immunopathol* 77: 4–13
21 Bach JF (1994) Insulin-dependent diabetes mellitus as an autoimmune disease. *Endocrine Rev* 15: 516–542
22 Kroemer G, Martinez C (1991) Cytokines and autoimmune disease. *Clin Immunol Immunopathol* 61: 275–295
23 Feldman M, Brennan FM, Chanty D, Haworth C, Turner M, Katsikis P, Londer M, Abney E, Buchan G, Barrett K et al (1991) Cytokine assays: Role in evaluation of the pathogenesis of autoimmunity. *Immunol Rev* 119: 105–123
24 Mossman TR (1991) Cytokine secretion patterns and crossregulation of T cell subsets. *Immunol Res* 10: 183–188
25 van Bekkum DW (1994) Biology of acute and chronic graft-versus-host reactions: predictive value of studies in experimental animals. *Bone Marrow Transpl* 14 (Suppl 4): 51–55
26 Holoshitz J, Matitiau A, Cohen IR (1984) Arthritis induced in rats by cloned T lymphocytes responsive to mycobacteria but not to collagen type II. *J Clin Invest* 73: 211–215
27 Taurog J (1983) The cellular basis of adjuvant arthritis. II. Characterization of the cells mediating passive transfer. *Cellular Immunol* 80: 198–204
28 Makino S, Kunimoto K, Muraoka Y, Mizushima Y, Katagiri K, Tochino Y (1980) Breeding of a non-obese, diabetic strain of mice. *Exp Anim* 29: 1–13
29 Nakhooda AF, Like AA, Chappel CI, Murray FT, Marliss EB (1977) The spontaneously diabetic Wistar rat. Metabolic and morphologic studies. *Diabetes* 26: 100–112
30 Leiter E (1998) The NOD mouse: A model for insulin-dependent diabetes. In: Shevach EM, Coico R (eds): *Current protocols in immunology*, Vol. 3. John Wiley & Sons, Inc, New York, Section 15.9
31 Kikutani H, Makino S (1992) The murine autoimmune diabetes model: NOD and related strains. *Adv Immunol* 51: 285–322
32 Shieh D-C, Cornelius J, Winter W, Peck A (1993) Insulin dependent diabetes in the NOD mouse model. 1. Detection and characterization of autoantibody bound to the surface of pancreatic beta cells prior to development of the insulitis lesion in prediabetic NOD mice. *Autoimmunity* 15: 123–135
33 Bach JF (1995) Insulin-dependent diabetes mellitus as a beta-cell targeted disease of immunoregulation. *J Autoimmunity* 8: 439–463
34 Coleman DL (1980) Acetone metabolism in mice: increased activity in mice heterozygous for obesity genes. *Proc Natl Acad Sci USA* 77: 290–293
35 Wicker LS, Todd JA, Peterson LB (1995) Genetic control of autoimmune diabetes in the NOD mouse. *Annu Rev Immunol* 13: 179–200
36 Ikegami H, Makino S, Yamato E, Kawaguchi Y, Ueda H, Sakamoto T, Takekawa K,

Ogihara T (1995) Identification of a new susceptibility locus for insulin-dependent diabetes mellitus by ancestral haplotype congenic mapping. *J Clin Invest* 96: 1936–1942
37 Lo D (1996) Transgenic and knockout models of autoimmunity: building a better disease? *Clin Immunol Immunopathol* 79: 96–104
38 Rossini AA, Handler ES, Mordes JP, Greiner DL (1995) Human autoimmune diabetes mellitus: Lessons from BB rats and NOD mice-Caveat Emptor. *Clin Immunol Immunopathol* 74: 2–9
39 Hutchings P, Rosent H, O'Reilly LA, Simpson E, Gordon S, Cooke A (1990) Transfer of diabetes in mice prevented by blockade of adhesion-promoting receptor on macrophages. *Nature* 348: 639–642
40 Wong S, Guerder S, Visintin I, Reich E-P, Swenson KE, Flavell RA, Janeway CA (1995) Expression of the co-stimulator molecule B7-1 in pancreatic beta-cells acclerates diabetes in the NOD mouse. *Diabetes* 44: 326–329
41 Katz JD, Benoist C, Mathis D (1995) T helper cell subsets in insulin-dependent diabetes. *Science* 268: 1185–1188
42 Serreze, DV, Hamaguchi K, Leiter EH (1993) Immunostimulation circumvents diabetes in NOD/Lt mice. *J Autoimmunity* 2: 759–776
43 Fox CJ, Danska JS (1997) IL-4 expression at the onset of islet inflammation predicts nondestructive insulitis in nonobese diabetic mice. *J Immunol* 158: 2414–2424
44 Mueller R, Bradley LM, Krahl T, Sarvetnick N (1997) Mechanism underlying counterregulation of autoimmune diabetes by IL-4. *Immunity* 7: 411–418
45 von Herrath MG, Oldstone MBA (1997) Interferon-γ is essential for destruction of β cells and development of insulin-dependent diabetes mellitus. *J Exp Med* 185: 531–539
46 Grewal IS, Grewal KD, Wong FS, Picarella DE, Janeway CA, Flavell RA (1996) Local expression of transgene encoded TNFα in islets prevents autoimmune diabetes in nonobese diabetic (NOD) mice by preventing the development of auto-reactive islet-specific T cells. *J Exp Med* 184: 1963–1974
47 McSorely SJ, Soldera S, Malherbe L, Carnaud C, Locksley RM, Flavell RA, Glaicherhaus N (1997) Immunological tolerance to a pancreatic antigen as a result of local expression of TNFα by islet β cells. *Immunology* 7: 401–409
48 Serreze DV, Chapman HD, Varnum DS, Hanson MS, Reifsnyder PC, Richard SD, Fleming SA, Leiter EH, Shultz, LD (1997) B lymphocytes are essential for the initiation of T cell-mediated autoimmune diabetes: analysis of a new "speed congenic" stock of NOD.Ig mu null mice. *J Exp Med* 184: 2049–2053
49 Dallas-Pedretti A, McDuffie M, Haskins K (1995) A diabetes-associated T-cell autoantigen maps to a telomeric locus on mouse chromosome 6. *Proc Natl Acad Sci USA* 92: 1386–1390
50 Martin R, McFarland HF (1995) Immunological aspects of experimental allergic encephalomyelitis and multiple sclerosis. *Crit Rev Clin Lab Sci* 32: 121–182
51 Hafler DA, Weiner HL (1995) Immunological mechanisms and therapy in multiple sclerosis. *Immunol Rev* 144: 75–107

52 Scolding NJ, Zajicek JP, Wood N, Compston DS (1994) The pathogenesis of demyelinating disease. *Prog Neurobiol* 43: 143–173
53 Sriram S, Carroll L, Fortin S, Cooper S, Ranges G (1988) *In vitro* immunomodulation by monoclonal anti-CD4 antibody: II. Effect on T cell response to myelin basic protein and experimental allergic encephalomyelitis. *J Immunol* 141: 464–468
54 Schmidt S, Linington C, Zipp F, Sotgiu S, de Waal Malefyt R, Wekerle H, Hohlfeld R (1997) Multiple sclerosis: comparison of the human T-cell response to S100 beta and myelin basic protein reveals parallels to rat experimental panencephalitis. *Brain* 120: 1437–1445
55 Barron JJ, Madri N, Ruddle G, Hashim G, Janeway CA (1993) Surface expression of α4-integrin by CD4 T cells is required for their entry into brain parenchyma. *J Exp Med* 177: 57–68
56 Renno T, Krakowski M, Piccirillo C, Lin J, Owens T (1995) TNFα expression by resident microglia and infiltrating leukocytes in the central nervous system of experimental allergic encephalomyelitis: regulation by TH1 cytokines. *J Immunol* 154: 944–953
57 Godiska R, Chantry D, Dietsch GN, Gray PW (1994) Chemokine expression in murine experimental allergic encephalomyelitis. *J Neuroimmunol* 58: 167–176
58 Karpus WJ, Kennedy KJ (1997) MIP-1α and MCP-1 differentially regulate acute and relapsing autoimmune encephalomyelitis as well as TH1/TH2 lymphocyte differentiation. *J Leuk Biol* 62: 681–687
59 Gladue RP, Carroll L, Milici AJ, Pettipher ER, Salter ED, Contillo L, Showell H (1996) Inhibition of leukotriene B4-receptor interaction suppresses eosinophil infiltration and disease pathology in a murine model of experimental allergic encephalomyelitis. *J Exp Med* 183: 1893–1898
60 Ferber IA, Brocke S, Taylor-Edwards C, Ridgway W, Dinisco C, Steinman L, Dalton D, Fathman CG (1996) Mice with a disrupted IFN-gamma gene are susceptible to the induction of experimental autoimmune encephalomyelitis (EAE). *J Immunol* 156: 5–7
61 Rose LM, Richards TL, Peterson J, Petersen R, Alvord EC (1997) Resolution of CNS lesions following treatment of experimental allergic encephalomyelitis in macaques with monoclonal antibody to the CD18 leukocyte antigen. *Multiple Sclerosis* 2: 259–266
62 Goverman J, Woods A, Larson L, Weiner LP, Hood L, Zaller DM (1993) Transgenic mice that express a myelin basic protein-specific T cell receptor develop spontaneous autoimmunity. *Cell* 72: 551–560
63 Wang LY, Fujinami RS (1997) Enhancement of EAE and induction of autoantibodies to T-cell epitopes in mice infected with a recombinant vaccinia virus encoding myelin proteolipid protein. *J Neuroimmunol* 75: 75–83
64 Ichikawa M, Johns TG, Adelmann M, Bernard CC (1996) Antibody response in Lewis rats injected with myelin oligodendrocyte glycoprotein derived peptides. *Int Immunol* 8: 1667–1674
65 Hartung HP, Rieckmann P (1997) Pathogenesis of immune-mediated demyelination in the CNS. *J Neural Transmission* 50: 173–181
66 Devaux B, Enderlin F, Wallner B, Smilek DE (1997) Induction of EAE in mice with

recombinant human MOG and treatment of EAE with a MOG peptide. *J Neuroimmunol* 75: 169–173
67 Kerlero de Rosbo N, Mendel I, Ben-Nun A (1995) Chronic relapsing experimental autoimmune encephalomyelitis with a delayed onset and an atypical clinical course induced in PL/J mice by myelin oligodendrocyte glycoprotein (MOG)-derived peptide: preliminary analysis of MOG T cell epitopes. *Eur J Immunol* 25: 985–993
68 Ding M, Wong JL, Rogers NE, Ignarro LJ, Voskuhl RR (1997) Gender differences in inducible nitric oxide production in SJL/J mice with experimental autoimmune encephalomyelitis. *J Neuroimmunol* 77: 99–106
69 Berger T, Weerth S, Kojima K, Wekerle H, Lassmann H (1997) Experimental autoimmune encephalmyelitis: the antigen specificity of T lymphocytes determines the topography of lesions in the central and peripheral nervous system. *Lab Invest* 76: 355–364
70 Bernard CC, Johns TG, Slavin A, Ichikawa M, Ewing C, Liu J, Bettadapura J (1997). Myelin oligodendrocyte glycoprotein: a novel candidate autoantigen in multiple sclerosis. *J Mol Med* 75: 77–88
71 Lafaille JJ, Keere FV, Hsu AL, Baron JL, Haas W, Raine CS, Tonegawa S (1997) Myelin basic protein specific T helper 2 (Th2) cells cause experimental autoimmune encephalomyelitis in immunodeficient hosts rather than protect them from the disease. *J Exp Med* 186: 307–312
72 Cannella B, Gao YL, Brosnam C, Raine CS (1996) IL-10 fails to abrogate experimental autoimmune encephalomyelitis. *J Neuroscience Res* 45: 735–746
73 Shaw MK, Lorens JB, Dhawan A, DalCanto R, Tse HY, Tran AB, Bonpane C, Eswaran C, Eswaran SL, Brocke S, Sarvetnick N, Steinman L, Nolan GP, Fathman CG (1997) Local delivery of interleukin 4 by retrovirus-transduced T lymphocytes ameliorates experimental autoimmune encephalomyelitis. *J Exp Med* 185: 1711–1714
74 Liblau R, Steinman L, Brocke S (1997) Experimental autoimmune encephalomyelitis in IL-4 deficient mice. *Int Immunol* 9: 799–803
75 Conboy IM, DeKruyff RH, Tate KM, Cao ZA, Moore TA, Umetsu DT, Jones PP (1997) Novel genetic regulation of T helper 1 (Th1/Th2) cytokine production and encephalogenecity in inbred mouse strains. *J Exp Med* 185: 439–451
76 Smith T, Hewson AK, Kingsley CL, Leonard JP, Cuzner ML (1997) Interleukin-12 induces relapse in experimental allergic encephalomyelitis in the Lewis rat. *Am J Pathol* 150: 1909–1917
77 Segal BM, Stevach EM (1996) IL-12 unmasks latent autoimmune disease in resistant mice. *J Exp Med* 184: 771–775
78 Gladue RP, Laquerre AM, Magna HA, Carroll LA, O'Donnell M, Changelian PS, Franke AE (1994) *In vivo* augmentation of IFNγ with a rIL-12 human /mouse chimera: pleiotropic effects against infectious agents in mice and rats. *Cytokine* 6: 318–328
79 Leonard JP, Waldburger KE, Goldman SJ (1996) Regulation of experimental autoimmune encephalomyelitis by interleukin-12. *Ann NY Acad Sci* 795: 216–226
80 Korner H, Lemckert FA, Chaudhri G, Etteldorf S, Sedgwick JD (1997) Tumor necrosis factor blockade in actively induced experimental autoimmune encephalomyelitis pre-

vents clinical disease despite activated T cell infiltration to the central nervous system. *Eur J Immunol* 27: 1973–1981

81 Taupin V, Renno T, Bourbonniere L, Peterson A., Rodriguez M, Owens T (1997) Increased severity of experimental autoimmune encephalomyelitis, chronic macrophage/microglial reactivity, and demyelination in transgenic mice producing tumor necrosis factor-alpha in the central nervous system. *Eur J Immunol* 27: 905–913

82 Sun D, Hu X, Liu X, Whitaker JN, Walker WS (1997) Expression of chemokine genes in rat glial cells: the effect of myelin basic protein reactive encephalogenic T cells. *J Neurosci* Res 48: 192–200

83 Miyagishi R, Kikuchi S, Takayama C, Inoue Y, Tashiro K (1997) Identification of cell types producing RANTES, MIP-1 alpha, and MIP-1 beta in rat experimental autoimmune encephalomyelitis by *in situ* hybridization. *J Neuroimmunol* 77: 17–26

84 Rollins BJ (1997) Chemokines. *Blood* 90: 909–928

85 Sabelko KA, Kelly KA, Nahm MH, Cross AH, Russell JH (1997) Fas and Fas ligand enhance the pathogenesis of experimental allergic encephalomyelitis, but are not essential for immune privilege in the central nervous system. *J Immunol* 159: 3096–3099

86 Waldner H, Sobel RA, Howard E, Kuchroo VK (1997) Fas- and FasL-deficient mice are resistant to induction of autoimmune encephalmyelitis. *J Immunol* 159: 3100–3103

87 Clements JM, Cossins JA, Wells GM, Corkill DJ, Helfrich K, Wood LM, Pigott R, Stabler G, Ward GA, Gearing AJ, Miller KM (1997) Matrix metalloproteinase expression during experimental autoimmune encephalmyelitis and effects of a combined matrix metalloproteinase and tumor necrosis factor-alpha inhibitor. *J Neuroimmunol* 74: 85–94

88 Stinissen P, Raus J, Zhang J (1997) Autoimmune pathogenesis of multiple sclerosis: role of autoreactive T lymphocytes and new immunotherapeutic strategies. *Crit Rev Immunol* 17: 33–75

89 Poliak S, Mor F, Conlon P, Wong T, Ling N, Rivier J, Vale W, Steinman L (1997) Stress and autoimmunity: the neuropeptides corticotropin-releasing factor and urocortin suppress encephalomyelitis via effects on both the hypothalmic-pituitary adrenal axis and the immune system. *J Immunol* 158: 5751–5756

90 Bolton C, O'Neill JK, Allen SJ, Baker D (1997) Regulation of chronic relapsing experimental allergic encephalomyelitis by endogenous and exogenous glucocorticoids. *Int Archives Allergy Imm* 114: 74–80

91 Kobayashi Y, Kawai K, Ito K, Honda H, Sobue G, Yoshikai Y (1997) Aggravation of murine experimental allergic encephalomyelitis by administration of T-cell receptor gamma-delta-specific antibody. *J Neuroimmunol* 73: 169–174

92 Martin R, McFarland H (1996) Experimental immunotherapies for multiple sclerosis. *Sem Immunopathol* 18: 1–24

93 Gerritse K, Deen C, Fasbender M, Ravid R, Boersma W, Claassen E (1994) The involvement of specific anti-myelin basic protein antibody-forming cells in multiple sclerosis immunopathology. *J Neuroimmunol* 49: 153–159

94 Merelli E, Bedin R, Sola P, Barozzi P, Mancardi GL, Ficarra G, Franchini G (1997)

Human herpes virus 6 and human herpes virus 8 DNA sequences in brains of multiple sclerosis patients, normal adults, and children. *J Neurology* 244: 450–454

95 Yu M, Nishiyama A, Trapp BD, Tuohy VK (1996) Interferon-beta inhibits progression of relapsing-remitting experimental autoimmune encephalomyelitis. *J Neuroimmunol* 64: 91–100

96 Brod SA, Nelson LD, Khan M, Wolinsky JS (1997) IFN-beta 1β treatment of relapsing multiple sclerosis has no effect on CD3-induced inflammatory or counterregulatory anti-inflammatory cytokine secretion *ex vivo* after nine months. *International J Neuroscience* 90: 135–144

97 Weiner HL, Friedman A, Miller A, Khoury SJ, al-Sabbagh A, Santos L, Sayeh M, Nussenblatt RB, Trentham DE, Hafler DA (1994) Oral tolerance: immunologic mechanisms and treatment of animal and human organ-specific autoimmune diseases by oral administration of autoantigens. *Ann Rev Immunol* 12: 809–837

98 Drake CG, Rozzo SJ, Vyse TJ, Palmer E, Kotzin BL (1995) Genetic contributions to lupus-like disease in (NZB × NZW)F_1 mice. *Immunol Rev* 144: 51–74

99 Kotzin BL (1997) Susceptibility loci for lupus: A guiding light from murine models. *J Clin Invest* 99: 557–558

100 Steinberg AD, Huston DP, Taurog JD, Cowdery JS, Raveche ES (1981) The cellular and genetic basis of murine lupus. *Immunol Rev* 55: 121–154

101 Peng SL, Craft J (1996) T cells in murine lupus: propagation and regulation of disease. *Molec Biol Rep* 23: 247–251

102 Vyse TJ, Kotzin BL (1996) Genetic basis of systemic lupus erythematosus. *Cur Opin Immunol* 8: 843–851

103 Corna D, Morigi M, Facchinetti D, Bertani T, Zoja C, Remuzzi G (1997) Mycophenolate mofetil limits renal damage and prolongs life in murine lupus autoimmune disease. *Kid Internat* 51: 1583–1589

104 Serreze DV, Leiter EH (1994) Genetic and pathogenic basis of autoimmune diabetes in NOD mice. *Cur Opin Immunol* 6: 900–906

105 Mueller R, Sarvetnick N (1995) Transgenic/knockout mice- tools to study autoimmunity. *Cur Opin Immunol* 7: 799–803

106 Renold AE, Porte D, Shafrir E (1988) In: E Shafrir, AE Renold (eds): *Frontiers in diabetes research: Lessons from animal diabetes II.* Libbey, London, 3–5

107 Allison AC, Lafferty KJ, Fliri H (eds) (1993) *Immunosuppressive and antiinflammatory drugs.* Annals of New York Academy of Sciences, vol 696

108 St. Georgiev V, Yamaguchi H (eds) (1993) *Immunomodulating drugs.* Annals of New York Academy of Sciences, vol 685

109 Przepiorka D, Sollinger H (eds) (1995) *Recent developments in transplantation medicine. Volume I. New immunosuppressive drugs.* Physician and Scientist Publishing Co., Inc., Illinois

Transplantation

Charles G. Orosz[1], M. Elaine Wakely[1], Ginny L. Bumgardner[2] and Elora J. Weringer[3]

[1]Departments of Surgery, Pathology & Medical Microbiology/Immunology, The Ohio State University, Columbus, Ohio 43210-1228, USA; [2]Department of Surgery, The Ohio State University, Columbus, Ohio 43210-1228, USA; [3]Department of Immunology, Central Research Division, Pfizer Inc., Box 1189, Eastern Point Road, Groton, CT 06340, USA

Cell-mediated alloimmunity

Basic pathobiology

When tissues are transplanted, they become rapidly inflamed due to surgical and mechanical tissue damage, ischemia and reperfusion injury [1, 2]. These antigen-independent inflammatory processes are transient. They will resolve and the transplanted tissues will be accepted, unless infiltrating T cells encounter foreign peptides at the graft site. Under such conditions, the T cells promote a severe, antigen-dependent enhancement of local inflammation. This *in vivo* T cell response, known as acute allograft rejection, is associated with prominent local cytokine production [3, 4], widespread pro-inflammatory activation of vascular endothelia [1, 3], as well as the intense leukocytic infiltration, and development of graft-reactive, cytolytic T cells (CTL) that has traditionally been associated with the acute loss of graft function (reviewed in [5, 6]).

The antigen-dependent phase of acute graft rejection occurs when T cells productively engage appropriate "foreign" peptides, i.e., alloantigens or xenoantigens. Productive engagement requires two sets of signals, one delivered by antigen via the T cell receptor (TcR), the other delivered by costimulatory molecules via their receptors on T cells. Foreign proteins are invisible to T cells, but they become visible when their peptide fragments are displayed to TcR by either MHC class I or MHC class II molecules (reviewed in [7]). The requisite costimulation can be delivered through any of several ligand/receptor systems [8], including CD28/B7 [9], CD40/CD40L [10], and various integrin systems like LFA-1/ICAM-1 [11] and VLA-4/VCAM-1 [12]. Foreign peptide recognition by T cells is best promoted by antigen-presenting cells (APC), which present both MHC/antigen complexes and costimulatory molecules. The most effective APC include dendritic cells, macrophages, and B cells [13].

In Vivo Models of Inflammation, edited by D.W. Morgan and L.A. Marshall

The APC can be derived either from the recipient or the donor of the graft [5]. Hence, T cells can recognize alloantigens displayed by either self MHC molecules (low frequency response) or foreign MHC molecules (high frequency response) [14]. Regardless, the consequence of T cell activation is the production of numerous pro-inflammatory cytokines.

Mechanisms available for study

Animal models of acute graft rejection are extremely useful for basic studies regarding mechanisms of inflammation and for pharmacological investigations regarding the control of inflammation. Table 1 gives examples of several new agents and the models in which they are active. (The reader is referred to Table 7 of the previous chapter by Weringer et al. which describes the molecular targets and the mechanisms of actions of these agents.) The utility of these models derives from (a) their inherent ability to dissociate features of antigen-independent inflammation (isografts/autografts) from antigen-dependent inflammation (allografts/xenografts), (b) the large body of accumulated knowledge regarding many of the key elements of the response (T cell behavior, MHC structure/function, APC function, pro-inflammatory endothelial behavior), (c) a well-defined clinical and experimental histopathologic process, and (d) the ease with which graft-induced inflammation can be induced, localized and accessed for study. Thus, acute rejection models provide tools for studies on more than just the *in vivo* behavior of graft-reactive T cells. They permit the study of inflammatory endothelial behavior in the presence or absence of antigen-activated T cells [1, 19]. They have excellent, but unappreciated potential for studies on mechanisms of leukocyte homing, extravasation and migration at sites of inflammation. They have application for studies on macrophage pro-inflammatory activity *in vivo* [20]. In each of these situations, acute rejection models are useful for pharmacological studies and drug screening [21, 22]. They also have significant potential for studies in the rapidly developing area of gene therapy [23, 24]. Finally, the unusual degree of integration between basic research and clinical transplantation provides acute rejection models with additional practical advantages.

Applications

There are several established experimental models of acute rejection, each with distinct advantages and disadvantages, depending on the needs and interests of the investigator. These include skin grafts, cardiac grafts, renal grafts, pancreatic islet grafts, and sponge matrix allografts.

*Table 1 - Immunosuppressive agents in experimental organ transplant models**

	CsA	FK506	Rap	15-DSG	MMF	Breq
Mouse						
skin	yes	ND	yes	ND	ND	ND
cardiac	ND	yes	yes	ND	yes	
kidney	yes	ND	ND	ND	ND	ND
GVHD	ND	yes	yes	ND	ND	ND
bone marrow	ND	yes	ND	yes	ND	ND
Rat						
skin	yes	yes	ND	yes	ND	ND
cardiac	yes	yes	yes	yes	yes	yes
kidney	yes	ND	yes	yes	ND	yes
GVHD	yes	yes	yes	yes	ND	ND
liver	yes	yes	ND	ND	ND	yes
pancreas	yes	yes	yes	yes	ND	ND
Primate - Cynomolgus						
cardiac	yes	ND	yes	yes	yes	yes
kidney	ND	ND	yes	yes	ND	ND
liver/pancreas	ND	yes	ND	ND	ND	ND
Primate - Rhesus						
kidney	yes	ND	ND	yes	ND	ND
Primate - Baboon						
kidney	ND	yes	yes	ND	ND	ND
pancreas	yes	ND	ND	ND	ND	ND

**Source of information: [15–18].*
CsA, cyclosporin A; Rap, rapamycin; 15-DSG, 15-deoxyspergualin; MMF, mycophenolate mofetil; breq, brequinar sodium; ND, not done or not reported.

Skin grafts

Of these tissue grafts, skin grafts are easiest to perform. The technique can be mastered quickly and multiple transplants/day can be performed. However, skin allografts induce acute rejection that is extremely intense and very difficult to influence with immunosuppressive drugs, even those that effectively block cardiac or renal allograft rejection [25, 26]. Thus, skin allografts represent the worst case scenario for acute rejection and are not advised for use in the screening of new immunosuppressive drugs. This may be related to their high content of dendritic (Langerhans) cells, which are potent activators of T cells [13, 27]. Further, skin grafts must be neovascularized, an ill-defined and little-studied angiogenic process whereby the vasculature of the recipient and the graft develop linkages. This process, which takes 2–3 days, delays the onset of inflammation and immunity, and complicates the utility of skin grafts for studies on early mechanisms of inflammation or immunity. Finally, skin grafts have no physiologic functions that can be quantitatively monitored, and end points are basically interpretive assessments of scab formation. While skin grafts may have relatively limited application as a model for studies on inflammation or immunity, they may have much more utility for studies on agents that promote neovascularization and/or wound healing. Also, recent advances in the understanding of xenograft rejection mechanisms [28] should facilitate studies on the use of xenogeneic skin grafts for burn therapy. Cardiac and renal allografts are more widely used models of acute rejection.

Cardiac allografts

Although cardiac grafts are technically easier than renal grafts, both organ transplants require sophisticated technical skills and long periods of training. A trained individual can perform, at most, three to four transplants/day. Cardiac grafts are commonly performed in both mice and rats. They are placed heterotopically in the abdomen, where their function can be easily and routinely monitored by transabdominal palpation. However, they are physiologically compromised by the surgical technique. They are attached to the recipient blood supply in such a way that blood flow to the graft is partial and retrograde. The entire myocardium is perfused, but only the right ventricle receives and pumps blood. The left ventricle never fills, and often develops an organizing clot that can influence local histologic events. The native heart of the recipient remains intact and functional (an excellent control for some studies), so the grafted heart is not compromised by systemic effects of its own developing malfunction. While cardiac allografts are rejected in most murine strain combinations, rejection rates and efficiencies can be highly variable, even among transplants mismatched for both MHC class I and II molecules [29–31], so strain combination can be an important variable in experimental design. Nevertheless, the cardiac isograft/allograft system is an excellent model for studies on various aspects of antigen-independent vs. antigen-dependent inflammation. Studies in this model have demonstrated that isografts and allografts develop different patterns of endothelial activation [1, 3], cytokine production [3, 4] and chemokine production [32, 33], and apoptosis expression [34], in addition to the better appreciated differences in patterns of leukocytic infiltration and tissue destruction. Each of these parameters can be used to monitor and dissect the effects of new immunosuppressive drugs. Indeed, the cardiac allograft model is used for pre-clinical drug studies, but its full potential in this capacity has yet to be realized.

Renal allografts

Renal allografts are slightly more difficult than cardiac grafts, due mainly to the need to attach the fragile ureter. They are less commonly performed than cardiac grafts, and more commonly performed in rats than mice. Among mice, there are clear strain-related differences in rejection rate, ranging from rapid to none, so strain combination is an important experimental variable [31]. Renal graft function is slightly more difficult to monitor, and requires assessment of blood chemistry or urine output. The renal graft is placed intra abdominal and is physiologically normal. The recipient can either be dependent or independent on graft function, depending on whether the native kidneys are nephrectomized. In general, renal grafts are similar to heart grafts in their utility for drug studies and other studies on transplant-associated inflammatory processes.

Within the last 10 years, it has become apparent that selected pharmacological agents can cause prolonged or indefinite acceptance of cardiac or renal allografts. These include selected anti-T cell reagents, such as anti-CD8 and/or anti-CD4 monoclonal antibodies (mAb) [35–37], anti-CD3 [38, 39] and anti-CD2 mAb [40], and CTLA4Ig [26], selected anti-adhesion molecule antibodies, such as anti-LFA-1/anti-ICAM-1 mAb [41] and anti-VCAM-1 mAb [42], and selected anti-APC antibodies, such as anti-CD40/anti-CD40 ligand mAb [43]. Allograft acceptance is also promoted by pre-transplant perfusion with donor leukocytes, which is most effective when performed in combination with various immunosuppressive agents like anti-CD4 mAb [44] or CTLA4Ig [45]. This expands the utility of cardiac and renal allografts into studies on the inflammatory, immune and pharmacological mechanisms associated with the onset of induced immunologic tolerance in adult animals. Interestingly, allograft tolerance has many similarities to the oral tolerance that interferes with the development of experimental allergic encephalomyelitis (EAE) in mice [46], suggesting that fundamental immunologic mechanisms of adult-induced tolerance exist and might eventually be harnessed to treat various autoimmune diseases. Such tolerogenic mechanisms are readily studied in renal and cardiac allograft models.

Pancreatic islet allografts

The prevalence of diabetes has fostered an interest in experimental pancreas or pancreatic islet transplantation. Experimental islet transplants are more common than pancreas transplants, but considerably less common than cardiac or renal transplants. However, islet transplants have some interesting features that deserve consideration. The general approach is to place isolated islets under the kidney capsule in a diabetic mouse or rat. Blood glucose levels are then used to monitor graft function. The transplant procedure is relatively easy. The chief limitation is islet cell preparation, which takes time and resources. Like skin grafts, islet grafts must be neovascularized, which complicates their use in studies on early inflammation and immunity. Further, the neovascularization process in islet grafts and skin grafts is somewhat different. The neovascularization process in skin grafts serves to knit together graft and recipient vasculature, whereas neovascularization of islet grafts sequesters the grafted cells behind a wall of recipient endothelia. This effectively shifts the host-graft interface away from the vascular endothelia, where it exists for the other acute rejection models. Finally, isolated islets are poor APC, so there is a predominant use of recipient APC for islet rejection responses [47]. This is also unusual among acute rejection models. Nevertheless, much has been learned about the inflammatory and immune responses that develop in islet isografts and allografts [48], and they can be an effective model for studies of new immunosuppressive drugs.

Sponge matrix allografts

For many years, sponge matrix allografts were used to study acute rejection [49, 50], but the model is rarely used now. In this model, small pieces of polyurethane sponge are loaded with allogeneic leukocytes and placed subcutaneously in the recipient. Over the next 2 weeks, the sponge implant induces a prominent inflammatory/foreign body response, coating it with a fibrous, neovascularized capsule and creating a cavernous interstitial space within it. During this time, the sponges are infiltrated by large numbers of leukocytes, which can be easily obtained for study by explanting and squeezing the sponge. The retrieved leukocytes are quite similar in subset distribution and function to infiltrating leukocytes from organ allografts [51, 52]. The fluid from the sponge is rich in mediators of inflammation and immunity [53, 54]. The fibrous capsule is rich in pro-inflammatory endothelia and could be used for migration and adhesion studies [55, 56]. This overlooked experimental system is highly versatile, since virtually any antigenic stimulus could be localized within the sponge. It is particularly well suited to drug studies involving localized vs. systemic deposition of drugs to study their influence on inflammatory/immune responses [57, 58]. In an interesting derivation of this model, polyurethane sponge can be replaced with a collagen sponge. These cause less mechanical agitation to the graft site and minimizes the foreign body response. Further, collagenase digestion, rather than squeezing, can be used to retrieve cells for further study. The primary disadvantage of sponge grafts for acute rejection studies is their lack of a physiologic function that could be monitored as an index of acute rejection. The contribution of the concurrent foreign body response must also be considered, but this can often be assessed by evaluation of responses in sponge matrix isografts. In general, sponge or collagen matrix grafts are the least technically demanding of all the graft models and have significant application for studies on the pharmacological control of *in vivo* pro-inflammatory and antigen recognition processes, yet the model remains largely undeveloped for these purposes.

Graft-vs-host disease (GVHD)

Graft-vs-host disease (GVHD) represents the final practical model of acute rejection. In general, allogeneic leukocytes are transferred into immunocompromised mice, where, unchecked by a host immune system, they aggressively target host tissues for destruction, especially the skin and gut. This disease is lethal, due predominantly to diarrhetic fluid loss from the damaged gut. This model is of importance to the field of bone marrow transplantation, where GVHD is a major clinical complication. However, the GVHD model also represents the easiest, most straightforward model with which to demonstrate acute rejection-like immune responses [59–61], and test therapeutic strategies [62]. As such, it represents a simple and often-overlooked model for studies with drugs that influence mechanisms of allosensitization, tissue inflammation and leukocyte migration.

Related pathologies and relevant human diseases

The vascularized heart, renal, and nonvascularized islet transplant models have been used by transplant immunologists to study *in vivo* cell mediated immune responses to alloantigens. However, these models have the potential for study of specific autoimmune diseases and viral diseases which target the heart, kidney, or pancreatic islets such as autoimmune myocarditis, autoimmune nephritis, viral myocarditis, virally mediated glomerulonephritis, and diabetes. A number of clinical disease states such as systemic vasculitis or autoimmune diseases such as systemic lupus erythematosus (SLE) or immune complex diseases involve damage to endothelial beds. Vascularized heart and renal transplant models offer the opportunity to study the pathophysiology of the disease state upon a specific organ endothelial bed (heart and/or kidney). Using a more global approach, these models can be used for *in vivo* study of non-organ specific autoimmune disease and viral infections. In these cases, autoantigens and viral antigens replace alloantigens as the antigenic stimulus which initiates the host cellular and humoral immune responses. In other words, the *in vivo* host response to the "X" autoantigen or viral antigen can be studied using the heart, renal, or islet allograft as a vehicle for *in vivo* presentation of the "X" antigen. In this way, these models can be used to study such nontransplant disease states in an analogous way to their use for the study of transplant immune responses. Current technology which allows genetic manipulation of the "donor" organ prior to transplant and the availability of a number of hosts with specific immunologic defects or diseases mimicking clinical diseases provides an opportunity to utilize these transplant models to investigate inflammatory immune responses to the organs under specific "controlled" host and donor organ conditions.

All these transplant models when performed with an isograft allow investigations regarding ischemia reperfusion injury as well as the efficacy of pharmacologic agents and preservation solutions in mitigating ischemia/reperfusion injury. The kidney transplant model for example can be used to examine inflammatory processes which influence renal function after revascularization for renal artery stenosis or after autotransplant of a kidney which required "bench" repair of a hilar aneurysm. The kidney transplant model could also be used to study pathophysiologic processes which occur with progressive chronic ischemia from atherosclerotic vascular occlusion.

Allograft tissue remodeling

Basic pathobiology

In addition to acute rejection, a second pathologic response can develop in allografts. Unlike acute rejection, which reflects a pathologic inflammatory response driven by allosensitized T cells, chronic rejection reflects a pathologic tissue remodeling response. The arterial structures develop an occlusive neointima, the hallmark of this disease, and the interstitium becomes increasingly fibrotic [63]. This tissue remodeling is apparently initiated by vascular trauma that occurs in the peri-transplant period due to ischemia and reperfusion injury [64]. In rodent isografts, the remodeling process takes about a year to develop [65]. In allografts, tissue remodeling develops much more rapidly, and may be apparent as soon as 30 days post-transplant [66]. Of course, for allografts to survive more than 2 weeks, the recipients must usually be immunosuppressed to subvert acute rejection. The few exceptions to this, which involve selected strain combinations or grafted tissues, will also develop remodeled tissues with time. In rodent allograft models, tissue remodeling has been observed after treatment with a variety of immunosuppressive agents, including anti-CD4 mAb [67], CTLA4-Ig [68] and others.

The accelerated tissue remodeling observed in accepted allografts is promoted by ill-defined components of T cell allosensitization that persist, despite sufficient immunosuppression to block acute rejection. The accepted allografts display continuous, low grade inflammation, i.e. vascular endothelial activation and infiltration by T cells and macrophages [67]. Recipients of accepted allografts often make high levels of graft-reactive alloantibodies [67, 69]. Such antibodies are potent accelerators of allograft tissue remodeling [70]. However, other unidentified elements must also be operative, since accelerated remodeling occurs in the accepted allografts of B cell knock-out mice (C.G. Orosz, unpublished data).

Mechanisms available for study

Animal models of chronic graft rejection are relatively new. Yet, it is intuitively obvious that they have great potential for basic studies regarding the expression and interplay between physiologic, inflammatory and immunologic mechanisms of pathologic vascular remodeling. Although remodeled graft vasculature has some similarities to atherosclerotic lesions [71], quite similar vascular lesions are associated with autologous vein graft restenosis [72], balloon catheter injury [73] and vascular access failure [74]. These are serious and common clinical problems that offer new opportunities for drug development, and pre-clinical drug testing should be facilitated by animal models of chronic rejection. The exact mechanism of graft neointimal formation remains unclear, but apparently involves the accumulation of smooth muscle cells and leukocytes within the sub-endothelial space. The current working hypothesis [71] suggests that the smooth muscle cells are recruited from the arterial media after they differentiate from a contractile to a synthetic phenotype under the influence of ill-defined immune or inflammatory mediators. Upon arrival, they secrete large amounts of new matrix material, which results in an increasingly occlusive arterial neointima.

Interstitial fibrosis represents a second feature of chronic rejection that can be studied in these animal models. It was originally thought that fibrosis was a secondary, down-stream effect of ischemia due to vascular occlusion, but recent evidence indicates that fibrosis in accepted grafts can occur independently of vascular remodeling [75]. Hence, chronic rejection models offer the opportunity to study the physiologic, inflammatory or immunologic mechanisms that work alone or together to produce interstitial fibrosis. Tissue fibrosis is associated with a wide range of pathologic conditions outside of transplantation, and chronic rejection models have significant, untapped potential for studies into the fibrotic process. Again, these models would be especially useful for studies on the pharmacologic control of fibrosis.

Applications

The primary advantage of chronic rejection models for studies on vascular occlusion and fibrosis is the ease with which physiologic, inflammatory and immunologic contributions can be studied either alone or in combination. The kidney graft model is especially useful in this regard [76, 77]. To study the effects of physiologic insufficiency in rodents, one kidney can be removed. The other can be left alone, or its renal artery can be temporarily clamped or partially ligated. To study the effects of peri-transplant trauma and transient inflammation, renal isografts can be transferred into rodent recipients (nephrectomized or non-nephrectomized), or renal allografts can be transferred into SCID rodent recipients. To study additional immune contributions, renal allografts can be transferred into immunosuppressed rodents. To study immune contributions in relative isolation, renal allografts can be placed in SCID recipients that are reconstituted with immunocompetent splenocytes at some late period post-transplant when peri-transplant inflammation has subsided. This remarkable utility is unique to the chronic rejection models, and provides these models with a significant advantage for mechanistic or pharmacologic studies on processes of vascular occlusion or tissue fibrosis. Again, the chief disadvantages of the chronic rejection models are the surgical skill required to perform the tissue transfers, and the protracted period (> 1 month) necessary to develop the tissue pathologies.

Few other allograft systems are as versatile as renal grafts for these studies, including heart allografts and pancreatic islet allografts. Occlusive vascular pathology and interstitial fibrosis can develop rapidly in accepted cardiac allografts [66]. Vascular pathology is confined to a small number of large and medium-sized vessels, which are apparently affected at random along their length and throughout the heart, making quantitation of responses difficult [78]. Accepted islet allografts are not associated with local vascular remodeling, although they develop prominent interstitial fibrosis [79]. Thus, this model clearly dissociates the immune fibrotic process from fibrosis that may be consequent to concurrent upstream vascular occlusion.

The study of occlusive vascular remodeling has been facilitated by the recent development of aortic and carotid graft models [80, 81]. Like accepted organ grafts, vascular allografts develop prominent occlusive neointima within 30 days of transplantation. These vessels are very large and relatively uncomplicated tissues, which makes them relatively easy to study. The conduit function of vascular grafts is somewhat insensitive to local immunity and inflammation, and vascular graft recipients are rarely immunosuppressed. The resulting uninhibited alloimmune responses contributes significantly to the remodeling process in vascular grafts, since isografts develop similar pathology much more slowly [82]. In addition to prominent neointima, these vascular allografts demonstrate significant disruption of medial and adventitial components and prominent interstitial inflammation [82]. Vascular allograft recipients can be immunosuppressed with common immunosuppressants like cyclosporine [83] or CTLA4Ig [80], which minimizes tissue disruption and slows the neointimal formation. However, neointimal formation may not be completely blocked, but merely decelerated toward the much slower rate of development that is observed in vascular isografts. Pharmacologic studies must take this into account. In general, drugs that affect immunity would be expected to decelerate, but not block vascular remodeling in these models. Other drugs that interfere with the poorly understood, basic vascular remodeling response that underlies this pathology need to be developed. Some experimental immunosuppressive drugs may actually perform both functions. For example, rapamycin [84] and leflunomide [85], which are potent inhibitors of T cell activation, may also interfere with smooth muscle cell differentiation promoted by inflammatory mediators. Thus, these drugs may directly interfere with neointimal formation in ways that are somewhat independent of alloimmune responses.

Sponge or collagen matrix grafts have been completely overlooked as models of tissue remodeling, despite the fact that they rapidly induce a neovascularized, fibrotic capsule. This makes them amenable for studies of either fibrosis or neovascularization. The primary advantage of the model is its technical simplicity. Further, the sponge can be loaded at any time with modifying agents, ranging from alloantigens and/or leukocyte populations, through immune mediators or their antagonists, to pharmacologic agents. There are few obvious disadvantages to the model, except that it is relatively unknown.

Relevant human diseases

Chronic rejection models which investigate the pathophysiology contributing to intimal hyperplasia as well as tissue fibrosis results in basic information contributing to a better understanding of the relationships between inflammation, immunity and tissue repair/remodeling. These investigations have application for the study of physiologic and immune processes which contribute to intimal hyperplasia and thrombotic vascular occlusion occurring in human disease conditions such as thrombosis of hemodialysis vascular access conduits, thrombosis of coronary artery bypass grafts, restenosis occurring after angioplasty procedures on coronary, renal, or peripheral blood vessels, and in general to atherosclerotic disease.

Evaluation of common experimental models of transplantation

Skin transplantation

The original rodent model of experimental transplantation involved skin grafts, as described in the classic paper by Medawar and Billingham [86]. This model has been widely used since, with many alterations and innovations. Steinmuller et al. [87] have provided a detailed description of an effective version of the skin graft technique that is clear and useful. There are several advantages to the skin graft model. Proficiency with the grafting technique can be easily acquired after about five practice grafts. Bilateral skin grafts from two different donors can be easily performed on the same recipient, which is useful for experimental controls. Finally, for a given transplant procedure and degree of histoincompatibility, skin graft rejection times are very reproducible.

In theory, the transplant procedure is straightforward. A graft bed is prepared on the recipient animal and a similarly-sized fragment of skin from the donor animal is attached to this bed in a manner that permits healing and engraftment. A graft bed is easily prepared by excising a dorsal patch of recipient skin, which is cut down to, but not through, the richly vascularized superficial fascia immediately over the panniculus carnosus. This layer of fascia contains the main arteries, veins, and lymphatics of the skin. The most widely used sources of skin for murine grafts are ear, tail, or trunk skin. Thin grafts heal and engraft better than thick grafts, providing an advantage for ear and tail skin. When ear skin is used, one ear yields sufficient tissue for two grafts. The ear must be pulled apart and the cartilage gently scraped off with a scalpel. Tail skin can be peeled from the bone/cartilage, cut into one cubic centimeter pieces and flattened dermis side down on saline saturated filter paper. For trunk skin grafts, ventral skin is preferable to the significantly thicker dorsal skin. In all cases, the donor skin fragment is trimmed to a size slightly smaller than the graft bed, and placed dermis side down on the graft bed. The graft can be sutured or stapled in place, or they can be held in place by bandaging.

Applying a bandage is not trivial, and secure bandaging may be the most difficult aspect of this technique, since care must be taken to avoid asphyxiating the host. In mice, the graft can be held in place with an adhesive bandage that is stapled into place with a stainless steel wound clip. In rats, a dressing is kept in place over the graft by applying Nexaband around the perimeter of 2×2 gauze. Roller gauze is wrapped around the entire chest, followed by adhesive tape. Appropriate pressure dressing facilitates healing and revascularization and minimizes nonspecific necrosis. Bandages can be removed 5–6 days after transplantation, allowing subsequent evaluation of graft acceptance or rejection. Accepted grafts remain pink and pliant, rejecting grafts become darkened due to circulation failure and exhibit grossly visible breakdown of surface epidermis leading to scab formation.

The rate of skin graft rejection varies with degree of major and minor histocompatibility disparities between the graft donor and recipient. For major histocompatibility disparities, grafts of similar size and thickness reject in roughly the same time frame, about 10 days post-transplant. For minor histocompatibility disparities, skin graft rejection times can vary. One factor is the source of skin, which may reflect the number of Langerhans cells (fewest in tail skin as compared to the most in belly skin) and other antigen presenting cells in the epidermis [88].

Heterotopic cardiac transplantation

The standard method for heterotopic transplantation of primarily vascularized heart grafts was originally developed in the rat by Abbott et al. [89] and adapted to the mouse by Corry, Winn and Russell [30]. In general, an intact heart is excised from an adult donor and placed in the peritoneum of the recipient, where it is anastomosed to the aorta and vena cava using microsurgical techniques. The technical skill required for this procedure is its chief limitation. The donor heart can be excised with relatively little difficulty. A butterfly thoracotomy is performed to achieve adequate exposure of the heart in the donor animal. Following systemic heparinization, the supradiaphragmatic vena cava, right and left superior vena cava and pulmonary veins are ligated and the heart removed and placed in iced, lactated Ringer's solution during the peri-transplant period.

Placement of the donor heart in the recipient is more demanding. Using 25× magnification, the donor ascending aorta is anastomosed end-to-side to the recipient abdominal aorta and the donor pulmonary artery is sutured to the recipient inferior vena cava using running 10–0 microvascular nylon sutures. Total ischemia time, which constitutes the interval from aortic ligation in the donor to coronary reperfusion in the recipient, should be no more than 30 min. The microsurgical expertise and technical reproducibility takes several months to achieve, particularly in mice. Complications due to technical failure include exsanguinating hemorrhage and hind limb paralysis. Generally, there is little variation in technical complications from strain to strain, but graft thrombosis has been frequently (up to 50%) observed within 24 h of transplantation when normal BALB/c and BALB SCID mice are used as recipients, even when isografts are performed. When the technique has been mastered, overall operative mortality is less than 5%. If these technical difficulties can be overcome, the murine cardiac transplant model has significant advantages, due to the availability of many congenic, transgenic, and knockout strains that can facilitate investigations.

The unusual location and the method of attachment of these cardiac grafts deserve comment. The graft recipient continues to rely on its native heart for effective cardiovascular function, and the transplanted heart probably has little hemodynamic effect on the graft recipient. Recipient blood flows in a retrograde manner through the donor aorta and is shunted into the donor coronary arteries by the closed aortic valve. After perfusing the myocardium, blood enters the coronary sinus, and traverses the right atrium. Blood is ejected from the right ventricle through the donor pulmonary artery and into the recipient's inferior vena cava. Since circulation to the left side of the heart is virtually bypassed, ventricular atrophy tends to occur over time, which can reduce the size of accepted heart grafts by as much as 30–60%. It is not clear how the unusual physiologic conditions in the transplanted heart contribute to the various pathologies that develop in this experimental system.

Cardiac graft survival is generally monitored by transabdominal palpation and /or electrocardiography (ECG). Palpable cardiac impulses are often graded on a 0 to 4+ scale by at least two independent observers. Decline of impulse (2+) is commonly characterized by palpable graft enlargement and firmness. Persons performing palpation should keep in mind that, as indicated above, long-term cardiac grafts reduce in size as early as 14 days post-transplant and can become very small between 60 and 90 days. Cardiac graft rejection is defined as the complete cessation of impulse (0+) and confirmed visually after laparotomy [90]. In some experimental systems, a transient decline (2+) and recovery (3–4+) of palpable impulses have been observed. This phenomenon appears to be unrelated to the presence or absence of immunosuppression or to the degree of MHC incompatibility [90, 91], and its etiology and significance are unknown.

Some investigators recommend combining palpation with ECG as a more reliable method of assessing graft function. An estimated 17% of allografts judged to be 1+ by palpation have been reported to lack electrical activity [92]. To monitor grafts by ECG, 27-gauge needles attached to ECG leads are placed subcutaneously in the left and right axillae and in the right groin, and the ECG of the recipient's native heart is recorded. The axillary leads are then repositioned on the anterior abdominal wall on either side of the cardiac graft and its ECG is recorded. Although reliable, ECG measurements require anesthetization of the graft recipient, which is disadvantageous due to the risk of anaesthetic complications. For many studies, palpation is an adequate indicator of graft rejection. With practice, the distinction can be made between a palpable contraction of the graft versus an impulse caused by aortic systole that is transmitted through a non-functional graft. This is facilitated by gentle compression of the graft between thumb and forefinger during palpation. Such graft compression should only be performed when graft impulses are declining, since compression could damage the heart and initiate graft thrombosis or thrombolytic embolism.

It should be noted that heterotopic cardiac allograft survival in mice is strain-dependent. Some strain combinations spontaneously accept cardiac allografts, while others vigorously and reliably reject allografts within 7–10 days of transplantation [29]. The biologic basis for this high degree of variability remains conjectural. However, it is very important to consider this phenomenon when initiating or evaluating studies that employ heterotopic cardiac allografts.

Kidney transplantation

A widely adopted method for rat kidney transplantation was introduced by Lee [93] and adapted to mice by Skoskiewicz et al. [94]. In general, the donor kidney, ureter and parts of the aorta, vena cava and bladder are excised and anastomosed to the recipient aorta, vena cava and bladder using microsurgical techniques. Preparation of the graft recipient requires dissection of short segments of the abdominal aorta and vena cava, and loosely applying two ligatures proximal and distal to the sites of intended anastomoses. The recipient's left kidney is removed. To obtain the graft, the left kidney, ureter, and bladder of the donor are mobilized. Proximal and distal ties to the renal artery are placed around the donor aorta to enable *in situ* perfusion with lactated Ringer's solution. The perfused kidney, with a patch of aorta, a patch of vena cava, and a segment of bladder, are removed and placed in iced, lactated Ringer's solution until needed. For transplantation, the donor and recipient aorta and vena cava are connected by end-to-side anastomoses, the ligatures are released, and perfusion of the transplant begins. The donor ureter with its attached patch of bladder is passed behind the vas deferens (in males) to avoid kinking, and the small bladder patch is then sutured to an opening in the dome of the recipient bladder. Contralateral nephrectomy may be added to the procedure, if graft function is to be monitored. Renal graft function is usually monitored via blood chemistry, i.e. blood urea nitrogen (BUN) and creatinine levels, but radioactive inulin clearance, intravenous pyelography (IVP), scintigraphic imaging of MHC class II antigen induction [95] have all been used for this purpose. However, monitoring recipient survival remains the easiest way to assess kidney graft function.

Kidney transplantation is considerably more difficult than heart transplantation, particularly in mice. The novice is initially confronted with long ischemia times, which tend to result in irreversible renal damage with subsequent graft dysfunction. Additional difficulties include donor ureter and bladder edema, necrosis and blockage. Dissection of the donor organ is made more complicated for the kidney than the heart because the adrenal and spermatic or ovarian vessels must be ligated, the very delicate renal artery and vein must be dissected free of surrounding tissues, and the bladder and ureter must be freed of surrounding inguinal and abdominal region tissues. The preparation of the recipient involves exposing larger segments of the vena cava and aorta, as well as bilateral nephrectomy. The anastomoses are somewhat more difficult because the donor vessels are both smaller and more fragile. In addition, the bladder anastomosis extends surgery time. Newer modifications of the mouse kidney transplant procedure, which make use of suprarenal aorta and vena cava cuffs, have been reported to reduce ischemia time by as much as 10 min [96, 97].

Like cardiac allografts, the acute rejection of kidney allografts is strain-specific, and many MHC-disparate allografts are spontaneously accepted. Further, the rate and incidence of renal allograft rejection may be highly variable among individuals within a given strain combination [31, 94]. This severely complicates the use of renal allografts for studies on the mechanisms or pharmacologic control renal allograft.

Aortic transplantation

Aortic transplantation has recently become an attractive experimental model for the study of the vascular remodeling processes that are associated with the chronic rejection of various organ allografts. Similar vascular remodeling occurs in both organ allografts and aortic allografts, but the aorta is much larger than the arteries of organ grafts, thus facilitating the study of events that occur specifically within the intima, the media and the adventitia. In addition, the surgical technique is relatively easy to perform in rats [98] and mice (J. Shelby, personal communication), especially for individuals who are proficient at either cardiac or kidney transplantation.

The primary disadvantage of this model is that graft function (patency) cannot be easily monitored after the transplant operation.

The aortic transplant procedure requires merely the isolation and excision of a length of the donor aorta, and its anastomosis to the abdominal aorta of the recipient. To minimize graft ischemia time, the recipient should be prepared first. To do this, the recipient's aorta is clamped caudal to the renal arteries and cephalad to the bifurcation. If an end-to-end anastomoses is to be performed, the recipient aorta must be divided to allow the insertion of the aortic graft. If an end-to-end anastomosis is planned, the lumbar arteries between the ties are gently occluded with a cool ophthalmic cautery, and two successive aortotomies are made.

To obtain donor aorta, some investigators systemically heparinize the donor; while others do not. For orthotopic transplantation, the donor abdominal aorta can be transected. For heterotopic transplantation, the lengthier thoracic aorta can be used, but all of the intercostal arteries must first be ligated or cauterized. It is divided distal to the left subclavian artery and cephalad to the diaphragm and placed in iced, lactated Ringer's solution. The desired length of the aortic segment depends whether it is to be inserted end-to-side as a loop or end-to-end as a conduit. The latter surgery is more easily performed in rats than in mice, and may be preferable due to considerations regarding flow and shear stress associated with the loop model. In either case, when blood flow is resumed, the graft should inflate and pulsate.

An experienced microsurgeon can complete the entire aortic transplant procedure within 30 to 40 min. Graft thrombosis and hind limb paralysis are common complications of this technique. Looped aortic grafts appear to have a higher risk for complications, presumably due to their increased potential for blood flow stasis. However, the looped grafts provide more tissue for histological analysis, and the potential for stasis can be significantly reduced by ligation of the native aorta between the anastomoses.

Another version of this procedure is the technically more demanding technique of carotid artery transplantation model in mice [99]. In this model, a carotid loop from the donor is sutured end-to-side to the carotid artery of the recipient. Two such grafts can be performed per recipient. It is essential to perform control carotid isografts when carotid allografts are under study because both the donor and recipient carotid arteries must undergo a considerable amount of mechanical stretching and pulling to gain adequate access, and this can contribute significantly to vascular remodeling.

Pancreatic islet transplantation

Pancreas transplantation has long been envisioned as a therapy for diabetes. However, pancreas transplantation in rodents is technically impossible, and investigators have resorted to the transfer of isolated pancreatic islets. Diabetic rodents can be obtained in two ways. One involves the use of spontaneously diabetic rat or mouse strains, like the BB rat and the NOD, but this limits the investigator to the vagaries of immune responses in these strains. The other is to induce diabetes in a preferred rat or mouse strain by treatment with streptozotocin or alloxan. This strategies have provided effective experimental models for the study of Type I diabetes and its control by islet transplantation. Indeed, transplantation of purified pancreatic islets have been shown to prevent hyperglycemia in both rats and mice, although by stringent criteria, the metabolic status is not completely restored [100].

Islet cells are first isolated from the donor(s) pancreata by collagenase digestion and Ficoll purification. Islets with a smooth surface and no attached exocrine tissue are hand-picked under a dissecting microscope following a second Ficoll gradient separation. A green filter is used to differentiate the islets from contaminating lymph nodes [101]. It appears that some non-islet components of pancreas can operate to increase graft immunogenecity. Thus, many stringent purification and culture methods have been proposed to reduce islet immunogenecity and prolong graft survival. Such methods include culturing islets at 37° C in an atmosphere of 95% O_2 for extended periods of time [102], immunomagnetic cell separation [103], and fluorescence-activated cell sorting [104]. However, late failure of islet isografts, and even autografts, has been reported in both rodents and humans (unpublished observations, [105]).

For islet cell transplantation, the islet cells are usually placed in a well-vascularized site, such as the spleen, liver or kidney capsule after recipients reach a stable blood glucose level of > 400 mg/dl. Depending on the strain combination, additional purification measures, and experimental protocol used, the injection of roughly 200 to 400 islets can restore the recipient to a euglycemic state. Reportedly, the renal subcapsular space provides better growth conditions for mouse islet cells than does the liver or spleen [106].

In islet graft recipients, blood sugar levels are used to monitored graft function at regular intervals. Islet graft rejection is denoted when the recipient displays a non-fasting blood sugar of more than 300 mg/dl or more than 250 mg/dl for three consecutive days. In studies involving prolongation of islet graft survival, a nephrectomy of the islet-bearing kidney at the end of the experiment is often performed to verify that hyperglycemia will resume in the absence of the islet graft.

Sponge matrix transplantation

The sponge matrix transplant has evolved as an *in vivo* model of allograft rejection for investigators who lack the resources required to perform experimental organ allografts. This model was also popularized because it facilitated the study of graft infiltrate cells. In organ allograft models, there are significant technical problems with the study of graft infiltrating cells caused by the enzymatic digestion necessary to release the infiltrating cells from the organ graft. This is a traumatic process that can compromise the viability and behavior of infiltrating leukocytes.

In this experimental model, small pieces of polyurethane sponge, which serve as an artificial graft matrix, are inserted subcutaneously into mice or rats. There, they induce a foreign body inflammatory response that involves fibrotic encapsulation and neovascularization of the sponge. If allogeneic cells are incorporated into the sponge matrix, alloantigen-specific immune responses become detectable at the graft site within 10–14 days. The sponges can be removed and simply squeezed to release the infiltrating cells. These cells display many of the immunologic reactions that are displayed from infiltrate cells derived from rejecting organ allografts [107]. For sponge allografts, certain procedural items are important. The polyurethane sponge cylinders, approximately 10×13 mm, can be cut with a cork borer from polyurethane foam sheets obtained at a local millinery shop. These cylinders must be thoroughly cleaned and sterilized [107]. They can be coated and/or infused with donor leukocytes by either of two methods. In the original method, they are implanted into the peritoneal cavity of the donor mice 48 h after an i.p. injection of protease peptone, then harvested 24 h later for implantation in the recipient. This is accomplished through a single middorsal incision in the skin of the ventrolateral thorax that is subsequently closed with wound clips. Alternatively, 3.5×10^6 donor splenocytes can be injected into sponges after they have been placed subcutaneously, but prior to wound closure. The latter is an easier method that is less prone to infection, a problem associated with gut trauma during peritoneal loading of the sponges. The sponges must be removed from the recipient within 21 days, or they become too fibrosed to work with. Usually, they are harvested within 10–14 days.

Graft-versus-host disease

Graft-vs-host disease occurs in immunocompromised rodents that are repopulated with allogeneic leukocytes. The host can be genetically immunocompromised, like SCID mice, or can be made immunologically incompetent by procedures that deplete T cells, such as gamma irradiation. Interestingly, the severity of GVHD can vary from strain to strain and species to species. In some rodent strains, GVHD can be induced in one direction, but not in the reciprocal direction.

One protocol for GVHD induction in mice [108] involves a multiple minor histocompatibility disparate strain combination, B10.BR (H-2^k) donor to CBA/J (H-2^k) recipient. The CBA/J recipients are given 1100 cGy (^{137}Cs source) lethal radiation in two doses separated by 3 h. The irradiated recipients are then injected intravenously with 1×10^6 naive B10.BR splenic T cells, isolated by nylon-wool purification, together with 5×10^6 T cell depleted bone marrow cells from the same donor mouse strain. This protocol yields 60% mortality by day 75 post injection.

The severity of systemic GVHD is generally indicated by the presence of alopecia, skin erythema, dramatic weight loss, and diarrhea. GVHD can be validated histologically by examining sections of skin, intestine, liver and spleen. Unfortunately, there is considerable variability between species, particularly when evaluating skin pathology [109, 110]. Mouse GVHD histopathology is more readily monitored in the oral mucosa and liver instead of skin.

References

1 Pelletier RP, Morgan CJ, Sedmak DD, Miyake K, Kincade PW, Ferguson RM, Orosz CG (1993) Analysis of inflammatory endothelial changes, including VCAM-1 expression in murine cardiac grafts. *Transplantation* 55: 315–320

2 Takada M, Nadeau KC, Shaw GD, Marquette KA, Tilney NL (1997) The cytokine-adhesion molecule cascade in ischemia/reperfusion injury of the rat kidney. Inhibition by a soluble P-selectin ligand. *J Clin Invest* 99: 2682–2690

3 Morgan CJ, Pelletier RP, Hernandez CJ, Teske DL, Huang EH, Ohye RG, Orosz CG, Ferguson RM (1993) Alloantigen-dependent endothelial phenotype and lymphokine mRNA expressions in rejecting murine cardiac allografts. *Transplantation* 55: 919–923

4 Dallman MJ, Larsen CP, Morris PJ (1991) Cytokine gene transcription in vascularised organ grafts: analysis using semiquantitative polymerase chain reaction. *J Exp Med* 174: 493–496

5 Vanbuskirk A, Pidwell D, Adams PW, Orosz CG (1997) Transplantation immunology. *JAMA* 278: 1993–1999

6 Rosenberg AS, Singer A (1992) Cellular basis of skin allograft rejection: An *in vivo* model of immune-mediated tissue destruction. *Annu Rev Immunol* 10: 333–358

7 Krensky, A.M. (1997): The HLA system, antigen processing and presentation. *Kidney International* 58: S2–S7

8 Sharpe AH (1995) Analysis of lymphocyte costimulation *in vivo* using transgenic and 'knockout' mice. *Curr Opin Immunol* 7: 389–395

9 Bluestone JA (1996) Costimulation and its role in organ transplantation. *Clin Transpl* 10: 104–109

10 Cayabyab M, Phillips JH, and Lanier LL (1994) CD40 preferentially costimulates activation of CD4+ T lymphocytes. *J Immunol* 152: 1523–1531

11 van Seventer GA, Shimizu Y, Horgan KJ, Ginther Luce, GE, Webb DS, and Shaw S (1991) Remote T cell co-stimulation via LFA-1/ICAM-1 and CD2/LFA-3: demonstrated with immobilized ligand/mAb and implication in monocyte-mediated co-stimulation. *Eur J Immunol* 21: 1711–1718

12 Damle NK, Aruffo A (1991) Vascular cell adhesion molecule 1 induces T-cell antigen receptor-dependent activation of CD4+ T lymphocytes. *Proc Natl Acad Sci* 88: 6403–6407

13 Sundstrom JB, Ansari AA (1995) Comparative study of the role of professional versus semiprofessional or nonprofessional antigen presenting cells in the rejection of vascularized organ allografts. *Transplant Immunol* 3: 273–289

14 Liu Z, Sun YK, Xi YP, Maffei A, Reed E, Harris P, Suciu-Foca N (1993) Contribution of direct and indirect recognition pathways to T cell alloreactivity. *J Exp Med* 177: 1643–1650

15 Cramer DV, Podesta L, Makowa L (eds) (1994) *Handbook of animal models in transplantation research.* CRC Press, Boca Raton

16 Perico N, Remuzzi G (1997) Prevention of transplant rejection. Current treatment guide and future developments. *Drugs* 54: 533–570

17 Przepiorka D, Sollinger H (eds) (1995) *Recent developments in transplantation medicine. Volume I. New immunosuppressive drugs.* Physician and Scientist Publishing Co., Inc., Illinois

18 Shoker AS (1997) Kidney transplant immunosuppressants of today. 2. Transplant immunosuppressant agents. *Drugs Today* 33: 221–236

19 Turunen JP, Mattila P, Halttunen J, Hayry P, Renkonen R (1992) Evidence that lymphocyte traffic into rejecting cardiac allografts is CD11a- and CD49d-dependent. *Transplantation* 54: 1053–1058

20 Agarwal A, Kim Y, Matas AJ, Alam J, Nath KA (1996) Gas-generating systems in acute renal allograft rejection in the rat. Co-induction of heme oxygenase and nitric oxide synthase. *Transplantation* 61: 93–98

21 Bishop DK, Li W (1992) Cyclosporin A and FK506 mediate differential effects on T cell activation *in vivo*. *J Immunol* 148: 1049–1054

22 Cramer DV, Chapman FA, Jaffee BD, Jones EA, Knopp M, Hreha-Eiras G, Makowka L (1992) The effect of a new immunosuppressive drug, Brequinar Sodium, on heart, liver, and kidney allograft rejection in the rat. *Transplantation* 53: 303–308

23 Lew AM, Bradley JL, Silva A, Coligan JE, Georgiou HM (1996) Secretion of CTLA4Ig by SV40 T antigen-transformed islet cell line inhibits graft rejection against the neoantigen. *Transplantation* 62: 83–89

24 Qin L, Chavin KD, Ding Y, Tahara H, Favaro JP, Woodward JE, Suzuki T, Robbins PD, Lotze MT, Bromberg JS (1996) Retrovirus-mediated transfer of viral IL-10 gene prolongs murine cardiac allograft survival. *J Immunol* 156: 2316–2323

25 De Fazio SR, Masli S, Gozzo JJ (1996) Effect of monoclonal anti-CD4 and anti-CD8 on skin allograft survival in mice treated with donor bone marrow cells. *Transplantation* 61: 104–110

26 Pearson TC, Alexander DZ, Winn KJ, Linsley PS, Lowry RP, Larsen CP (1994) Transplantation tolerance induced by CTLA4-Ig. *Transplantation* 57: 1701–1706

27 Guery J-C, Adorini L (1995) Dendritic cells are the most efficient in presenting endogenous naturally processed self-epitopes to class II-restricted T cells. *J Immunol* 154: 536–544

28 Takahashi T, Saadi S, Platt JL (1997) Recent advances in the immunology of xenotransplantation. *Immunol Res* 16: 273–297

29 Madsen JC, Morris PJ, Wood KJ (1997) Immunogenetics of heart transplantation in rodents. *Transplant Rev* 11: 141–150

30 Corry RJ, Winn HJ, Russell PS (1973):Heart transplantation in congenic strains. *Transplant Proc* 5: 733–735
31 Zhang Z, Zhu L, Quan D, Garcia B, Ozcay N, Duff J, Stiller C, Lazarovits A, Grant D, Zhong R (1996): Pattern of liver, kidney, heart, and intestine allograft rejection in different mouse strain combinations. *Transplantation* 62: 1267–1272
32 Russell ME, Adams DH, Wyner LR, Yamashita Y, Halnon NJ, Karnovsky MJ (1993) Early and persistent induction of monocyte chemoattractant protein 1 in rat cardiac allografts. *Proc Natl Acad Sci USA* 90: 6086–6090
33 Fairchild RL, VanBuskirk AM, Kondo T, Wakely ME, Orosz CG (1997) Expression of chemokine genes during rejection and long-term acceptance of cardiac allografts. *Transplantation* 663: 1807–1812
34 Bergese SD, Klenotic SM, Wakely ME, Sedmak DD, Orosz CG (1997) Apoptosis in murine cardiac grafts. *Transplantation* 63: 320–325
35 Mottram PL, Pietersz GA, Smyth MJ, Purcell LJ, Clunie GJA, McKenzie IF (1993) Evidence that an anthracycline-anti-CD8 immunoconjugate, idarubicin-anti-LY-2.1, prolongs heart allograft survival in mice. *Transplantation* 55: 484–490
36 Bushell A, Morris PJ, Wood KJ (1995) Transplantation tolerance induced by antigen pretreatment and depleting anti-CD4 antibody depends on $CD4^+$ T cell regulation during the induction phase of the response. *Eur J Immunol* 25: 2643–2649
37 Madsen JC, Peugh WN, Wood KJ, Morris PJ (1987) The effect of anti-L3T4 monoclonal antibody treatment on first-set rejection of murine cardiac allografts. *Transplantation* 44: 849–852
38 Mottram PL, Han W-R, Murray AG, Mandel TE, Pietersz GA, McKenzie IFC (1997) Idarubicin-anti-CD3: a new immunoconjugate that induces alloantigen-specific tolerance in mice. *Transplantation* 64: 684–690
39 Nicolls MR, Aversa GG, Pearce NW, Spinelli A, Berger MF, Gurley KE, Hall BM (1993) Induction of long-term specific tolerance to allografts in rats by therapy with an anti-CD3-like monoclonal antibody. *Transplantation* 55: 459–468
40 Chavin KD, Qin L, Lin J, Woodward JE, Baliga P, Bromberg JS (1993) Combination anti-CD2 and anti-CD3 monoclonal antibodies induce tolerance while altering interleukin-2, interleukin-4, tumor necrosis factor, and transforming growth factor-beta production. *Ann Surg* 218: 492–503
41 Isobe M, Yagita H, Okumura K, Ihara A (1992) Specific acceptance of cardiac allograft after treatment with antibodies to ICAM-1 and LFA-1. *Science* 255: 1125–1127
42 Orosz CG, Bergese SD, Huang EH, VanBuskirk AM (1995) Immunologic characterization of murine cardiac allograft recipients with long-term graft survival due to anti-VCAM-1 or anti-CD4 monoclonal antibody therapy. *Transplant Proc* 27: 387–388
43 Larsen CP, Elwood ET, Alenander DZ, Ritchie SC, Hendrix R, Tucker-Burden C, Cho HR, Aruffo A, Hollenbaugh D, Linsley PS, Winn KJ, Pearson TC (1996) Long-term acceptance of skin and cardiac allografts after blocking CD40 and CD28 pathways. *Nature* 381: 434–438
44 Pearson TC, Madsen JC, Larsen CP, Morris PJ, Wood KJ (1992) Induction of trans-

plantation tolerance in adults using donor antigen and anti-CD4 monoclonal antibody. *Transplantation* 54: 475–483

45 Pearson TC, Alexander DZ, Hendrix R, Elwood ET, Linsley PS, Winn KJ, Larsen CP (1996) CTLA4-Ig plus bone marrow induces long-term allograft survival and donor-specific unresponsiveness in the murine model. *Transplantation* 61: 997–1004

46 Khoury SJ, Hancock WW, Weiner HL (1992) Oral tolerance to myelin basic protein and natural recovery from experimental autoimmune encephalomyelitis are associated with downregulation of inflammatory cytokines and differential upregulation of transforming growth factor beta, interleukin 4, and prostaglandin E expression in the brain. *J Exp Med* 176: 1355–1364

47 Coulombe M, Yang H, Guerder S, Flavell RA, Lafferty KJ, Gill RG (1996) Tissue immunogenicity: the role of MHC antigen and the lymphocyte costimulator B7-1. *J Immunol* 157: 4790–4795

48 Gill RG, Coulombe M, Lafferty KJ (1996) Pancreatic islet allograft immunity and tolerance: the two-signal hypothesis revisited. Immunol Rev 149: 75–96

49 Robert P, Hayry P (1976) Effector mechanisms in allograft rejection. I. Assembly of "sponge matrix" allografts. *Cell Immunol* 26: 160–167

50 Ascher N, Hoffman R, Chen S, Simmons R (1980) Specific and non-specific infiltration of sponge matrix allografts by specifically sensitized cytotoxic lymphocytes. *Cell Immunol* 52: 38–47

51 Orosz CG, Zinn NE, Sirinek L, Ferguson RM (1986) *In vivo* mechanisms of alloreactivity. 1. Frequency of donor reactive CTL in sponge matrix allografts. *Transplantation* 41: 75

52 Sirinek LP, Zinn NE, Ferguson RM, Orosz CG (1986) *In vivo* mechanisms of alloreactivity. III. Development of donor-specific antibody in sponge matrix allografts. *Transplantation* 41: 349–356

53 Langrehr JM, Dull KE, Ochoa JB, Billiar TR, Ildstad ST, Schraut WH, Simmons RL, Hoffman RA (1992) Evidence that nitric oxide production by *in vivo* allosensitized cells inhibits the development of allospecific CTL. *Transplantation* 53: 632– 640

54 Morgan CJ, Hernandez CJ, Ward JS, Orosz CG (1993) Detection of cytokine mRNA *in vivo* by PCR: problems and solutions. *Transplantation* 56: 437–443

55 Bishop D., Sedmak DD, Leppink DM, Orosz CG (1990) Vascular endothelial differentiation in sponge matrix allografts. *Hum Immunol* 28: 128–133

56 Bishop DK, Jutila MA, Sedmak DD, Beattie MS, Orosz CG (1989) Lymphocyte entry into inflammatory tissues *in vivo*. Qualitative differences of high endothelial venule-like vessels in sponge matrix allografts vs isografts. *J Immunol* 142: 4219–4224

57 Sirinek LP, Zinn NE, Ferguson RM, Orosz CG (1987) Use of sponge matrix allografts for concurrent monitoring of immunologic and pharmacologic events at a graft site. *Transplantation* 44: 161–164

58 Bishop DK, Li W, Chan SY, Ensle, RD, Shelby J, Eichwald EJ (1994) Helper T lymphocyte unresponsiveness to cardiac allografts following transient depletion of CD4-positive cells. Implications for cellular and humoral responses. *Transplantation* 58: 576–584

59 Norton JA, Sloane JP, Al-Saffar N, Haskard DO (1991) Vessel associated adhesion molecules in normal skin and acute graft-versus-host disease. *J Clin Pathol* 44: 586–591
60 De Wit D, Van Mechelen M, Zanin C, Doutrelepont JM, Velu T, Gerard C, Abramowicz D, Scheerlinck JP, De Baetselier P, Urbain J et al (1993) Preferential activation of Th2 cells in chronic graft-versus-host reaction. *J Immunol* 150: 361–366
61 Harning R, Pelletier J, Lubbe K, Takei F, Merluzzi VJ (1991) Reduction in the severity of graft-versus-host disease and increased survival in allogeneic mice by treatment with monoclonal antibodies to cell adhesion antigens LFA-1 alpha and MALA-2. *Transplantation* 52: 842–845
62 Helene M, Lake-Bullock V, Bryson JS, Jennings CD, Kaplan AM (1997) Inhibition of graft-versus-host disease. Use of a T cell-controlled suicide gene. *J Immunol* 158: 5079–5082
63 Solez K (1994) International standardization of criteria for histologic diagnosis of chronic rejection in renal allografts. *Clin Transpl* 8: 345–350
64 Schmid C, Heemann U, Tilney NL (1996) Retransplantation reverses mononuclear infiltration but not myointimal proliferation in a rat model of chronic cardiac allograft rejection. *Transplantation* 61: 1695–1699
65 Tullius SG, Hancock WW, Heemann UW (1994) Long-term kidney isografts develop functional and morphological changes which mimic those of chronic allograft rejection. *Ann Surg* 220: 425–432
66 Russell PS, Chase CM, Winn HJ, Colvin RB (1994) Coronary atherosclerosis in transplanted mouse hearts. 1. Time course and immunogenetic and immunopathological considerations. *Am J Pathol* 144: 260–274
67 Orosz CG, Wakely E, Sedmak DD, Bergese SD, VanBuskirk AM (1997) Prolonged murine cardiac allograft acceptance: characteristics of persistent active alloimmunity after treatment with gallium nitrate versus anti-CD4 monoclonal antibody. *Transplantation* 63: 1109–1117
68 Sayegh MH, Zheng X.-G, Magee C, Hancock WW, Turka LA (1997) Donor antigen is necessary for the prevention of chronic rejection in CTLA4Ig-treated murine cardiac allograft recipients. *Transplantation* 64: 1646–1650
69 Hancock WH, Whitley WD, Tullius SG, Heemann UW, Wasowska B, Baldwin WMI, Tilney NL (1993) Cytokines, adhesion molecules, and the pathogenesis of chronic rejection of rat renal allografts. *Transplantation* 56: 643–650
70 Russell PS, Chase CM, Winn HJ, Colvin RB (1994) Coronary atherosclerosis in transplanted mouse hearts. II. Importance of humoral immunity. *J Immunol* 152: 5135–5141
71 Hayry P, Mennander A, Raisanen-Sokolowski A, Ustinov J, Lemstrom K, Aho P, Yilmaz S, Lautenschlager I, Paavonen T (1993) Pathophysiology of vascular wall changes in chronic allograft rejection. *Transplant Rev* 7: 1–20
72 Davies MG, Hagen PO (1994) Structural and functional consequences of bypass grafting with autologous vein. *Cryobiology* 31: 63–70
73 Thyberg J, Blomgren K, Hedin U, Dryjski M (1995) Phenotypic modulation of smooth muscle cells during the formation of neointimal thickenings in the rat carotid artery after

balloon injury: An electron-microscopic and stereological study. *Cell Tissue Res* 281: 421–433

74 Windus DW (1993) Permanent vascular access: a nephrologist's view. *Am J Kidney Dis* 21: 457–471

75 Armstrong AT, Strauch AR, Starling RC, Sedmak DD, Orosz CG (1997) Morphometric analysis of neointimal formation in murine cardiac grafts. III. Dissociation of interstitial fibrosis from neointimal formation. *Transplantation* 64: 1198–1202

76 Mackenzie HS, Tullius SG, Heemann UW, Azuma H, Rennke HG, Brenner BM, Tilney NL (1994) Nephron supply is a major determinant of long-term renal allograft outcome in rats. *J Clin Invest* 94: 2148–2152

77 Tullius SG, Tilney NL (1995) Both alloantigen-dependent and independent factors influence chronic allograft rejection. *Transplantation* 59: 313–318

78 Armstrong AT, Strauch AR, Starling RC, Sedmak DD, Orosz CG (1997) Morphometric analysis of neointimal formation in murine cardiac allografts. II. Rate and location of lesion development. *Transplantation* 64: 322–328

79 Squifflet JP, Sutherland DE, Morrow CE, Monda L, Field MJ, Najarian JS (1984) The postoperative course and quantitative aspects of rat islet and segmental pancreas isografts. J Surg Res 36: 578–587

80 Sun H, Subbotin V, Chen C, Aitouche A, Valdivia LA, Sayegh MH, Linsley PS, Fung JJ, Starzl TE, Rao AS (1997) Prevention of chronic rejection in mouse aortic allografts by combined treatment with CTLA4-Ig and anti-CD40 ligand monoclonal antibody. *Transplantation* 64: 1838–1856

81 Shi CW, Lee W-S, He Q, Zhang D, Fletcher DL, Newell JB, Haber E (1996) Immunologic basis of transplant-associated arteriosclerosis. *Proc Natl Acad Sci USA* 93: 4051–4056

82 Kolb F, Heudes D, Mandet C, Plissonnier D, Osborne-Pellegrin M, Bariety J, Michel JB (1996) Presensitization accelerates allograft arteriosclerosis. *Transplantation* 62: 1401–1410

83 Bernucci P, Lepidi S, diGioia C, Fiore F, Pietromarchi A, Ioppolo A, Cavallaro A, Gallo P (1995) Does cyclosporin A have any effect on accelerated atherosclerosis in absence of graft rejection? Pathologic and morphometric evaluation in an experimental model. *J Heart Lung Transplant* 14: 1187–1196

84 Marx SO, Jayaraman T, Go LO, Marks AR (1995) Rapamycin-FKBP inhibits cell cycle regulators of proliferation in vascular smooth muscle cells. *Circ Res* 76: 412– 417

85 Xiao F, Chong A, Shen J, Yang J, Short J, Foster P, Sankary H, Jensik S, Mital D, McChesney L, Koukoulis G, Williams JW (1995) Pharmacologically induced regression of chronic transplant rejection. *Transplantation* 60: 1065–1072

86 Billingham RE, Medawar PB (1951) The technique of free skin grafting in mammals. *J Exp Biol* 28: 385–402

87 Steinmuller D (1984) Immunochemical techniques, skin grafting. In: DiSabato G, Langone JJ, van Vunakis H (eds): *Methods in enzymology*, vol. 108. Academic Press, New York, 20–28

88 Odling KA, Halliday GM, Muller HK (1987) Enhanced survival of skin grafts depleted of Langerhans's cells by treatment with dimethylbenzanthracene. *Immunology* 62: 379–385

89 Abbott CP, Lindsey EJ, Creech O, DeWitt CW (1964) A technique for heart transplantation in the rat. *Arch Surg* 89: 645–652

90 Orosz CG, Wakely E, Bergese SD, VanBuskirk AM, Ferguson RM, Mullet D, Apseloff G, Gerber N (1996) Prevention of murine cardiac allograft rejection with gallium nitrate: comparison with anti-CD4 mAb. *Transplantation* 61: 783–791

91 Stepkowski SM, Raza-Ahmad A, Duncan WR (1987) The role of class I and class II MHC antigens in the rejection of vascularized heart allografts in mice. *Transplantation* 44: 753–759

92 Superina RA, Peugh WN, Wood KJ, Morris PJ (1986) Assessment of primarily vascularized cardiac allografts in mice. *Transplantation* 42: 226–227

93 Lee S (1998) An improved technique of renal transplantation in the rat. *Surgery* 61: 771–779

94 Skoskiewicz M, Chase C, Winn JH, Russell PS (1973) Kidney transplants between mice of graded immunogenetic diversity. *Transplant Proc* 5: 721–725

95 Isobe M (1993) Scintigraphic imaging of MHC class II antigen induction in mouse kidney allografts; a new approach to noninvasive detection of early rejection. *Trans Int* 6: 263–269

96 Kalina SL, Mottram PL (1993) A microsurgical technique for renal transplantation in mice. *Aust NZ J Surg* 63: 213–216

97 Zhang Z, Schlachta C, Duff J, Stiller C, Grant D, Zhong R (1995) Improved techniques for kidney transplantation in mice. *Microsurgery* 16: 103–109

98 Mennander A, Tiisala S, Halttunen J, Yilmaz S, Paavonen T, Hayry P (1991) Chronic rejection in rat aortic allografts: an experimental model for transplant arteriosclerosis. *Arterio & Thromb* 11: 671–680

99 Shi C, Russell ME, Bianchi C, Newell JB, Haber E (1994) Murine model of accelerated transplant arteriosclerosis. *Circ Res* 75: 199–207

100 Shi CL, Taljedal IB (1996) Dynamics of glucose-induced insulin release from mouse islets transplanted under the kidney capsule. *Transplantation* 62: 1312–1318

101 Gotoh M, Maki T, Kiyoisumi T, Satomi S, Monaco AP (1985) An improved method for isolation of mouse pancreatic islets. *Transplantation* 40: 437–438

102 Coulombe M, Gill RG (1994) Tolerance induction to cultured islet allografts. II. The status of antidonor reactivity in tolerant animals. *Transplantation* 57: 1201–1207

103 Davies JE, Winoto-Morbach S, Ulrichs K, James RFL, Robertson GSM (1996) A comparison of the use of two immunomagnetic microspheres for secondary purification of pancreatic islets. *Transplantation* 62: 1301–1306

104 Gray DW, Gohde W, Carter N, Heiden T, Morris PJ (1989) Separation of pancreatic islets by fluorescence-activated cell sorting. *Diabetes* 38 (Suppl. 1): 133–135

105 Hering BJ, Browatzki CC, Schultz AO, Bretzel RG, Federlin K (1994) Islet transplant registry report on adult and fetal allografts. *Transplant Proc* 26: 565–568

106 Mellgren A, Schnell-Landstrom AH, Petersson B, Andersson A (1986) The renal subcapsular site offers better growth conditions for transplanted mouse pancreatic islet cells than liver or spleen. *Diabetologia* 29: 670–672
107 Orosz CG, Zinn NE, Sirinek L, Ferguson RM (1986) *In vivo* mechanisms of alloreactivity. I. Frequency of donor-reactive cytotoxic T lymphocytes in sponge matrix allografts. *Transplantation* 41: 75–83
108 Krenger W, Snyder KM, Byon CH, Falzarano G, Ferrara JLM (1995) Polarized type 2 alloreactive CD4+ and CD8+ donor T cells fail to induce experimental acute graft-versus-host disease. *J Immunol* 155: 585–593
109 Beschorner WE, Tutschka P, Santos GW (1982) The sequential morphology of acute graft versus host disease in the rat radiation chimera. *Clin Immunol Immunopathol* 22: 203–224
110 Sullivan KM, Storb R, Buckner CD, Fefer A, Fisher L, Weiden PL, Witherspoon RP, Applebaum FR, Banaji M, Hansen J, et al (1989) Graft-versus-host disease as adoptive immunotherapy in patients with advanced hemtatologic neoplasms. *N Engl J Med* 320: 828–834

Transgenics

David S. Grass

DNX Transgenic Sciences, 301 B College Road East, Princeton, NJ 08540, USA

Introduction

With the explosion of genetic information becoming available in large part due to the Human Genome Project, there is a need to determine the function and role of previously unknown genes. Much can be learned from the comparison of the sequences of these new genes to those of previously characterized genes, and from computer modeling to infer the structure of specific gene products. *In vitro* assay systems can also often provide information on the function of gene products. These assays, as well as cell culture systems, have the advantage of being relatively high throughput and low cost. However, these systems lack the complexity of the whole organism. The inability to study the interaction of different organs and/or cell types limits the information that can be derived from these technologies. *In vivo* experiments, while more costly, can provide information that the above-mentioned systems cannot approach, due to the fact that they provide the metabolic, physiologic and pathologic complexity absent in these other systems. Transgenic technology can be thought of as two individual technologies, gene addition and gene modification (Tab. 1), which allow researchers to perform genetic engineering to investigate the roles of specific genes during development and in various disease states. Gene addition often involves the overexpression of genes, while gene targeting is most often used to create null mutations or "knock outs". Characterizing and studying the resulting animals can often contribute great insights into the role that individual genes may play in normal physiology and in various disease states. In addition, this technology provides the potential to create new *in vivo* disease and metabolic models.

In this chapter, transgenic technology will be reviewed, focusing on the use of this technology in mice to study gene function and to produce disease models. Several examples of the use of gene addition and gene modification will be discussed that illustrate their utility in inflammation research, specifically in the areas of inflammatory mediators, adhesion molecules, receptors, major histocompatability molecules, and enzymes.

In Vivo Models of Inflammation, edited by D.W. Morgan and L.A. Marshall

Table 1 - Technology: A comparison of gene addition versus gene modification

Gene addition	Gene modification
• Requires regulatory and structural sequences	• Regulatory sequences not required
• Transgene integration in the mouse genome is random	• Positive and negative selection performed
• Gene introduced by pronuclear microinjection	• Gene introduced through transfection of ES cells
	• Mouse produced by injection of ES cells into blastocyst

Background

Gene addition

Introducing an exogenous gene or genes into the genome of an animal is most often accomplished by pronuclear microinjection (Fig. 1), although other methods, such as infection of embryos with retroviral vectors, have also been successfully performed. Pronuclear microinjection has been used to produce transgenic animals in several species, such as mice, rats, rabbits, chickens, goats, sheep, cows, pigs, and fish [1].

To perform this technique (for review, see [2]), fertilized eggs are isolated at the single cell stage from donor females. Prior to the fusing of the pronuclei, the male pronucleus, which is larger and easier to visualize than the female pronucleus, is injected with 100 to 200 copies of the transgene fragment. The injected embryos are then transferred into the fallopian tube of a pseudopregnant recipient female and allowed to develop to term. In mice, pseudopregnant recipient females are generated by mating the recipient females with vasectomized male mice. Recipient mice not mated with vasectomized males will not be receptive to transferred embryos. The resulting pups, referred to as potential founder mice, are screened for the transgene integration event by analyzing the DNA from tail biopsies of the pups. Those mice that are determined by this analysis (PCR or Southern blot) to have an integrated copy (or copies) of the transgene fragment are mated to non-transgenic mice. Those

Figure 1
Gene addition by pronuclear microinjection. From "Access to Transgenic Technologies" by Taconic, 1997. Used by permission.

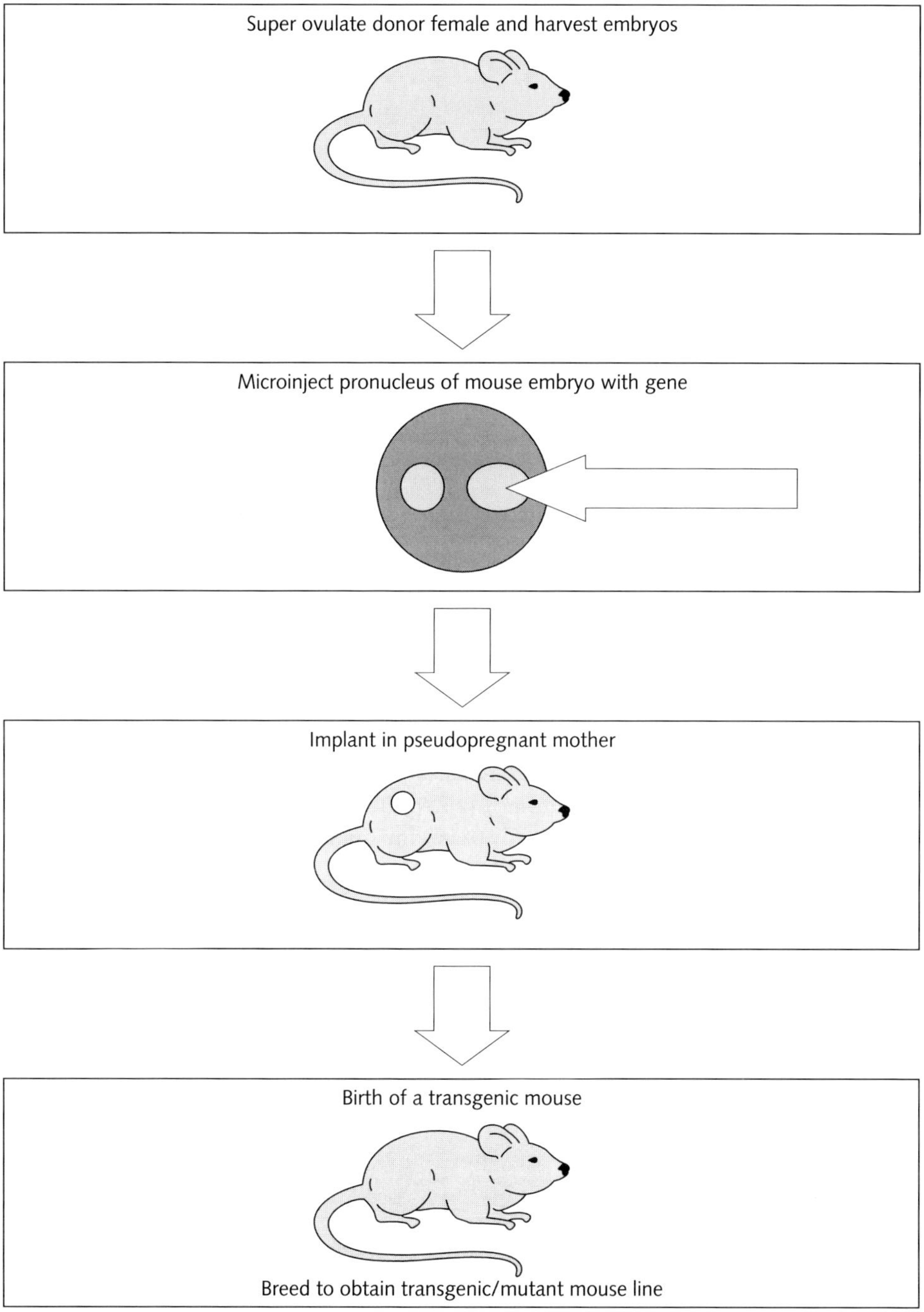
Super ovulate donor female and harvest embryos
Microinject pronucleus of mouse embryo with gene
Implant in pseudopregnant mother
Birth of a transgenic mouse
Breed to obtain transgenic/mutant mouse line

mice in which the integrated transgene fragment can be found in the germ cells can pass the transgene locus to their progeny, establishing a transgenic line. Transgenic lines can be established in a majority of the cases, approximately 90% of the time. Often, for ease of future breeding or to increase the expression of the transgene, mice from the same transgenic line are intercrossed to generate mice homozygous for the transgene locus. In some cases, homozygous mice cannot be generated, because the transgene insertion event can also be an insertional mutation event that can either "knock out" or otherwise affect the expression of an endogenous gene.

The transgene fragment can be thought of as consisting of structural sequence and regulatory sequence. The structural sequences code for the amino acids making up the protein product of the gene. The regulatory sequences dictate the tissue or cell location of expression, the timing of expression, and the level of expression. These sequences can be located several kilobases upstream or downstream with respect to the coding sequences of the gene and/or in the introns of the gene. In addition, regulatory sequences from one gene can be fused to structural sequences of a heterologous gene. This can be done to direct the expression of a gene product in cells where it is not ordinarily expressed or at times when it is not normally expressed. The site of transgene integration in the genome is thought to be random. Because of this, each individual transgenic founder animal is considered unique, and can be used to generate a transgenic line that has the transgene locus at a unique chromosomal location. In general, the transgene fragment integrates into a particular chromosomal location in a head to tail array. In addition, the number of copies of the gene integrating into a particular chromosomal locus is variable. In addition to the regulatory sequences, the location of the transgene locus in the genome of the animal and the number of copies of the transgene integrated can play a role in mediating expression levels, timing of expression, and expression in specific tissues or cell types. LCR (locus control region) sequences and MAR (matrix attachment region) sequences are additional regulatory sequences often placed on 5' of the promoter/enhancer elements to insulate the gene from the effects of the chromosomal insertion site [3, 4]. Both of these types of regulatory sequences are thought to establish an "open" chromatin conformation, allowing the gene to be accessible to the transcriptional machinery. In general, DNA fragments containing large amounts of 5' and 3' flanking sequence in addition to all of the introns are more likely to contain the regulatory sequences necessary for efficient, temporally and spatially correct expression of the gene. These transgene fragments are also less likely to be affected by the chromosomal location of integration.

Historically, the gene addition technology has been used to map regulatory sequences (Tab. 2). This has been done by fusing the putative regulatory sequence with a heterologous reporter gene and characterizing the expression of the reporter gene in the mice. Multiple putative promoter sequences can be rapidly screened this way. Gene addition has also been used to study the phenotype caused by overexpressing or ectopically expressing a gene. Often this phenotype is abnormal and

Table 2 - Utility of gene addition

- Map regulatory sequences
- Phenotypic effects of overexpression or ectopic expression – often results in disease model
- Phenotypic effects of dominant mutations
- Cell ablation studies – studies of specific cell types or lineages
- "Knock down" expression (antisense or ribozyme)

resembles that of a human pathological condition. In addition, gene addition has been used to determine the effect of the expression of a mutated allele. It is important to note that the utility of gene addition is limited to the study of dominant mutations, however, because the presence of the endogenous wild type allele precludes studying the effects of recessive mutations. Gene addition has also been used to perform specific cell ablation studies [5]. In these types of studies, a cell type specific promoter is fused to a toxin gene, such as the diptheria toxin A gene. Cells expressing this gene do not survive, so the affect of knocking out a specific cell type or cell lineage can be studied.

Often, depending on the promoter and the gene chosen for the study, the expression of the transgene is lethal to the developing embryo. To get around this limitation, investigators have utilized inducible promoter sequences. Several inducible systems have been reported, such as the tetracycline inducer or repressor systems, and the ecdysone inducible system (for review, see [6]). Often, a limitation of these systems is that it is difficult to achieve low constitutive levels of expression.

More recently, gene addition has been performed to develop mice that express antisense copies of a particular gene, or a ribozyme designed to cleave a particular endogenous mRNA [7, 8]. The goal of these experiments is to "knock down" the expression of an endogenous gene either by interfering with or digesting the mRNA. This technology has the potential to augment gene targeting technology to lower the expression of an endogenous gene (see the next section) in a more high throughput way because gene addition experiments require less extensive genomic information and less time and resources to perform than gene targeting experiments. However, animals generated using this technology will not serve as the perfect models for any specific genetic mutation, due to the fact that the "knock down" in expression comes not from a mutation in the gene of interest, but the effect of the added "gene" on the expression of the endogenous gene. In addition, these antisense and ribozyme transgenic experiments depend on the ability to express the transgene in every cell in the animal or in the specific targeted tissue. If expression is not achieved in every organ or cell type targeted, there will be a knockdown of expression in only a fraction of the cells and the results could be very misleading. Thus, identification of ubiquitous promoter systems will be necessary to maximize the utility of these approaches.

Gene targeting

Modification of an endogenous gene is accomplished by homologous recombination in embryonic stem (ES) cells (Fig. 2, and for review, see [2]). This technology allows researchers to modify genes in defined ways and study the consequences *in vivo*. When exogenous DNA is transfected into ES cells, it can recombine with the endogenous targeted genes in these cells, substituting the sequence containing the mutation for the endogenous sequence.

ES cells are isolated from the inner cell mass of a developing mouse embryo at the blastocyst stage. If carefully grown in tissue culture, they can be manipulated without losing their potential to contribute to the germline when they are either injected back into a blastocyst or aggregated with eight-cell morulae embryos to produce chimeric mice. Because these cells can be propagated in cell culture, a targeting vector can be introduced into the ES cells by transfection, electroporation, or microinjection, and low probability homologous recombination events can be selected for. The targeting vector requires 5' and 3' regions of homology, surrounding the targeted change. (For review of targeting vectors, see [9].) In contrast to the gene addition technology discussed above, for gene targeting, the vector does not require a promoter for the gene of interest, because the endogenous promoter sequences will serve to drive the expression from the new allele that will be generated. However, the homologous recombination event occurs at a much lower frequency than the frequency of random integration. Therefore, the targeting vector is designed to select for the low frequency event. Usually, this involves both positive and negative selection. Positive selection is often accomplished by the expression of a selectable gene marker, often the neomycin resistance gene driven by the phosphoglycerate kinase (PGK) promoter. The selectable marker is often placed in the targeting construct such that it disrupts the coding sequence of the mouse gene. After the ES cells are transfected with the targeting vector, positive selection is performed, and in the case of the neomycin gene, only colonies resistant to the antibiotic G418 can grow. Negative selection is often performed by placing the HSV-*tk* gene onto the 5' and/or 3' end of the vector [10]. If a homologous recombination event occurs, the *tk* sequence will be lost, whereas it will not be lost if the vector is randomly inserted into the genome. The cells are grown under selection with the nucleoside analog ganciclovir. Because ganciclovir is converted by the HSV-*tk* into a cytotoxic derivative, only if the vector has integrated by homologous recombination, thus losing the *tk* sequence, will the cells be resistant to the drug.

Figure 2
Gene targeting by homologous recombination. From "Access to Transgenic Technologies" by Taconic, 1997. Used by permission.

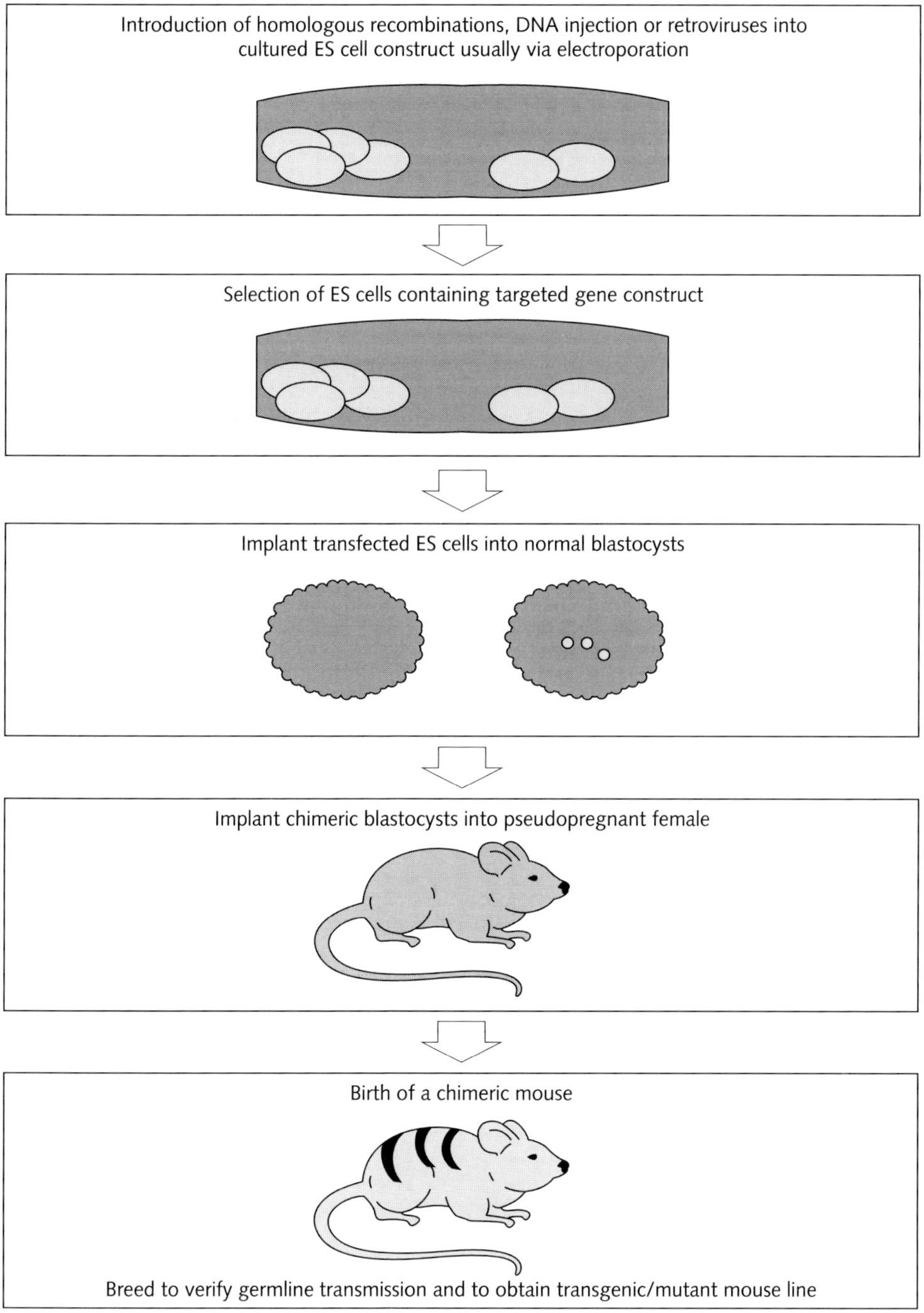
Introduction of homologous recombinations, DNA injection or retroviruses into cultured ES cell construct usually via electroporation
Selection of ES cells containing targeted gene construct
Implant transfected ES cells into normal blastocysts
Implant chimeric blastocysts into pseudopregnant female
Birth of a chimeric mouse
Breed to verify germline transmission and to obtain transgenic/mutant mouse line

Once an ES cell line with the targeted mutation is selected, the ES cells are either injected into the blastocyst cavity or aggregated with an eight cell morulae embryo (often from a mouse strain that differs from the strain that the ES cells were derived from) and transferred into the uterus of a pseudopregnant recipient female to complete development. In order to easily determine the degree of chimerism, the host blastocyst (or morula) strain and the donor ES cell strain are selected such that they do not have the same coat color. The chimeric mouse that is born is made of cells from the ES cell line as well as cells from the host embryo. Chimeric mice can then be identified by looking for a combination of the coat colors between the host and the ES cell strain. The chimeric mice are then bred with mice from the host strain. Mice derived from the germ cells of the ES cell line can be determined by coat color and may be heterozygous for the recombined allele. These mice are genotyped to determine whether they carry the modified allele. The larger the ES cell contribution to the chimera, the more likely it is that the progeny of the chimeric mouse will be derived from the ES cell line and transmit the targeted mutation. Mice homozygous for the modified allele are then produced by interbreeding mice heterozygous for this allele.

In contrast to gene addition, gene modification technology has been used to answer a different set of questions (Tab. 3). One application has been to produce structural mutations. If a particular gene defect is thought to produce a specific phenotype in humans, a mutation can be produced in the homologous mouse gene in an attempt to create a model for the disease. These mutations need not be dominant, because, unlike the case of gene addition, the change is in the endogenous gene and the mice can be bred to homozygosity for the targeted allele. However, the most common application of this technology has been to create null mutations. These experiments have been very useful in determining the role of many genes during development and in the adult animal. Often, however, homozygous null mutations are phenotypically normal, perhaps due to the ability of other gene products to compensate for the loss of gene function. In addition, other homozygous null mutations are often not viable. While this suggests an important role during embryonic development, these experiments are not informative with respect to the role of the gene during adulthood and possibly during disease conditions, which is the major issue for an investigator performing research on inflammation. Because of this, more advanced methods for gene targeting have been developed. These methods give investigators the ability to create tissue-specific and conditional mutations. One method being used currently involves the Cre-loxP system [11]. By placing lox P sites in the targeting vector flanking a DNA sequence, this sequence can be deleted if the Cre enzyme is present. If the Cre gene product is only produced in a particular tissue or cell type, the mutation is only produced in that cell type. If expression of the Cre gene product is under control of an inducible promoter, the mutation can be produced by inducing Cre expression, perhaps in the adult animal. These experiments are usually performed by producing transgenic mice (by gene addition)

Table 3 - Utility of gene modification

• Phenotypic effect of null mutations
• Phenotypic effect of recessive mutation
• "Knock in" of a human allele

expressing Cre under control of a specific promoter, and breeding them with mice in which a targeted mutation was produced by inserting lox P sites flanking a critical area in a targeted gene. The Cre-loxP system can also be utilized to perform "knock in" experiments, in which a different structural allele is substituted for the wild type mouse allele. A specific human gene can be substituted for its mouse homologue using this technology.

The use of transgenic technology in inflammation research

In the following section, recent examples of the utility of transgenic technology in inflammation research will be discussed. The examples will be cataloged by the type of molecule that was expressed or knocked out. Often, inflammatory conditions are induced in transgenic animals by some of the more traditional techniques to better determine the role of a specific gene. Examples of this will be shown.

Inflammatory mediators

Transgenic mice expressing human TNFα were generated by Keffer et al. in 1991 [12]. A construct utilizing the human TNFα gene encompassing the entire gene and 0.6 kb of 5' flank but substituting the human α-globin 3' untranslated flank for the TNFα 3' flank was used. This exchange of the heterologous 3' untranslated flanking sequence for the homologous 3' flanking sequence served to deregulate the expression of the TNFα transgene, and proved that the TNFα 3' untranslated flanking sequence has an important regulatory element involved in correct TNFα expression. Mice expressing the 3' modified transgene developed chronic inflammatory arthritis, and can be used as a model for investigating the pathogenesis and possible treatments for arthritis. Interestingly, when treated with antibodies to the type I IL-1 receptor, these mice no longer develop the arthritic phenotype, suggesting that IL-1 acts downstream of TNFα in the inflammatory progression towards arthritis [13].

Gene targeting has also been utilized to gain insights into the role of TNFα. Mice deficient for TNFα [14] are phenotypically normal developmentally and develop lymph nodes and Peyer's patches. However, they do not form B cell follicles or organized follicular dendritic cell networks and germinal centers in the spleen, suggesting that TNFα is critical in regulating the organization and development of splenic follicular structure and the maturation of the humoral immune response. These mice were challenged by the hapten 2,4,6-trinitrobenzene sulfonic acid (TNBS), which can induce a chronic intestinal inflammation that has some similarity to Crohn's disease in humans [15]. These mice did not develop any significant colitis in response to this TNBS challenge. When TNFα overexpressing mice were challenged in the same assay, the colitis was more severe than what was seen in wild type animals. This data supported work that determined that antibodies to TNFα could ameliorate the colitis in TNBS induced wild type mice, and supported the notion that TNFα would be a good target for inhibition in Crohn's disease.

Many neurodegenerative disorders, including Alzheimer disease, AIDS dementia complex, viral and bacterial meningitis, and stroke, can be characterized by inflammation and the expression of IL-6 in the brain. To study the role of IL-6 in CNS disorders, transgenic mice were generated in which IL-6 was overexpressed in the brain under the regulatory control of the glial fibrillary acidic protein promoter, which targets expression to astrocytes [16]. These mice develop tremor, ataxia, seizure, and runting. Neurodegeneration, astrocytosis, and angiogenesis were the neuropathological effects of IL-6 expression in these animals. Behavioral studies with these mice indicated an age- and dose-related avoidance learning deficit [17]. This learning deficit corresponded with the neuropathological changes, and established the link between inflammatory neurodegeneration, learning impairment, and IL-6 expression. These results suggest that proinflammatory cytokines such as IL-6 can play an important role in the pathogenesis of neurodegenerative diseases such as AIDS and Alzheimer disease and suggest that these mice will be a useful model in studying these diseases.

Adhesion molecules

Intercellular adhesion molecule 1 (ICAM-1) is a member of the family of cellular adhesion molecules. Cellular adhesion molecules are present on the surface of endothelial cells as well as inflammatory cells and are involved in the accumulation of lymphocytes at sites of tissue injury. ICAM-1 is expressed on vascular endothelial cells. It binds to a family of integrins, including CD11b/CD18 (Mac-1) and CD11a/CD18 (LFA-1), on lymphocytes. A null allele of ICAM-1 was created using gene targeting technology. These mice were essential for demonstrating that ICAM-1 expression is important in the recruitment of neutrophils in response to thioglycollate injection (model of peritonitis) [18]. In addition, these mice had a significantly diminished response when contact hypersensitivity was elicited with 2,4-dinitrofluorobenzene (DNFB). The maximal ear swelling was reduced 74%. More recently, these ICAM-1 knock-out mice have been tested in a model in which radiation is used to induce pulmonary inflammation in the mouse [19]. These experiments have confirmed a requirement for ICAM-1 in the recruitment of lymphocytes due to radiation induced inflammation.

Receptors

The overexpression of receptor molecules has often been used to gain insights into the inflammatory process. Epidermal keratinocytes have been shown to produce IL-1α and IL-1β, IL-1 receptor antagonist (IL-1ra) and both the IL-1 type I and IL-1 type II receptors (IL-1R1 and IL-1R2). Because of this, they are thought to be a good model for studying the IL-1 mediated inflammatory conditions. To study the role of the IL-1R2, Rauschmayr et al. targeted the expression of this receptor to the basal keratinocytes using the human keratin 14 promoter [20]. Inflammation in these mice was then induced by topical application of the phorbol ester, PMA. IL-1 has previously been shown to play a prominent role in this inflammatory model, which is characterized by epidermal hyperplasia, inflammatory cell infiltration, and vascular leakage. In both an acute (vascular permeability after PMA application and subsequent Evan's blue dye injection) and a more chronic (PMA ear painting three times within a week) model, there was a significant inhibition of the inflammatory response in the transgenic mice compared to the non-transgenic control mice. These results showed that IL-1R2 can act as a local IL-1 antagonist. These mice provide a useful model to determine the role of IL-1 in different types of skin inflammation.

Chemokines are a large group of secreted proteins that regulate multiple aspects of inflammation and host defense, including leukocyte chemotaxis, hematopoiesis, and angiogenesis. Chemokines are broken down into two groups, CXC and CC, based on whether or not there is a single amino acid between the first two of four conserved cysteines. CC chemokines target monocytes, lymphocytes, basophils and eosinophils, but rarely target neutrophils. In contrast, CXC chemokines target mostly neutrophils and T cells. Gao and colleagues used gene targeting technology to create a null mutation in the CC chemokine receptor 1 (CCR1) gene [21]. This receptor is expressed in neutrophils, lymphocytes, monocytes and eosinophils, and binds several CC chemokines, including macrophage inflammatory protein (MIP)-1a. Since there are other CCR genes known whose chemokine and leukocyte specificities overlap with those of CCR1, these experiments were performed to determine whether CCR1 played a role distinct from these other receptors. The results indicated that CCR1 deficient mice were normal when raised in specific pathogen free environments. However, in response to MIP-1a, neutrophils from these mice did not mobilize into the peripheral blood and had higher mortality rates when challenged with *Aspergillus fumigatus*, a fungus controlled primarily by neutrophils. In addition, when synchronous egg-induced granulomas were induced in these mice by intravenous injection of *Schistosoma mansoni* eggs, there was a reduction of 40% in the size of lung granulomas compared to wild type mice. These mice should be an important tool in determining the role of CCR1 and have already shown that CCR1 has nonredundant functions in inflammatory conditions.

Major histocompatability complex proteins

Autoimmune diseases are often associated with the expression of specific major histocompatability complex (MHC) alleles. The expression of the HLA DR1 allele is strongly associated with the incidence of rheumatoid arthritis. Rheumatoid arthritis is an autoimmune disease, and although the antigen(s) which initiates the disease has not yet been determined, it is known that many patients have an autoimmune response to type II collagen (CII). Rosloniec et al. produced transgenic mice expressing a chimeric (human/mouse) HLA-DR1 gene [22]. These mice were used to determine whether the HLA-DR1 gene product was capable of mediating an immune response to the human CII antigen. Upon immunization with human CII, the transgenic mice developed a severe inflammatory arthritis, whereas the non-transgenic mice did not. In addition, antigen presentation assays determined that there was a strong DR1 restricted T and B cell response to human CII in the transgenic mice. The data show that the HLA DR1 molecule is able to confer susceptibility to arthritis by presenting antigenic determinants from the human CII protein. These mice will be a useful model for studying DR1 related arthritis.

Enzymes

The collagen-induced experimental model of human rheumatoid arthritis utilizes the DBA/1 mouse strain. Most ES cell lines used for performing gene targeting experiments are derived from the 129 mouse strain. In order to develop a system in which the role of different genes could be assessed in collagen-induced arthritis model, ES cells from the DBA/1 strain were developed by Griffiths and colleagues so that gene targeting could be performed in this strain. Using these DBA/1 ES cells, a null mutation in the 5-lipoxygenase-activating protein (FLAP) was generated [23]. FLAP is an integral membrane protein that is essential for leukotriene synthesis. As expected, in response to zymosan stimulation, a model of acute peritonitis, mice homozygous for this mutation were unable to produce leukotrienes, although substantial levels of prostaglandins were produced. When challenged in the collagen-induced arthritis model, the arthritis was significantly less severe in the FLAP deficient mice than the arthritis in wild type or FLAP heterozygous mice. These studies demonstrate the importance of FLAP and leukotriene production in acute and chronic inflammation.

Group II PLA_2 has been implicated in inflammatory processes in both man and other animals. Transgenic mice expressing human group II PLA_2 were produced using a 6.2 kb genomic fragment encompassing the entire gene as well as 1.6 kb of

5' untranscribed flanking sequence and 350 bases of 3' untranscribed flanking sequence [24]. Expression was found in several tissues, with the most abundant expression in the liver, lung, skin, and kidney. Serum PLA_2 activity levels were approximately eightfold higher than nontransgenic littermates. The group II transgenic mice exhibited epidermal and adnexal hyperplasia, hyperkeritosis, and almost total alopecia. The chronic epidermal hyperplasia and hyperkeritosis is similar to that seen in a variety of dermatopathies, including psoriasis. However, unlike what is seen with these dermatopathies, no significant inflammatory-cell influx was observed in the skin of these animals, or in any other tissues examined. More recently, these mice have been shown to have an increased susceptibility to arthritis when bred with human TNFα transgenic mice, and an increased susceptibility to LPS induced shock, suggesting that PLA_2 can exacerbate inflammatory conditions [25]. Consistent with this, there is growing evidence that PLA_2 plays a role in atherosclerosis, a condition which exhibits many of the characteristics of a chronic inflammatory condition. The group II PLA_2 transgenic mice have been found to have significantly decreased levels of HDL cholesterol and to be highly susceptible to atherosclerosis compared to non-transgenic mice after 12 weeks on either a high fat diet or a chow diet [26, 27]. In addition, it was determined that HDL from the PLA_2 mice was unable to prevent LDL oxidation and in fact was proinflammatory in an *in vitro* assay in which human aortic endothelial cells and human smooth muscle cells were co-cultured with LDL. These mice will be a good model for studying the inflammatory process and the role of group II PLA_2 in that process.

Summary

Transgenic technology has played an important role in inflammation research, helping to determine the role specific genes play in inflammatory conditions. As the pool of newly identified genes increases rapidly due to the worldwide efforts to characterize and sequence the human genome, the number of genes implicated in inflammatory conditions will also increase. The availability of gene addition and gene modification technologies and the relative similarity of the mouse and human physiology compared with simpler species where genetic techniques are available make the mouse a powerful system to study the mechanism of inflammatory conditions. This technology should help in our understanding of human inflammatory processes and should help in the design of novel therapies for the treatment of the debilitating diseases associated with inflammatory processes.

Acknowledgements
I thank Mark E. Swanson, Satbir Kaur, and Emily B. Cullinan for useful discussions and critical reading of the manuscript.

References

1 Swanson ME, Grass DS, Ciofalo VB (1994) Transgenic and gene targeting technology in drug discovery. *Annual Rep Med Chem* 29: 265–274

2 Hogan B, Beddington R, Costantini F, Lacy E (1994) *Manipulating the mouse embryo: a laboratory manual*, second edition. Cold Spring Harbor Laboratory Press, New York

3 McKnight RA, Shamay A, Sankaran L, Wall RJ, Hennighausen L (1992) Matrix-attachment regions can impart position-independent regulation of a tissue-specific gene in transgenic mice. *Proc Natl Acad Sci USA* 89: 6943–6947

4 Kioussis D, Festenstein R (1997) Locus control regions: overcoming heterochromatin-induced gene activation in mammals. *Curr Opin Genet Dev* 7: 614–619

5 Palmiter RD, Behringer RR, Quaife CJ, Maxwell F, Maxwell IH, Brinster RL (1987) Cell lineage ablation in transgenic mice by cell-specific expression of a toxin gene. *Cell* 50: 435–443

6 Saez E, No D, West A, Evans RM (1997) Inducible gene expression in mammalian cells and transgenic mice. *Curr Opin Biotechnol* 8: 608–616

7 Valera A, Solanes G, Fernandez-Alvarez J, Pujol A, Ferrer J, Asins G, Gomis R, Bosch F (1994) Expression of GLUT-2 antisense RNA in β cells of transgenic mice leads to diabetes. *J Biol Chem* 269: 28543–28546

8 Larsson S, Hotchkiss G, Andang M, Nyholm T, Inzunza J, Jansson I, Ahrlund-Richter L (1994) Reduced β2-microglobulin mRNA levels in transgenic mice expressing a designed hammerhead ribozyme. *Nuc Acids Res* 22: 2242–2248

9 Hasty P, Bradley A (1993) Gene targeting vectors for mammalian cells. In: AL Joyner (ed): *Gene targeting: a practical approach*. IRL Press, Oxford, 1–31

10 Mansour SL, Thomas KR, and Capecchi MR (1988) Disruption of the proto-oncogene int-2 in mouse embryo-derived stem cells: a general strategy for targeting mutations to non-selectable genes. *Nature* 336: 348–352

11 Gu H, Marth JD, Orban PC, Mossmann H, Rajewsky K (1994) Deletion of a DNA polymerase β gene segment in T cells using cell type-specific gene targeting. *Science* 265: 103–106

12 Keffer J, Probert L, Cazlaris H, Georgopoulos S, Kaslaris E, Kioussis D, Kollias G (1991) Transgenic mice expressing human tumour necrosis factor: a predictive genetic model of arthritis. *EMBO J* 10: 4025–4031

13 Probert L, Plows D, Kontogeorgos G, Kollias G (1995) The type I interleukin-1 receptor acts in series with tumor necrosis factor (TNF) to induce arthritis in TNF-transgenic mice. *Eur J Immunol* 25: 1794–1797

14 Pasparakas M, Alexopoulou L, Episkopou V, Kollias G (1996) Immune and inflammatory responses in TNF alpha deficient mice: a critical requirement for TNF alpha in the formation of primary B cell follicles, follicular dendritic cell networks and germinal centers, and in the maturation of the humoral response. *J Exp Med* 184: 1397–1411

15 Neurath MF, Fuss I, Pasparakis M, Alexopoulou L, Haralambous S, Meyer zum

Buschenfelde K-H, Strober W, Kollias G (1997) Predominant pathogenic role of tumor necrosis factor in experimental colitis in mice. *Eur J Immunol* 27: 1743–1750
16 Cambell IL, Abraham CR, Masliah E, Kemper P, Inglis JD, Oldstone MBA, Mucke L (1993) Neurologic disease induced in transgenic mice by cerebral overexpression of interleukin 6. *Proc Natl Acad Sci USA* 90: 10061–10065
17 Heyser CJ, Masliah E, Samimi A, Cambell IL, Gold LH (1997) Progressive decline in avoidance learning paralleled by inflammatory neurodegeneration in transgenic mice expressing interleukin 6 in the brain. *Proc Natl Acad Sci USA* 94: 1500–1505
18 Sligh JEJ, Ballantyne CM, Rich SS, Hawkins HK, Smith CW, Bradley A (1993) Inflammatory and immune responses are impaired in mice deficient in intercellular adhesion molecule 1. *Proc Natl Acad Sci USA* 90: 8529–8533
19 Hallahan DE, Virudachalam S (1997) Intercellular adhesion molecule 1 knockout abrogates radiation induced pulmonary inflammation. *Proc Natl Acad Sci USA* 94: 6432–6437
20 Rauschmayr T, Groves RW, Kupper TS (1997) Keratinocyte expression of the type 2 interleukin 1 receptor mediates local and specific inhibition of the interleukin 1-mediated inflammation. *Proc Natl Acad Sci USA* 94: 5814–5819
21 Gao J-L, Wynn TA, Chang Y, Lee EJ, Broxmeyer HE, Cooper S, Tiffany HL, Westphal H, Kwon-Chung J, Murphy (1997) Impaired host defense, hematopoiesis, granulomatous inflammation and type 1-type 2 cytokine balance in mice lacking CC chemokine receptor. *J Exp Med* 185: 1959–1968
22 Rosloniec EF, Brand DD, Myers LK, Whittington KB, Gumanovskaya M, Zaller DM, Woods A, Altmann DM, Stuart JM, Kang AH (1997) An HLA-DR1 transgene confers susceptibility to collagen-induced arthritis elicited with human type II collagen. *J Exp Med* 185: 1113–1122
23 Griffiths RJ, Smith MA, Roach ML, Stock JL, Stam EJ, Milici AJ, Scampoli DN, Eskra JD, Byrum RS, Koller BH, McNeish JD (1997) Collagen-induced arthritis is reduced in 5-lipoxygenase-activating protein-deficient mice. *J Exp Med* 185: 1123–1129
24 Grass DS, Felkner RH, Chiang M-Y, Wallace RE, Nevalainen TJ, Bennett CF, Swanson ME (1996) Expression of human group II PLA_2 in transgenic mice results in epidermal hyperplasia in the absence of inflammatory infiltrate. *J Clin Invest* 97: 2233–2241
25 Chapdelaine JM, Ciofalo VB, Grass DS, Felkner R, Wallace RE, Swanson ME (1995) Human extracellular (type II) phospholipase A_2 (PLA_2) transgenic mice provide a tool to determine the role of PLA_2 in inflammatory conditions. *Arthritis and Rheumatism* 38: S293
26 deBeer FC, deBeer MC, van der Westhuyzen DR, Castellani LW, Lusis AJ, Swanson ME, Grass DS (1997) Secretory non-pancreatic phospholipase A_2: influence on lipoprotein metabolism. *J Lipid Res* 38: 2232–2239
27 Ivandic B, Castellani L, Wang X-P, Qiao JH, Mehrabian M, Navab M, Fogelman A, Grass DS, Swanson ME, deBeer MC, deBeer F, Lusis AJ (1998) Role of group II secretory PLA_2 in atherosclerosis: 1. Increased atherosclerosis and altered lipoproteins in transgenic mice expressing group IIa PLA_2; *in press*

Gene transfer technology

Karen M. Anderson, Sandhya S. Nerurkar and Michael R. Briggs

Department of Cardiovascular Pharmacology, SmithKline Beecham Pharmaceuticals, P.O. Box 1539, 709 Swedeland Road, King of Prussia, PA 19406, USA

Introduction

Since the initial descriptions in the early 1970s of methods to successfully generate recombinant DNA molecules [1, 2], recombinant DNA technology has advanced to the point that researchers now have the ability to routinely introduce DNA into the genome of animals to experimentally alter patterns of gene expression *in vivo*. This gene transfer capability provides a powerful tool with which to evaluate the function of genes within the intact animal and to study the interplay of genes in tissue and organ systems. Perturbation of the system by overexpressing wildtype and/or mutant forms of genes allows for an elucidation of the role of the test gene in whole animal biology. It is a critical component of the post-genome era of functional genomics. In addition, these technologies provide the opportunity, through genetic manipulation, to create or mimic human disease in animal models.

Exogenous genes are introduced into cells by two basic approaches [3, 4]: (1) *ex vivo* gene transfer, an indirect method in which autologous cells are removed from the host, genetically modified by inserting DNA sequences in culture, and then returned to the same subject by transplantation, or (2) *in vivo* gene transfer, a direct method by which genes are delivered into the test subject via specific vectors to allow for cellular uptake and gene expression.

Functional genes have been transferred successfully to both germ cells and somatic cells. Germ cell genetic transfer relies on *ex vivo* gene transfer methods and results in the stable expression of the foreign DNA sequence in both germ cells and somatic cells, and transmission of the new DNA sequence (transgene) to the animal's progeny as a simple Mendelian trait [5]. The technique most often used to generate transgenic animals which have gene additions is direct microinjection of one-cell fertilized embryos whereas gene deletions or knock-outs are produced by gene transfer in embryonic stem cells by specifically targeting the site of integration in the genome [5]. Somatic cell gene transfer involves the insertion of a functional gene into a cell other than a germ cell. Genes inserted into somatic cells are not passed on to future generations and are rarely expressed long-term. Thus, we refer to ani-

In Vivo Models of Inflammation, edited by D.W. Morgan and L.A. Marshall

mals resulting from somatic cell gene transfer as "transient transgenics," and others have referred to them as "somatic transgenics" [6, 7].

Generation of germ cell transgenic animals, most commonly mice, is expensive both in terms of resources and time. Aside from the greater scientific issues of identifying and cloning a gene and successfully engineering a recombinant construct to introduce that gene into the germ cells, development of a transgenic strain takes many months and involves the cumbersome tasks of monitoring animal breeding, animal colony maintenance and genotyping. Nevertheless, the use of germ cell transgenic animals has increased dramatically during the last decade and has contributed substantially to our knowledge of the regulation of gene expression, developmental biology, immunology and cancer, among other fields. However, lethality during embryonic or neonatal development often confounds studies of gene function and prevents examination of the consequences of dysregulated gene expression in the context of the adult organism where most forms of acquired diseases are manifest [8]. In addition, some human diseases (e.g. atherosclerosis) do not have naturally occuring murine or rodent equivalents and therefore are difficult to model. In spite of these limitations, overexpression or knock-out of genes related to such diseases in mice or other rodents may still yield valuable information regarding the relevance of a gene product to a particular human disease. An advantage of somatic cell gene transfer is that it can be done at any stage of the animal's development and it therefore will allow for determination of gene product function and regulation independent of stage-specific effects on gene products [8]. In addition, since breeding is not involved, somatic gene transfer can be done in a broader range of animal species up to and including primates and it yields information more quickly.

The purpose of this chapter is to review the currently available methods used to generate transient trangenics as a means to develop animal models of human disease. The rate-limiting step in successful application of gene transfer technology to animals *in vivo* is the ability to deliver the experimental gene to the appropriate target cell or tissue [4, 9]. Efficient delivery of genetic material to cells or tissues generally requires delivery vehicles – so-called vectors – that encapsulate the gene and guide it to the target cells [9, 10]. There are two general categories of delivery vehicles, viral vectors and non-viral vectors [11]. This chapter reviews the vectors most commonly used for somatic gene transfer, introduces recently developed or emerging vector technology, and details applications of vector-mediated gene transfer to development of new animal models of disease. For those readers unfamiliar with gene transfer technology, a brief discussion of the general principles of gene transfer is included.

Principles of gene transfer

Gene transfer into host somatic cells involves the introduction of foreign DNA or gene sequences (transgenes) into the cells. Introduction of DNA into cells is generally referred to as transduction. Virally-mediated introduction of genetic material to cells is referred to as infection; all non-viral methods are referred to as the process of transfection. The major steps and tools involved in transfering DNA into somatic cells are shown in Figure 1, and the principle methods available for introduction of genes into cells and tissues are summarized in Table 1. After successful identification and cloning of a particular gene, standard techniques of molecular biology are used to insert the gene sequence into an expression vector plasmid. Within the plasmid are sequences that signal the origin of replication, promoter sequences which drive expression of the transgene under favorable conditions, and nucleotide sequence which conveys antibiotic resistance to cells harboring the plasmid to facilitate selective growth and isolation of transformed cells in bacterial cultures, and amplification of the recombinant plasmid. A variety of commercially prepared plasmid vectors containing various antibiotic resistance genes and promoter elements are available. A further refinement of the method involves introduction of specific promoter sequences (i.e. tissue-selective or regulatable promoters) [12]. This allows for more directed expression of the gene in specific target tissues and allows for

Table 1 - Vectors for gene transfer

I. Non-viral vectors

A Chemical methods of gene transfer
1. Calcium phosphate coprecipitation
2. Lipofection via cationic liposomes
3. Receptor-mediated gene delivery via glycoprotein-polycation conjugates

B Physical methods of gene transfer
1. Direct injection of naked DNA plasmids
2. Electroporation
3. Bioballistic or particle acceleration gene delivery

II. Viral vectors

1. Retroviruses
   - (Lentiviruses (e.g. human immunodeficiency virus, HIV)
2. Adenoviruses (Ad5, Ad2)
3. Adeno-associates viruses (AAV5, parvovirus)
4. Herpes-simplex virus
5. Vaccina virus
6. Sindbis and Semliki virus and other RNA viruses

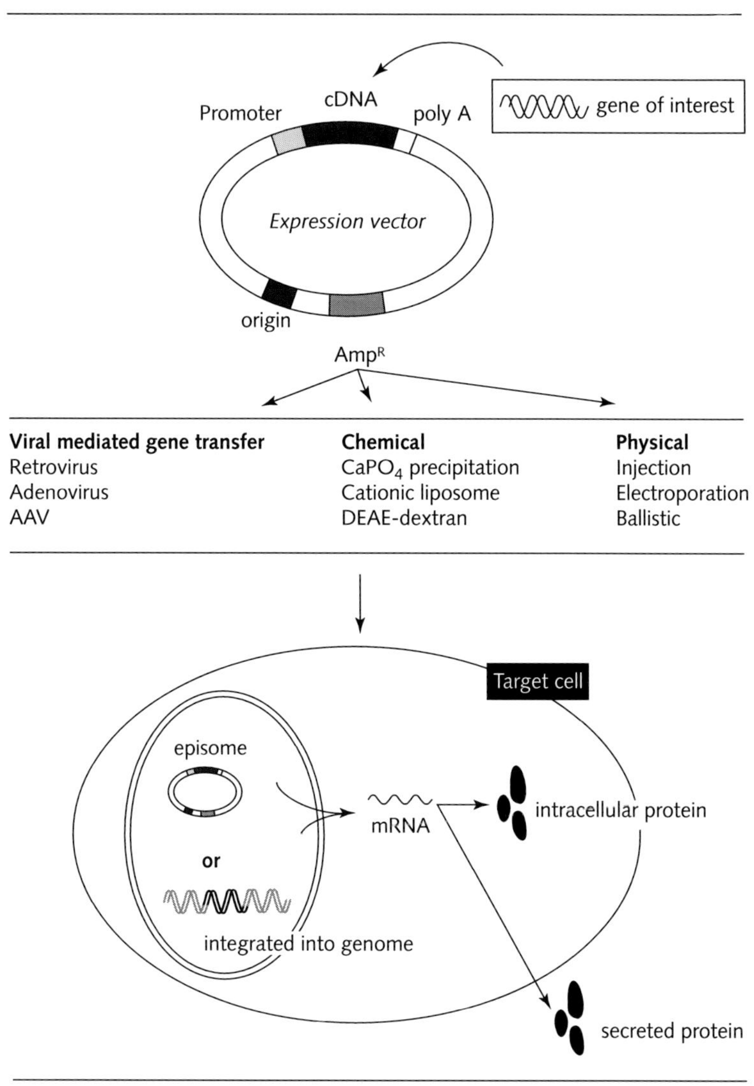

Figure 1
Somatic cell gene transfer
The gene of interest is cloned as a cDNA into an expression vector then transfected into target cells by any of the viral , chemical or physical methods. The vector then expresses the gene product as either an integrated part of the host genome or as an independent episome.

more controlled expression than the more global and constitutive expression often obtained with nonspecific promoters.

The next step entails introduction of the plasmid vector into the host cell. A number of methods have been used to achieve this step (detailed in the next section). In general, however, DNA enters cells by receptor mediated endocytosis or direct insertion by physical or chemical methods [3, 6]. Once inside the cell, the transformed DNA translocates to the nucleus. Once in the nucleus, the recombinant gene may integrate into the host genome, or it may exist as a separate and distinct extrasomal entity where it may or may not replicate [6, 10]. The new gene is transcribed into mRNA and translated into protein by host cell enzymes. Once the transgene is successfully introduced to the host cell and expressed, the resultant gene product may produce intracellular effects or extracellular effects, if a secreted protein. The phenotype associated with transgene expression provides a means to define the function of unknown genes and their novel or poorly characterized gene products, and more commonly, facilitates investigation of the physiologic consequences of overexpression of proteins with known activity and their relevance to disease.

Vectors for gene transfer

Introduction

Gene transfer requires the use of gene delivery vehicles, or vectors, that encapsulate or complex with the gene, guide it to the target cell and facilitate introduction of the DNA into the host cell [9, 10]. An optimal vector will have most of the following attributes [3, 11]: (1) exquisite tropism for the target cell to enable selective cellular uptake; (2) the ability to efficiently transduce nondividing as well as dividing cells; (3) the ability to carry a large functional gene "payload"; (4) selective integration within the genome to facilitate long-term expression without introduction of insertional mutagenesis; (5) transcriptional stabilizers; (6) gene expression regulators and modulators; (7) minimal cytotoxicity; (8) cell compartment tags; and (9) the ability to be manufactured readily and easily. Because no single such optimal vector yet exists, the Achilles heel of gene transfer remains an inability to both effect efficient gene delivery and sustained expression of the transgene [11]. Consequently, a new scientific discipline has emerged around vector technology and has resulted in the availability of numerous vectors and gene transfer techniques, each with differing efficiencies of transduction, and each with distinct advantages and disadvantages. [3, 4, 9–15]. Each of the vectors will likely have a role in gene transfer for individual applications that will be dependent on the target tissue, the size and number of genes to be delivered and the required duration of expression as determined by experimental design and objectives [13].

In general, there are two categories of vectors available for gene transfer (Tab. 1). The first comprises non-viral vectors, ranging from direct injection of naked DNA to complexes of DNA with polylysine or cationic lipids to allow genes to cross the cell membrane [3, 6, 7, 9, 11, 15]. The second category is viral vectors [3, 4, 6, 11, 15]. For the most part, viral vectors are more effective than non-viral vectors for achieving high-efficiency gene transfer but they have associated problems that render them less-than-optimal gene transfer vectors [4]. Details of the various vector technologies with their individual advantages and disadvantages are provided below.

Non-viral vectors for gene transfer

General characteristics of non-viral vector systems

Most non-viral methods of gene transfer rely on normal mechanisms used by mammalian cells for the uptake and intracellular transport of macromolecules [15]. Mammalian cells will take up DNA directly when it is in the form of a microprecipitate or when they are subjected to an electric current [16]. DNA can also be complexed with other macromolecules which condense it or enhance its ability to penetrate cells [9]. DNA complexes formed with lipids are called liposomes, whereas those formed with polycationic proteins are called molecular conjugates. Both chemical and physical techniques have been developed to facilitate non-viral tranfer of genes into cells (Tab. 1). Overall, the potential advantages of these techniques and non-viral vectors in general are increased safety and decreased immunogenicity. In some cases non-viral gene transfer methods represent a viable technical option for development of animal models of disease and the currently available techniques warrant review.

Non-viral vectors: chemical methods of gene delivery

Calcium phosphate coprecipitation exploits the natural ability of cells to take up DNA when it is in the form of a microprecipitate. This represents a very simple technique that is effective for delivering genes to a wide variety of cell types. However, the efficiency of gene transfer by this method is low and highly variable. In addition, the technique can be cytotoxic and it must be standardized for each cell type. To improve efficiency of gene transfer, DNA has been incorporated into a variety of lipid vesicle formulations generally know as liposomes [7] for delivery into cells by a process called lipofection. When plasmid DNA and the lipid particles are coincubated the lipid particles interact spontaneously with the DNA resulting in a liposome polycationic conjugate, and condense to form complexes in which nearly all the DNA is trapped in the aqueous interior [3, 6]. Liposome-DNA conjugates are simple to prepare and essentially have no limit to the size of the transgene that they can deliver. Fusion of liposomes with cell membranes enables delivery of the DNA to cells. Depending on the cell type, high transfection efficiency may be realized. Using this method, genes have been successfully transferred to a wide variety of target cells – both quiescent and dividing, and both *in vitro* and *in vivo* – with minimal toxicity [3, 6]. Thus, liposomal delivery of genes has promise as a useful technique in development of animal models. The duration of gene expression following successful lipofection is on the order of weeks to months [6]. Thus this approach is most suitable for studies of acute disease processes. Another means for transferring DNA to cells involves formation of complexes between glycoprotein-polycation conjugates and the plasmid DNA containing the transgene [3, 6]. The soluble DNA/glycoprotein-polycation complex is recognized by cell surface receptors on the target cell and enters the cell by receptor-mediated endocytosis. This is a highly efficient method of delivering large copy numbers of DNA to cells and, importantly by carefully designing the glycoprotein-conjugate, can result in recognition of the complex by specific cell surface receptors thereby conferring cellular selectivity. As is the case with liposome-mediated transfer, the DNA complex is delivered to endosomes and ultimately lysosomes where most of it is degraded. Thus expression, although it may be robust, occurs over a limited period of time.

An excellent example of the application of cationic lipid-mediated gene transfer technology in the laboratory to understand a disease process is found in the work of Nabel and colleagues [17-19], who demonstrated independent changes in vascular smooth muscle cell proliferation, vascular extracellular matrix elaboration and angiogenesis (vasa vasorum) following transfer of different growth factors in a vascular injury model of restenosis. Together with work from Isner and his colleagues [20] who demonstrated successful *in vivo* gene transfer to atherosclerotic rabbit arteries, it is apparent that liposomal-mediated gene transfer technology will be valuable in defining molecular mechanisms of vascular disease.

Non-viral vectors: physical methods of gene delivery

Direct injection of naked DNA plasmids is an extremely low efficiency method of gene transfer which has been used successfully to transfect skeletal and cardiac muscle which were refractory to transfection by other means [3, 7, 21]. The low efficiency of this method limits its application, but it has been sufficient for certain applications such as cellular secretion of potent growth factors encoded by the transgene [7]. The transgene is not integrated into the host cell DNA following injection. Thus, expression of the transgene is generally short-lived as it is rapidly degraded. Electroporation, or the formation of transient hydrophilic pores in cell membranes upon subjecting the cells to an electric current has become a popular method for delivering DNA into cultured cells that are difficult to transfect [3], and therefore is more relevant to studies with *ex vivo* gene transfer. Commercial portable electroporators are available which precisely control the voltage and other parameters important in electroporation of cells, thereby allowing optimization of DNA delivery with minimal host cell trauma. Bioballistic or particle acceleration gene delivery involves coating microparticles with DNA and forcefully injecting them into cells or tissues as a means of delivering the transgene [3]. This technique was originally developed to introduce genes into plant cells and, to our knowledge, has not gained acceptance as a routine method of gene transfer into animals or animal cells. Nevertheless, it is a simple and versatile technique and it reportedly can be used in both *ex vivo* and *in vivo* delivery of genetic material [3].

In general, most of these non-viral gene transfer approaches are characterized by poor transfection efficiency and transient expression of the transgene. Although liposomes and molecular conjugates increase the efficiency of *in vivo* gene transfer, they still produce mostly transient expression of the transgene. Thus, unless one is interested solely in examination of acute phenomena, these polycationic technologies still fall short of the optimal vector system desired for development of relevant animal models.

Viral vectors for gene transfer

General characteristics of viral vector systems

The initial conceptualization and use of recombinant versions of viruses as vectors represented an important breakthrough in the development of methodology to effect successful somatic cell gene transfer [9]. For the most part, viral vectors are more effective than non-viral vectors in achieving high-efficiency gene transfer [4]. As such, viral vectors can facilitate specific and efficient introduction of genes into differentiated cells that are generally refractory to chemical or physical non-viral transfection protocols [22]. Many viruses have evolved methods for infecting mammalian cells, and once they successfully enter the cell, viruses rely on the host cell biosynthetic pathways to produce viral DNA, RNA and protein. In this context viruses are a natural gene transfer vector [6]. To create recombinant viral vectors the life cycles of their naturally occurring counterparts are exploited. The challenge has been to disable the viruses so that they are no longer able to replicate without affecting their ability to shuttle their genome containing the transgene into the desired cell [9]. In this regard, the single most important advance in the development of viruses as gene transfer vectors was the generation of specialized cell lines (termed "packaging or helper cells") that permit the production and purification of high titres of replication-defective recombinant virus free of wild-type virus [15, 23]. The viruses used have been substantially attenuated and disabled, in principle, of most intrinsic pathogenic effects. However, during *in vivo* applications, the animal's immune system responds to the foreign invading virus and efforts to deliver genes in viral vectors have been confounded by these host responses [11]. Thus, immunogenicity of viral vectors is an issue which must be considered during design of experiments and during selection of a gene transfer protocol. Viruses currently used as vectors for gene transfer are listed in Table 1. In this chapter, we review viral vector systems based on three different virus groups: retroviruses, adenoviruses and adeno-associated viruses. Herpes-based vectors, which are not as commonly used as these other viruses are reviewed elsewhere [4, 13, 14].

Recombinant retroviral vectors for gene transfer

Genome and life cycle of wild type retrovirus

Retroviruses are enveloped single-stranded RNA viruses. Retroviruses derive their name from the fact that during their replication their RNA is converted to DNA, the reverse (i.e. retro) of the normal flow of genetic information [16]. This feature of retroviruses is unique among animal viruses. The retrovirus genome is composed of two identical RNA molecules and contains three open reading frames designated *gag*, *pol* and *env* which encode structural proteins, viral enzymes and envelope glycoproteins, respectively. The diploid virus particle (Fig. 2) contains the two RNA strands complexed with *gag* proteins, viral enzymes (including reverse transcriptase, ribonuclease H and endonucleases) and host tRNA molecules within a core structure of *gag* proteins. Surrounding and protecting this capsid is a lipid bilayer, derived from host cell membranes and containing viral envelope proteins, generating a viral particle of about 100–150 nm across. The *env* proteins bind to membrane-bound receptors on the host cell and initiate viral infection of the host cell via receptor-mediated endocytosis (Fig. 3). Following infection, the outer envelope is shed and the viral RNA is copied into DNA by reverse transcription by the polymerase encoded by the viral *pol* sequences. Second strand DNA synthesis generates a double stranded DNA molecule – the provirus – which is then randomly integrated into the host genome. The provirus is flanked at both ends by identical elements called long terminal repeats (LTR) which are required, along with the *pol* product integrase, for integration into the host genome. The LTRs contain all the necessary and sufficient transcriptional regulatory sequences required for expression of the viral genome. An additional genomic element of importance is the packaging sequence (ψ) which functions to distinguish the viral RNA from other RNAs in the cell, and which identifies the viral RNA for packaging into virions.

Generation of recombinant retroviral vectors

The overall objective of performing gene transfer is to deliver a functional gene into the nucleus of a target cell with the goal that the target cell express that gene. The

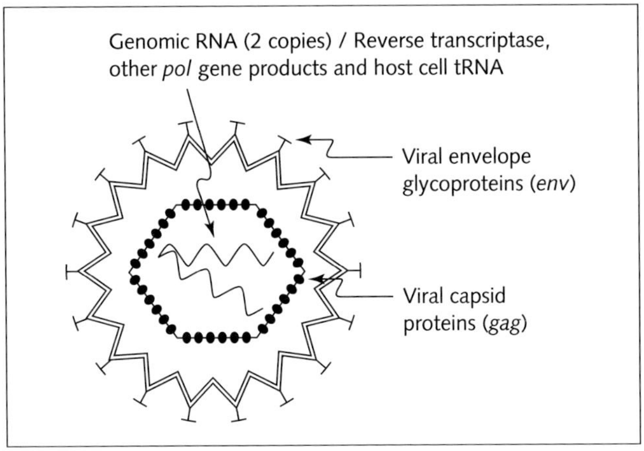

Figure 2
Schematic of key features of retrovirus.

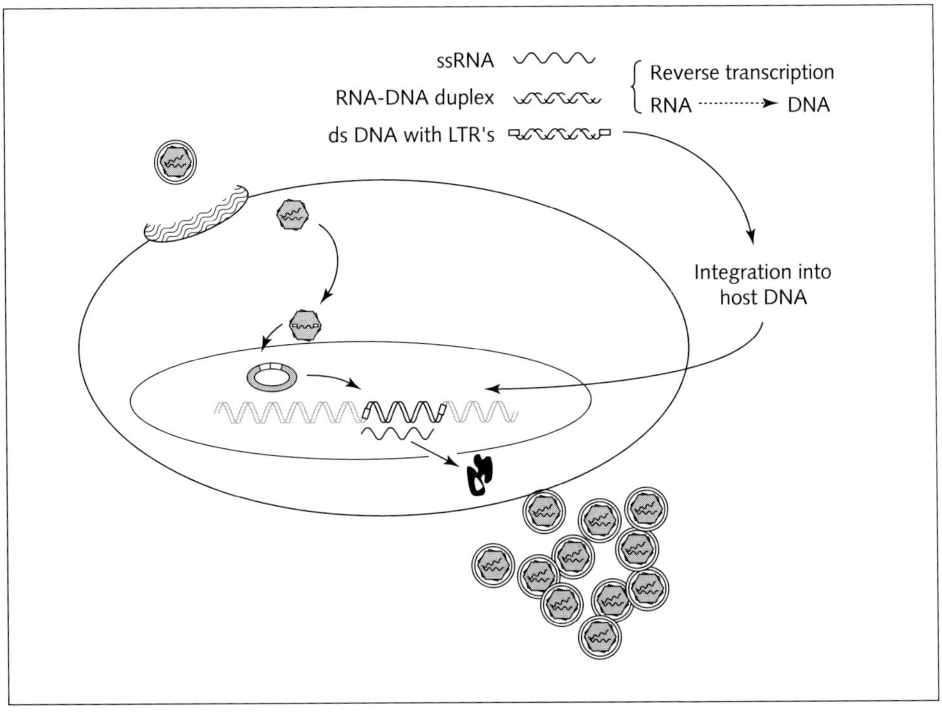

Figure 3
Wild type retrovirus life cycle.

intent of using viruses as the gene delivery vehicle is to take advantage of their ability to efficiently infect cells and deliver their genome to the host cell nucleus. Insertion of a mammalian DNA sequence (transgene) into the viral genome results in co-delivery of the transgene to the host cell nucleus where, under direction of an appropriate promoter for the target cell-type, it will be transcribed. It is not desirable that research subjects or patients experience symptoms of viral infection following exposure to viral vectors. Thus, a second objective to consider during construction of a recombinant viral vector is to render the virus replication-defective and to minimize its immunogenicity. Routine molecular techniques allow *in vitro* manipulation of the viral genome to effectively replace the viral genes involved in replication (thereby making a replication-defective particle) with the transgene (to enable gene transfer). Making the virus replication defective is not consistent with the need to purify high titers of the recombinant retroviral vector. Development of packaging cells which express the viral genes involved in replication and packaging in trans has largely overcome this apparent dilemma.

Figure 4 details schematically the steps required for generation of a recombinant retroviral vector and its subsequent use to create a transgenic cell. Step 1: a plasmid (helper plasmid) containing the retroviral genomic sequences encoding *gag*, *pol* and *env* but not the packaging sequence (ψ) is stably transfected into a suitable cell line, e.g., the human embryonic kindey 293 (HEK-293) cell line. This creates a cell line which constitutively expresses the products of *gag*, *pol* and *env*, i.e. the enzymes and structural proteins involved in viral replication and packaging. Despite production by these cells of viral gene products and viral RNA, the viral RNA will not be packaged into virions because it lacks the ψ packaging recognition sequence. These cells will package any subsequently introduced RNA that contains ψ and, therefore they are called packaging cells. Step 2: Subsequent transfection (stable or transient) into the packaging cell of a recombinant construct containing the cDNA of the gene of interest (transgene) in a viral genomic backbone which retains the ψ sequence, but from which the *gag*, *pol* and *env* genes have been deleted. Step 3: The packaging cells provide (in trans) *gag*, *pol* and *env* gene products (e.g. reverse transcriptase, viral proteases, integrase) which regulate replication and nuclear integration of the recombinant construct, encapsidation, viral RNA synthesis and infection. In the presence of these constitutively expressed proteins the recombinant construct translocates to the nucleus where it makes a viral RNA transcript. Step 4: Because this RNA carries the ψ packaging signal it is packaged into virions and shed by the packaging cell. A recombinant replication-defective virus which contains an RNA transcript encoded by the transgene has now been generated. Conditioned medium from these packaging cells containing viral particles from lysed cells may be used to isolate and concentrate the recombinant viral vector or may be used to infect target cells directly. Presently, packaging cell lines that produce titers of at least 10^7 infectious virus particles per ml are available [4]. Step 5: Infection of a target cell with the recombinant retroviral vector. The normal viral life cycle promotes integration

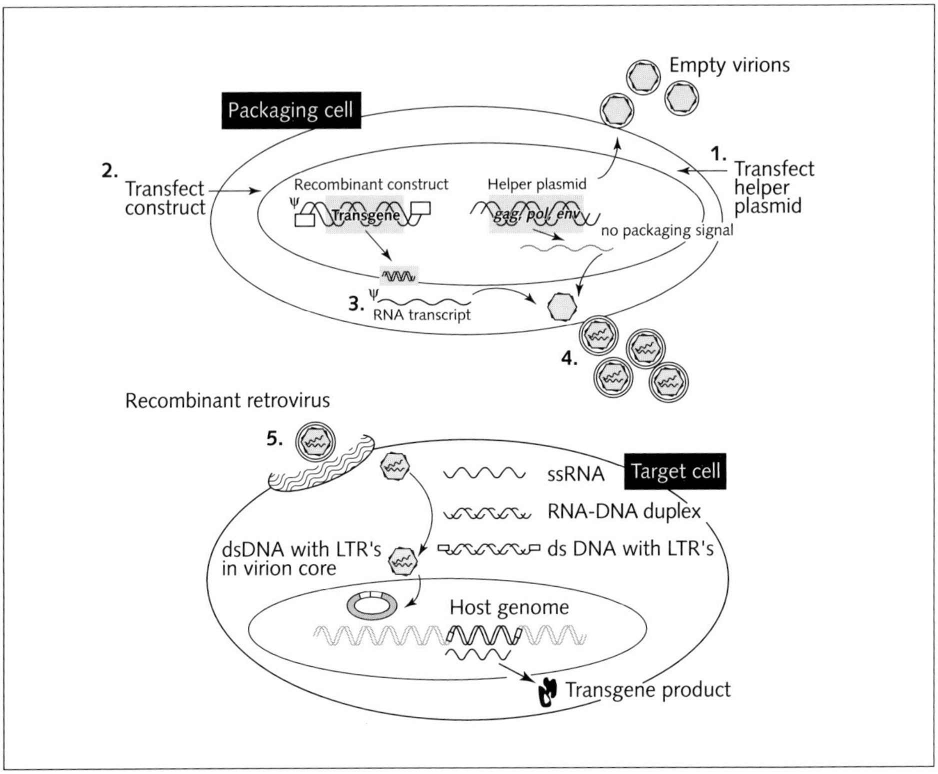

Figure 4
Generation of a recombinant retroviral vector and its subsequent use to create a transgenic cell.
Step 1: a helper plasmid containing the retroviral genomic sequences encoding gag, pol *and* env *but not the packaging sequence (ψ) is stably transfected into a suitable cell line to create a cell line which constitutively expresses the enzymes and structural proteins involved in viral replication and packaging. These cells will not package the viral RNA but will package any subsequently introduced RNA that contains ψ and, therefore they are called packaging cells. Step 2: Transfection of a recombinant construct containing the cDNA of the gene of interest (transgene) in a viral genomic backbone which retains the ψ sequence, but from which the* gag, pol *and* env *genes have been deleted. Step 3: In the presence of the trans* gag, pol *and* env *gene products the recombinant construct translocates to the nucleus where it makes a viral RNA transcript. Step 4: Because this RNA carries the ψ packaging signal it is packaged into virions and shed by the packaging cell. Step 5: Infection of a target cell with the recombinant retroviral vector. The normal viral life cycle promotes integration of the viral genome (including the transgene) into the host cell, and the machinery within the cell enables expression of the transgene. The target cell now expresses the new retrovirally-introduced gene but not infectious virus.*

of the viral genome (including the transgene) into the host cell, and the machinery within the cell enables expression of the transgene. The target cell now expresses the new retrovirally-introduced gene but not infectious virus.

Gene transfer via recombinant retrovirus vectors

The major advantage of retroviruses is that they integrate their viral genome into the host cell chromosomal DNA. Thus, the retrovirus genome becomes a permanent part of the host cell genome and it is expected that any foreign gene placed in the retrovirus also will be expressed in host cells [6, 9]. In addition, integration into the host genome should assure long-term stable expression [9]. Together with the high infection efficiency usually realized by retrovirus vectors, it appears that recombinant retroviral vectors fulfill two major requirements of an ideal vector [3, 11]. However, several features of retroviruses may limit their applications, particularly with regard to *in vivo* gene transfer [15]: (1) retroviral entry into cells is absolutely dependent on the existence of the appropriate viral receptor on the target cell and, therefore, problems encountered in transducing specific cell types may be due, at least in part, to the lack of expression of appropriate receptors; (2) replication of the target cell is necessary for proviral integration to occur; i.e. the cell cannot be terminally differentiated; (3) the retroviral particle is low titer and relatively labile in comparison to other viruses. Both the instability of retrovirus particles and the inability of retroviruses to integrate in nonreplicating cells accounts for most of the difficulties reported during attempts to use retroviral vectors for gene transfer *in vivo*; and, (4) retroviruses are species specific and while much work has been reported with rodent-based vector systems, their application in primate and/or human biology has been limited and usually requires at least P3 level biosafety containment protocols because of user safety concerns. Despite their limitations, retroviruses remain a popular vector option because they have been genetically optimized and evolved to efficiently insert, integrate, activate and express their own genetic material in mammalian target cells with very high efficiency [15]. Traditional retroviral vectors will, however, most likely continue to be limited primarily to *ex vivo* gene transfer procedures.

Interestingly, it was recently discovered that lentiviruses (which belong to the retrovirus family and include HIV) can infect a wide range of non-dividing cells and should, therefore, enable more diverse applications for retroviral vectors [4]. Lentiviruses achieved long-term (> 6 month) expression when injected into rodents [11]. Little is known about the possible immune problems associated with lentiviral vectors, but injection of 10^7 infectious units did not elicit a cellular (skin) immune response at the site of injection. Furthermore, there seems to be no potent antibody response. So, lentiviral vectors seem to offer an excellent opportunity to perform *in vivo* gene delivery with sustained expression. Unfortunately, lentiviral constructs have several deficiencies: they possess no greater coding capacity than other retroviral vectors (7–7.5 kb) [11], the recombinant construct remains relatively labile, and there are concerns about the ability to make truly replication defective and non-pathogenic HIV vectors. Further research will be required before these vectors become widely used.

Application of retrovirus-mediated gene transfer to animal model development

Retroviral vectors have been used successfully in numerous studies to develop animal models of human disease [6]. A critical limitation, however, to the use of retroviral vectors in animal models of disease is their inability to infect non-dividing cells, such as those that comprise muscle, brain, lung and liver tissue [11]. Thus, experiments with retroviruses generally rely on methods of *ex vivo* gene transfer which involves isolation of cells from the target tissue, maintenance and infection with the recombinant retroviral vector in culture, characterization to confirm cellular production of the transgene product, and transplantation back into the animal. Despite the technical demands of this approach, replication-defective retroviruses remain the vectors of choice for animal studies examining diseases of the lympho-hematopoietic system [24]. Examples of successful application of this approach include persistent, retroviral-mediated expression of adenosine deaminase in hematopoietic stem cells of mice, dogs, sheep and monkeys [24], and sustained abundant expression of factor IX in transplanted murine primary fibroblasts [11]. There is an extensive literature describing studies of this nature done in hematopoietic stem cells, immune cells, peripheral blood cells and tumor cells for both basic research and gene therapy (for reviews see [11, 15, 24–26]). There are examples of successful retroviral-mediated transfer of genes to cells *in vivo*. To produce a model of autoimmune vasculitis, a disease characterized by focal inflammation within arteries, Nabel et al. [27] introduced the human HLA-B7 gene into porcine femoral arteries by retroviral vectors and DNA liposome complexes. Expression of the HLA-B7 gene stimulated vascular inflammation characterized by adventitial mononuclear infiltration.

Recombinant adenovirus vectors for gene transfer

Genome and life cycle of wild type adenovirus

Compared to retroviral vectors, vectors based on adenoviruses have greater potential to successfully deliver genes *in vivo* because they can efficiently transfer genes into a wide variety of dividing and non-dividing cells. Human adenovirus is a lytic

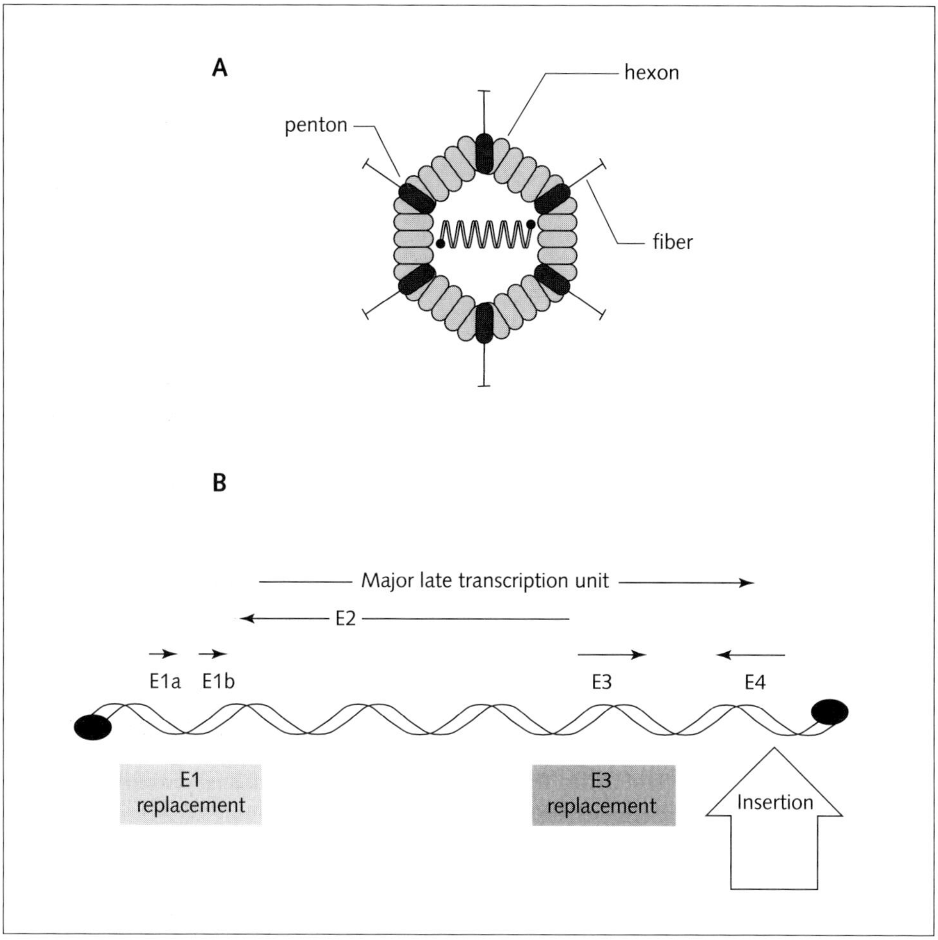

Figure 5
Adenovirus structure and cloning sites. (A) Important features of adenovirus for infection. (B) Usual cloning sites in "First Generation" recombinant adenovirus vectors. E1 replacement is the most common technique.

non-enveloped, icosahedral double-stranded DNA virus approximately 130 nm in size (Fig. 5A) [28]. To date, 49 serotypes of human adenovirus have been identified [29]. They are divided into six subgroups (A–F) on the basis of shared immunological and biochemical properties, and they are associated with a variety of acute infections, primarily respiratory, ocular and gastrointestinal [16]. Human adenoviruses types 2 and 5 of subgroup C were originally developed for *in vivo* gene transfer to human lung because of their presumed natural tropism to pulmonary epithelial cells [9], and they are now used extensively as gene transfer vectors [29]. The linear adenoviral genome (Fig. 5B) is approximately 36 kb in length and encodes multiple overlapping transcriptional units [28]. The genome of adenoviruses is larger and more complex than the genome of retroviruses. Adenovirus origins of replication are located at the ends of the genome within inverted terminal repeats (ITR) of about 100 bp. Covalently attached to each 5' end of the DNA is a 55 kDa protein of viral origin (terminal protein, TP). This protein directs the viral chromosome to specific sites of the nuclear matrix where viral transcription is initiated [30]. Expression of adenoviral genes occurs in two phases [16]: at early times after infection six blocks of genes are transcribed – E1a, E1b, E2a, E2b, E3 and E4 (Fig. 5B). E1a encodes proteins with the properties of a transcriptional regulator whereas E1b gene products are involved in both host cell shut-off and mRNA transport. Proteins encoded by the E2a and E2b regions are directly involved in viral DNA replication. E3 gene products are dispensable for growth in tissue culture, but are thought to be involved in modifying the infected cell such that it can escape immune surveillance by the host. The products of the E4 region are thought to be transcriptional activators, although their mode of action is less well defined. Late phase transcription predominantly initiates from the major late promoter. Protein products of the late genes are mainly components of the virion and proteins involved in viral particle assembly. In the vicinity of the left ITR is an indispensable cis-acting sequence – the packaging or ψ sequence-which is required for proper packaging of the progeny chromosomes during the terminal phase of the infection. The viral genome is surrounded by a virally encoded protein coat called the capsid (Fig. 5A). The capsid consists of three major subunits [12]: (1) the hexon which comprises the bulk of the coat; (2) the penton base; and (3) the penton fiber. Together, the penton base and the penton fiber are referred to as the penton complex. During infection, the fiber mediates initial binding of the virus to the Coxsackie virus receptor [31]; the penton base subsequently mediates viral particle internalization by interactions with αv integrins and subsequent receptor-mediated endocytosis in clathrin-coated pits. Thus, the penton complex is responsible for binding and internalization of the viral particle and, therefore, for viral tropism at the level of cell recognition. Following escape from lysosomes, viral DNA replication occurs in the cell nucleus. Although integration of adenoviral DNA sequences into chromosomal DNA of the target cell can occur, particularly at high multiplicities of infection in nonpermissive cells, or in replicating cells integration does not appear to be an integral part of the viral life cycle and rarely occurs [6, 15, 32]. Instead they are replicated as separate extrachromosomal entities (episome) in the nucleus of the host cell [11]. During wild-type lytic infection, the viral genome is replicated to several thousand copies per cell [28]. Replicated genomes rapidly associate with core proteins and are later packaged into capsids formed by self assembly of the major capsid proteins. The size of the adenoviral genome which can be packaged into nascent capsids is limited to approximately 105% of native genome length (~38 kb), and is an important consideration in the construction of recombinant adenoviral vectors. Likewise, inserts which result in much shorter genome sizes are also inefficiently packaged by the virus, but this problem can be overcome by inserting extra DNA "stuffer fragments" along with the gene of interest.

Generation of recombinant adenovirus vectors

Recombinant adenovirus vectors are easily constructed, analyzed and propagated using standard recombinant DNA and virological techniques [22, 28, 29]. Foreign genes have been inserted into recombinant adenoviral genomes as replacements for the E1 or E3 regions, or as insertions between the E4 region and the end of the adenovirus genome [28]. Vectors based on the insertion of foreign genetic material only into the right end of the adenovirus genome or in place of early region 3 are replication competent. However, because of the importance of the E1 gene products to viral gene expression and, ultimately replication of the viral genome, replacement of the E1 region by a foreign gene of interest results in a replication-defective virus [28]. While recombinant adenoviruses of each class have been used as gene transfer vectors, the replication-defective adenoviruses based on replacements of E1 are most commonly used. These E1 replacement adenoviral vectors are often referred to as first-generation vectors. Many of these first generation vectors have 2 kb deletions in the middle of E3 to provide additional cloning capacity beyond the usual 7–8 kb insertions accommodated by the adenovirus [32]. Expression of the transgene is driven by either the E1 promoter or more often by an exogenously inserted promoter-enhancer sequence such as the cytomegalovirus (CMV) early gene promoter. Replacement of E1, since it is required for the efficient expression of the remainder of the viral genome, generates a vector which can only be propagated in a complementary cell line which supplies the missing E1 (specifically E1a) functions in trans from an integrated copy of the appropriate fragment of the viral genome [29]. HEK-293 cells contain an integrated copy of the leftmost 12% of the adenovirus 5 genome and thus provide a functional E1a gene product in trans, and are routinely used as the complementary cell line for propagation of the recombinant adenovirus [28, 29]. Other cell lines capable of providing E1a in trans (e.g. 911 cells) may also be used [23]. Recombinant viruses produced in this manner can be stably propagated to very high titers and are readily purified free of contaminating viral proteins and empty capsids via CsCl density gradient ultracentrifugation for *in vivo* use [29]. Figure 6 diagrams the basic steps involved in the generation of a recombinant adenovirus. Briefly, the adenovirus genome is ligated into an expression vector (e.g. Ad 5 genome in pMJ17, available as a stock plasmid) while the gene of interest and its promoter-enhancer elements (if required) are ligated into a second plasmid with homology to the left end of the adenovirus genome. The two plasmids are then co-

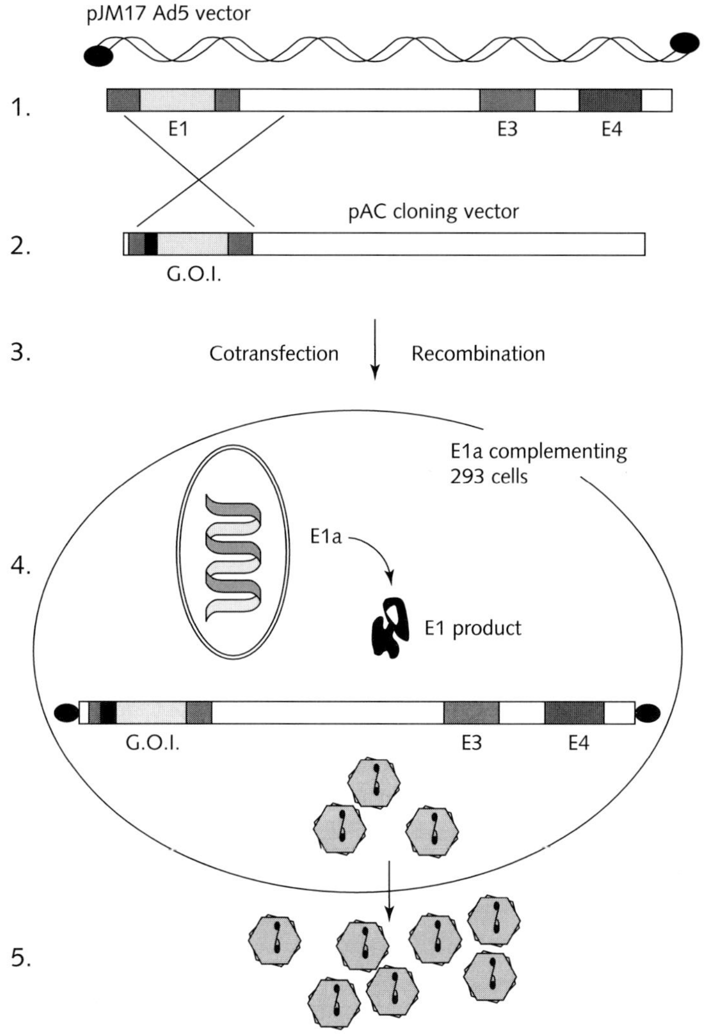

Figure 6
Recombinant adenovirus
1, 2: sub cloning srategy 3: Cotransfection of pJM17 and transgene-containing pAC vector leads to homologous recombination with gene of interest (G.O.I) 4: G.O.I. becomes inserted into Ad5 genome in place of E1 sequences leading to 5: recombinant adenovirus.

transfected into HEK-293 or another E1a complementary cell line for propagation. The transfected cells are incubated to allow homologous recombination between the tranfected DNA sequences and then identified by observation of plaque formation on soft agar plates. Highly purified and concentrated recombinant adenovirus constructs are then routinely obtained via several rounds of plaque purification on monolayers of 293 cells followed by ultracentrifugation. Detailed and specific protocols for generation of recombinant adenovirus by this method of homologous recombination are provided elsewhere [28].

Although easy to perform and widely used, this process of homologous recombination and plaque purificaton is lengthy. As such, alternative strategies have been developed to generate recombinant virus via more straight forward clonal techniques [30]. These include methods by which the recombinant adenovirus is obtained via *in vitro* ligation [28] and a newer simplified protocol which generates the recombinant adenovirus in *E. coli* rather than mammalian cells and which does not require plaque purification [33]. These newer methods show great promise to simplify and make more accessible the adenovirus technology, but have not as yet been widely applied.

Gene transfer via recombinant adenovirus vectors

Recombinant adenoviruses are highly efficient vectors for the transfer of foreign genes into cells. Compared to retroviral vectors, the single greatest advantage of adenovirus vectors is that they can infect both dividing and non-dividing cells. This property of adenoviruses enables infection of a broad range of mammalian cell types. Together with the relative ease of generating highly concentrated stocks, the efficient expression of recombinant gene products, and their high infection efficiency, this makes them the preferred gene delivery vector for many *in vivo* applications [9, 15, 28, 29]. Unfortunately, universal application of adenoviral vector-mediated gene transfer is thus far hampered by the fact that expression of the transgene is usually transient (weeks) and that expression after a second administration of vector is inefficient or impossible because of an elicited immune response [10]. In contrast to the retroviral vectors, long-term transgene expression can be achieved with adenovirus vectors if the recombinant adenovirus vectors are introduced into nude mice or if they are given together with immunosuppressive agents, thus indicating that immune recognition of the foreign viral particle and its components is responsible for the transient expression [11]. It appears that the host immune reaction includes both a cellular cytotoxic T lymphocyte response and a humoral response in which antibodies are directed against adenoviral proteins [9, 11]. Additional studies are underway to identify and characterize the proteins responsible for the immune response with the hope that their removal from the vector will enable longer transgene expression. The recombinant protein itself can be recognized as a foreign antigen if it is from a different species or introduced into a naïve animal such as a knockout animal (e.g. murine LDL receptor is recognized as foreign in the LDL receptor knockout mice). In spite of these limitations, current generation adenovirus vectors represent a very effective means for gene delivery and are gaining wide acceptance by the research community in the study of gene function via transient transgenic adult animals.

Application of adenovirus-mediated gene transfer to animal model development

The development of gene-transfer technology in whole animals has afforded unprecedented opportunities for investigators to research complex regulatory systems *in vivo* [5]. The successful development of adenovirus vectors which are easy to construct and which efficiently transfer genes of interest to a wide range of cells and tissues *in vivo* should allow routine use of this powerful technology in this emerging era of functional genomics. Recombinant adenovirus methodology, with its advantages of speed and relatively low cost, may provide important data to support longer term traditional transgenic or knockout animal experiments. One significant benefit of the transient transgenic technologies, in general, is the expression of a test gene in the tissue of interest in the adult animal, without the complicating factors of effects of the gene's expression (or absence thereof) during development. In addition, it is possible to use the recombinant adenovirus in many mammalian species including rodents, rabbits and monkeys whereas transgenic technology has been primarily limited to rodent species because of the time and cost constraints of obtaining germline transmission in larger animals.

Experiments in animal models have already demonstrated the potential of adenoviral vectors for achieving high-level gene transfer *in vivo*. For investigational purposes, the transient expression of adenoviral vector-mediated transgenes may be sufficient. In fact, depending on the particular questions being asked and the goal for the gene transfer experiment, transient expression may be an advantage. Recent functional *in vivo* tests of expression from recombinant adenoviruses have shown a half-life of about 6 weeks following intravenous and intramuscular administration into adult animals [32]. It is possible to see the effect of the transgene correlate with its increased expression in the model and then watch the phenotype return to normal as the transgene expression gradually diminishes, thereby giving a more dynamic picture of the function rather than simply obtaining constitutive expression as occurs with traditional transgenic animals.

Adenoviral-mediated gene transfer has proven a powerful tool for development of animal models. Several excellent examples include the successful delivery of genes to the vasculature [6, 29] and the heart [6, 34–37] to model cardiovasular disorders and to demonstrate the feasibility of adenovirus-mediated delivery of various genes for gene therapy approaches to hypercholesterolemia, thrombosis and restenosis [29, 38]. An exploding literature indicates exciting opportunities to use this technology to examine both normal physiology and pathology for a variety of organ systems.

Adeno-associated virus

A relative newcomer to the field, adeno-associated virus (AAV) is a simple, non-pathogenic single-stranded DNA virus [11]. The virion is icosahedral with a diameter of 20–30 nm, is non-enveloped, and contains a genome of about 5 kb [39]. AAV is a member of the Parvovirus family of human viruses [4]. There are five human serotypes (AAV 1–5) and they constitute the genus Dependovirus, so-named because of their unusual requirement for a helper virus coinfection to enable productive AAV infection to occur [32, 39]. Without a helper virus, such as adenovirus or herpes simplex virus, AAVs integrate into the host genome and remain dormant as a provirus [32]. AAVs are widely prevalent [32]. Their natural tropism is presumably the respiratory and gastrointestinal tracts, but all human cells tested *in vitro* have been successfully transduced [32].

The potential advantage of AAV vectors is that they appear capable of long-term expression in nondividing cells, possibly due to integration of the viral DNA into the host genome, and lower immunogenicity suggested to be due to its small size [26]. Thus, AAV may represent a gene transfer vector which blends the advantages of both retroviruses and adenoviruses into a vector system that could have applications in both *ex vivo* and *in vivo* gene transfer [9].

The AAV genome has been sequenced and is 4680 nucleotides (nt) long with 145 nt inverted terminal repeats (ITR) [32]. AAV contains only two open reading frames, rep and cap, encoding polypeptides important for replication and encapsidation, respectively. These two genes can be supplied in trans with only the ITRs required in cis for viral replication to occur [4]. Thus, a transgene with its appropriate regulatory sequences can be inserted between the two ITRs, and the recombinant viral vector generated by cotransfection into HEK-293 cells with a helper plasmid expressing *rep* and *cap*, and subsequent infection with a first-generation adenoviral vector [4]. Recombinant AAV vector is released only with lysis and death of the producer cells; and despite high viral particle concentration, the percentage of particles containing infectious functional vector may be quite small [25]. Recombinant virus must then be separated from any contaminating helper virus.

AAV vectors can accommodate only 3.5–4.0 kb of foreign DNA [11] and, as described above, preparation of the vector is laborious. Nevertheless, AAVs are becoming more popular as vectors in gene transfer experiments. The degree of infection of muscle, brain and liver cells with recombinant AAV is exceedingly high *in vivo*. Thus, AAV may be highly suitable for the delivery of genes to specific target cells *in vivo*, without inducing an immune response to the infected cells [4]. Importantly, use of AAV vectors in animal models has led to long-term expression of the transgene [39]. An example of the successful application of this technology is the long-term (> 6 months) production of therapeutic amounts of factor IX in liver and muscle of immunocompetent mice [11]. In order for AAVs to be used more routinely, technologies must be developed to enable less cumbersome production of the recombinant vector. To this end, efforts are underway to discover methods to produce AAV without the need for helper virus co-infection [40]. One goal is to achieve this in a stable producer cell line [25]. Ding et al. [41], recently described successful *in vitro* production of functional recombinant AAV from an extract of Hela cells previously infected with adenovirus-2 and transfected with AAV sequences. The titre obtained using this method was extremely low and it is necessary to optimize the method to increase the efficiency of packaging before this method can be broadly applied to produce recombinant AAV. Successful optimization and application of such techniques could, however, represent a breakthrough discovery in the field of gene transfer and, ultimately gene therapy.

Summary and conclusions

An exciting challenge of contemporary science is to define in detail the function of proteins in mediating or regulating their physiologic responses, and determining the potential role of proteins in the development and progression of disease. This is a particular challenge when one is faced with defining the function of the product of an unknown gene. A new term – functional genomics – has emerged to characterize the application of molecular technology to determination of protein function. One approach is the study of physiologic or pathophysiologic consequences of acute or chronic expression (or overexpression), or chronic inhibition of expression (knock-outs, antisense or dominant negatives) of the protein of interest in a whole animal model. An important tool allowing such an approach is gene transfer. Indeed, gene transfer technology has been successfully applied to the development of transgenic animals either overexpressing a gene or with a gene deleted (knock-outs). Gene transfer technology is now being routinely applied *in vivo* to produce transient transgenic older animals in an effort to better understand the function of certain proteins in adults when the protein's role in disease is more likely manifest.

The purpose of this chapter was to review the status of current methods by which genes are successfully transferred *in vivo* to elicit a functional physiologic response. As discussed, no single vector system yet exists which optimally combines highly efficient transfection, long term expression and easy preparation of the recombinant vector. Rather, numerous vectors exist that allow transfer of a gene *in vivo* to whole animals for development of animal models. Each vector system has its own advantages and disadvantages and the choice of which approach to use will depend on the objectives and design of a particular study. Future research will likely continue to focus on vector technology. To this end, research is already underway to increase the persistence of adenovirus vector gene expression by removing portions of the adenovirus genome to reduce its immunogenicity [42–45]. An extreme example of this is "gutless" or "gutted" vectors in which all of the viral genes are deleted, leaving only the inverted terminal repeats and the packaging sequence [42, 43]. These vectors have been used successfully to produce long-term expression *in vivo* [11]. They also have the added benefit of being able to accept larger inserts (up to 28 kb) of DNA. Accordingly, regulatory elements such as tissue specific promoters and genomic control elements can be accomodated to enable better targeting of expression to particular end organs, thus allowing one to better manipulate expression *in vivo*. Adeno-associated viruses apparently possess the advantages of both retroviruses and adenoviruses as vector systems for gene delivery. Despite their limitations of small coding capacity and difficult preparation, these vectors hold great promise for future application of gene transfer technology in the laboratory to establish relevant animal models. Other new approaches will rely on combinations of presently available vector systems in an attempt to blend the advantages of the component systems. One successful example involves combining liposomes and viral vectors [46].

Successful animal model development, in addition to affording the opportunity for defining the functions of proteins and unraveling the mysteries of the mechanisms by which proteins execute their functions, represents a critical step in establishing feasibility of gene transfer technology for potential therapeutic value (gene therapy). The current status of gene transfer technology has already yielded significant contributions to understanding the role of some proteins in animal development, normal physiology and disease. The future holds great promise and excitement as this and other molecular techniques are applied to the discovery of protein function in this new era of functional genomics.

References

1 Jackson DA, Symons RH, Berg P (1972) Biochemical method for inserting new genetic information into DNA of Simian Virus 40: circular SV40 DNA molecules containing lambda phage genes and the galactose operon of *Escherichia coli*. *Proc Natl Acad Sci USA* 69: 2904–2909

2 Cohen SN, Chang AC, Boyer HW, Helling RB (1973) Construction of biologically functional bacterial plasmids *in vitro*. *Proc Natl Acad Sci USA* 70: 3240–3244

3. Pappas MG (1996) The biotechnology of gene therapy. *Drug Development and Industrial Pharmacy* 22(8): 791–803

4 Robbins PD, Tahara H, Ghivizzani SC (1998) Viral vectors for gene therapy. *TIBTECH* 16: 35–40

5 Sigmund CD (1993) Major approaches for generating and analyzing transgene mice. An overview. *Hypertension* 22: 599–607

6 Nabel EG, Pompili VJ, Plautz GE, Nabel GJ (1994) Gene transfer and vascular disease. *Cardiovasc Res* 28(4): 445–455

7 Finkel T, Epstein SE (1995) Gene therapy for vascular disease. *FASEB J* 9: 843–851

8 Fishman GI (1998) Timing is everything in life. Conditional transgene expression in the cardiovascular system. *Circ Res* 82: 837–844

9 Wilson JM (1997) Vectors – shuttle vectors for gene therapy. *Clin Exp Immunol* 07 (Suppl 1): 31–32

10 Wilson J (1996) Adenoviruses as gene-delivery vehicles. *New Eng J Med* 334 (18): 1185–1187

11 Verma IM, Somia N (1997) Gene therapy – promises, problems and prospects. *Nature* 389: 239–242

12 Miller N, Vile R (1995) Targeted vectors for gene therapy. *FASEB J* 9: 190–199

13 Glorioso JC, DeLuca NA, Fink DJ (1995) Development and application of *Herpes simplex* virus vectors for human gene therapy. *Ann Rev Microbiol* 49: 675–710

14 Latchman DS (1994) *Herpes simplex* virus vectors for gene therapy. *Mol Biotechnol* 2 (2): 179–195

15 Mulligan RC (1993) The basic science of gene therapy. *Science* 260: 926–932

16 Kendrew J (ed) (1994) *The encyclopedia of molecular biology*. Blackwell Science, Cambridge

17 Nabel EG, Yang ZY, Plautz G, Forough R, Zhan X, Haudenschild CC, Maciag T, Nabel GJ (1993) Recombinant fibroblast growth factor-1 promotes intimal hyperplasia and angiogenesis in arteries *in vivo*. *Nature* 362: 844–846

18 Nabel EG, Shum L, Pompili VJ, Yang ZY, San H, Shu HB, Liptay S, Gold L, Gordon D, Derynk R (1993) Direct transfer of the transforming growth factor β1 gene into arteries stimulates fibrocellular hyperplasia. *Proc Natl Acad Sci USA* 90: 10759–10763

19 Nabel EG, Yang Z, Liptay S, San H, Gordon D, Haudenschild CC, Nabel GJ (1993) Recombinant platelet-derived growth factor β gene expression in porcine arteries induces intimal hyperplasia *in vivo*. *J Clin Invest* 91 (4): 1822–1829

20 LeClerc G, Gal D, Takeshita S, Nikol S, Weir L, Isner JM (1992) Percutaneous arterial gene transfer in a rabbit model. Efficiency in normal and balloon-dilated atherosclerotic arteries. *J Clin Invest* 90: 936–944

21 Gal D, Weir L, Leclerc G, Pickering JG, Hogan J, Isner JM (1993) Direct myocardial transfection in two animal models. Evaluation of parameters affecting gene expression and percutaneous gene delivery. *Lab Invest* 68 (1): 18–25

22 Gerard RD, Meidell RS (1993) Adenovirus-mediated gene transfer. *Trends Cardiovasc Med* 3: 171–177

23 Fallaux FJ, Kranenburg O, Cramer SJ, Houweling A, Van Ormondt H, Hoeben RC, Van Der Eb AJ (1996) Characterization of 911: A new helper cell line for the titration and

propagation of early-region-1-deleted adenoviral vectors. *Hum Gene Ther* 7: 215–222
24 Cournoyer D, Caskey CT (1993) Gene therapy of the immune system. *Ann Rev Immunol* 11: 297–329
25 Dunbar CE (1996) Gene transfer to hematopoietic stem cells: implications for gene therapy of human disease. *Ann Rev Med* 47: 11–20
26 Smith AE (1995) Viral vectors in gene therapy. *Ann Rev Microbiol* 49: 807–838
27 Nabel EG, Plautz G, Nabel GJ (1992) Transduction of a foreign histocompatability gene into the arterial wall induces vasculitis. *Proc Natl Acad Sci USA* 89 (11): 5157–5161
28 Gerard RD, Meidell RS (1995) Adenovirus vectors. In: BD Hames, D Glover (eds.): *DNA cloning: A practical approach: Mammalian systems*. Oxford University Press, Oxford, 285–307
29 Gerard RD, Collen D (1997) Adenovirus gene therapy for hypercholesterolemia, thrombosis and restenosis. *Cardiovasc Res* 35: 451–458
30 Yeh P, Perricaudet M (1997) Advances in adenoviral vectors: from genetic engineering to their biology. *FASEB J* 11: 615–623
31 Bergelson JM, Cunningham JA, Droguett G, Kurt-Jones EA, Krithivas A, Hong JS, Horwitz MS, Crowell RL, Finberg RW (1997) Isolation of a common receptor for Coxsackie B viruses and adenoviruses 2 and 5. *Science* 275 (5304): 1320–1323
32 Kremer EJ, Perricaudet M (1995) Adenovirus and adeno-associated virus mediated gene transfer. *Br Med Bull* 51: 31–44
33 He TC, Zhou S, daCosta LT, Yu J, Kinzler KW, Vogelstein B (1998) A simplified system for generating recombinant adenoviruses. *Proc Natl Acad Sci* 95: 2509–2514
34 Kypson AP, Peppel K, Akhter SA, Lilly R, Glower DD, Lefkowitz RJ, Koch WJ (1998) *Ex vivo* adenovirus-mediated gene transfer to the adult rat heart. *J Thorac Cardiovasc Surg* 115: 623–630
35 Koch WJ, Lefkowitz RJ, Milano CA, Akhter SA, Rockman HA (1998) Myocardial overexpression of adrenergic receptors and receptor kinases. *Adv Pharmacol* 42: 502–506
36 Barr E, Carroll J, Tripathy S, Kozarsky K, Wilson J, Leiden JM (1993) Effective catheter-mediated gene transfer into the heart using replication-defective adenovirus. *Gene Therapy* 1: 51–58
37 Peppel K, Koch WJ, Lefkowitz RJ (1997) Gene transfer strategies for augmenting cardiac function. *Trends Cardiovasc Sci* 7: 145–150
38 Kozarsky KF, Wilson JM (1995) Gene therapy of hypercholesterolemic disorders. *Trends Cardiovasc Med* 5: 205–209
39 Berns KI, Linden MR (1995) The cryptic life style of adeno-associated virus. *BioEssays* 17 (3): 237–245
40 XiaoX, Li J, Samulski RJ (1998) Production of high-titer recombinant adeno-associated virus vectors in the absence of helper adenovirus. *J Virology* 72 (3): 2224–2232
41 Ding L, Lu S, Munshi NC (1997) *In vitro* packaging of an infectious recombinant adeno-associated virus 2. *Gene Therapy* 4: 1167–1172
42 Kochanek S, Clemens PR, Mitani K, Chen H-H, Chan S (1996) A new adenoviral vector: replacement of all viral coding sequences with 28 kb of DNA independently express-

ing both full-length dystrophin and β-galactosidase. *Proc Natl Acad Sci USA* 93: 5731–5736

43 Schieder G, Morral N, Parks RJ, Wu Y, Koopmans SC, Langston C, Graham FL, Beaudet AL, Kochanek S (1998) Genomic DNA transfer with a high-capacity adenovirus vector results in improved *in vivo* gene expression and decreased toxicity. *Nature Genetics* 18: 180–183

44 Lusky M, Rittner K, Dieterle A, Dreyer D, Mourot B, Schultz H, Stoeckel F, Pavirani A, Mehtali M (1998) *In vitro* and *in vivo* biology of recombinant adenovirus vectors with E1, E1/E2A, or E1/E4 deleted. *J Virol* 72: 2022–2032

45 Amalfitano A, Hauser MA, Hu H, Serra D, Begy CR, Chamberlain JS (1998) Production and characterization of improved adenovirus vectors with the E1, E2b and E3 genes deleted. *J Virol* 72(2):926-933

46 Qiu C, De Young MB, Finn A, Dichek DA (1998) Cationic liposomes enhance adenovirus entry via a pathway independent of the fiber receptor and av-integrins. *Human Gene Therapy* 9: 507–520

Guidelines and regulations in animal experimentation

Kenneth N. Litwak and Howard C. Hughes

SmithKline Beecham Pharmaceuticals, LAS, UW2620, 709 Swedeland Rd., King of Prussia, PA 19406-0939, USA

Introduction

The welfare of animals used in biomedical research is a matter of importance to those who utilize the animals, as well as the general public. This is evidenced by the numerous laws and regulations currently governing the use of animals in effect, reflect this concern [1]. The purpose of this chapter is to provide a brief overview of these laws and regulations as they pertain to the use of animals in pharmaceutical and academic research. As most of the regulations affect animal care practices and as such are the responsibility of those who care for the animals (i.e. the attending veterinarian and the institution animal care and use committee), details for care practices will not be provided. We will, however, attempt to cover some of the responsibilities of the principal investigator who does animal based research. In addition, since these regulations are subject to change, we have tried to indicate sources where the most recent information additional sources sources and up to dated information can be obtained.

Animal experiments are subject to many different laws and regulations. At the Federal level, the primary these laws include the Animal Welfare Act, Health Research Extension Act (National Institutes of Health Policies), and the Food Drug and Cosmetic Act (Good Laboratory Practice Regulations) [2–4]. There are also a host of local and state laws, and International Treaties that deal with licencing, transportation and/or cruelty to animals [5, 6].

Federal laws

The Animal Welfare Act (AWA) was originally passed in 1966 as a pet theft protection Act and was subsequently amended in 1970, 1976, and 1985 to cover all facilities that use laboratory animals in any form of research, except those which exclusively use mice, rats, birds, and farm animals used for agricultural research [2]. The AWA falls under the jurisdiction of the Secretary of Agriculture and is administered by

In Vivo Models of Inflammation, edited by D.W. Morgan and L.A. Marshall

the Animal and Plant Health Inspection Service (APHIS). APHIS is responsible for issuing and enforcing regulation as well as setting minimum standards for housing and care. The 1985 revisions are those which have the most significant effect upon the investigator. These revisions call for (1) exercise for dogs, (2) animal care committees, (3) protocol review, (4) required training of scientists and care staff, (5) searching for alternatives, (6) minimizing pain and distress, (7) adequate veterinary care, and (8) programs to enhance the psychological well-being of non-human primates [2].

Under the AWA as well as NIH Policy, all investigators must submit a protocol to the institutional care and use committee (IACUC) for review and approval before an animal-based study may start [2]. The protocol should contain, at a minimum, a description of the animal portion of the study, evidence that the study is not unnecessarily duplicate, that alternatives to animals have been considered and, if the procedure produces more than slight or momentary pain, that alternatives to this procedure and/or adequate anesthesia, analgesia, or tranquilization are being used. A USDA inspector may now ask for evidence of the literature search including the source, key words and dates. The investigator must agree to permit on-site inspection of their study by the IACUC and attending veterinarian, and should a major change in the protocol occur, the IACUC will be notified. Should veterinary care be necessary, a veterinarian must be consulted. Furthermore, all members of the investigative team and the animal care staff must be adequately trained for the study as well.

Any facility conducting federally supported research or studies regulated under GLP's must also comply with the Policy on Humane Care and Use of Laboratory Animals [4]. This policy has evolved over the past twenty years and, most recently, was given a statutory foundation following the enactment of the Health Research Extension Act of 1985 [3]. Within the Policy there are several requirements. First, the institution must provide written proof that the facilities meet the requirements outlined in the Guide for the Care and Use of Laboratory Animals [7]. Second, the Policy mandates that there be an Institutional Animal Care and Use Committee (IACUC). Third, any grants submitted must address the number and species of animal used, rationale for animal use, description of use, procedures to minimize pain, and methods of euthanasia. Fourth, there must be records kept on file with animal welfare assurance and any documentation associated with the IACUC. For the most part, Federal agencies regulating animal based research have harmonized their guidelines and regulations. The only difference remaining between AWA regulations and NIH/GLP Policies is that the AWA does not regulate studies or care and use practices for rats and mice specifically bred for research while NIH policy does.

Resources

There are several widely used guidelines available to the scientific community detailing the use of laboratory animals. The most commonly cited resource is the "Guide for the Care and Use of Laboratory Animals" [7], which details most aspects of animal experimentation, from cage size to institutional monitoring to husbandry of the more commonly used species. The Guide also contains an extensive bibliography, categorized by topic, should there be any point that a researcher needs to know in greater detail. Other good references include the Biomedical Investigators Handbook [8], Animals and their Legal Rights [9], and State Laws Concerning the Use of Animals in Research [5], to name a few. Many resources on the laws and regulations are now available on-line for retrieval making access to them much easier (Tab. 1).

Conclusion

With these myriad rules and regulations from numerous levels of government, it is becoming increasingly difficult to ensure compliance with all laws. While there are agencies in place to help sort through the morass (i.e. the Interagency Research Animal Committee) [7], it is the researcher's responsibility to ensure that any research involving animals follows all pertinent laws. In summary, the researcher should provide the best animal care and environmental conditions possible and should be

Table 1 - Selected internet sites dealing with laboratory animals

Group	Internet address
Americans for Medical Progress	http://www.ampef.org/
American College of Laboratory Animal Medicine	http://chopin.osp.uh.edu/~rocky/aclam/hdg1055.htm
Institute of Laboratory Animal Research	http://www2.nas.edu/ilarhome/
Veterinary Organizations	http://netvet.wustl.edu/org/htm
Research Defense Society	http://www.uel.ac.uk/research/rds/
Foundation for Biomedical Research	http://www.fbresearch.org/
National Association for Biomedical Research	http://www.nabr.org/
Animal and Plant Health Inspection	http://www.aphis.usda.gov/
REAC	http://www.aphis.usda.gov:80/reac/
Scientists Center for Animal Welfare	http://www.scaw.com/
American Association of Laboratory Animal Scientists	http://www.aalas.org
American Society of Laboratory Animal Practitioners	http://www.aslap.org

aware of any practice which might exploit the animal [1]. These basic principles will make following the laws and regulations less difficult.

References

1 McPherson CW (1984) Laws, regulations and policies affecting the use of laboratory animals. In: JG Fox, JB Cohen, FM Loew (eds): *Laboratory animal medicine.* Academic Press, New York, 21–30

2 Animal Welfare Act (Title 7 U.S.C. 2131–2156), as amended by P.L. 99–198, December 23, 1986

3 US Department of Health and Human Services, National Institutes of Health, Office for Protection from Research Risks. *Public health service policy on human care and use of laboratory animals*, revised September, 1986, pursuant to Health Research Extension Act of 1985 (P.L. 99–158, November 20, 1985)

4 US Department of Health and Human Services, Food and Drug Administration (1994) *Good laboratory practice regulations: nonclinical laboratory studies* (21 CFR Part 58)

5 National Association for Biomedical Research (1991) *State laws concerning the use of animals in research*

6 Murphy RA, Rowan AN, Smeby R (eds) (1991) *Annotated bibliography on laboratory animal welfare.* Scientists Center for Animal Welfare, Greenbelt, MD

7 National Research Council, Institute of Laboratory Animal Resources (1996) *Guide for the care and use of laboratory animals.* National Academy Press, Washington, DC

8 Foundation for Biomedical Research (1987) *The biomedical investigator's handbook: for researchers using animal models.* Washington, DC

9 Animal Welfare Institute (1985) *Animals and their legal rights.* Washington, DC

Index